Image Processing
The Fundamentals

Image Processing

The Fundamentals

Maria Petrou
University of Surrey, Guildford, UK

Panagiota Bosdogianni
Technical University of Crete, Chania, Greece

JOHN WILEY & SONS, LTD
Chichester · New York · Weinheim · Brisbane · Singapore · Toronto

Other Wiley Editorial Offices

John Wiley & Sons, Inc., 605 Third Avenue,
New York, NY 10158-0012, USA

WILEY-VCH Verlag GmbH, Pappelallee 3,
D-69469 Weinheim, Germany

Jacaranda Wiley Ltd, 33 Park Road, Milton,
Queensland 4064, Australia

John Wiley & Sons (Asia) Pte Ltd, 2 Clementi Loop #02-01,
Jin Xing Distripark, Singapore 129809

John Wiley & Sons (Canada) Ltd, 22 Worcester Road
Rexdale, Ontario, M9W 1L1, Canada

Library of Congress Cataloging-in-Publication Data

Petrou, Maria
    Image Processing : the fundamentals / Maria Petrou, Panagiota
Bosdogianni.
        p.   cm.
    Includes bibliographical references.
    ISBN 0-471-99883-4 (alk. paper)
    1. Image processing—Digital techniques.   I. Bosdogianni,
Panagiota. II. Title.
    TA1637.P48   1999
    621.36 ' 7—dc21   99-32327
                                                       CIP

British Library Cataloguing in Publication Data
A catalogue record for this book is available from the British Library

ISBN   0-471-99883-4

Produced from PostScript files supplied by the authors.
Printed and bound in Great Britain by Bookcraft (Bath) Limited.
This book is printed on acid-free paper responsibly manufactured from sustainable forestry, in which at least two trees are planted for each one used for paper production.

To the Costases
in our lives

Contents

Preface

This book is the result of 11 years of teaching the subject to the students of Surrey University studying for an MSc degree in Signal Processing and Machine Intelligence.

As the subject of Machine Intelligence has mushroomed in recent years, so it has attracted the interest of researchers from as diverse fields as Psychology, Physiology, Engineering and Mathematics. The problems of Machine Intelligence may be tackled with various tools. However, when we have to perform a task using a computer, we must have in mind the language a computer understands, and this language is the language of arithmetic and by extension mathematics. So, the approach to solving the problems of Machine Intelligence is largely mathematical. Image Processing is the basic underlying infrastructure of all Vision and Image related Machine Intelligence topics. Trying to perform Computer Vision and ignoring Image Processing is like trying to build a house starting from the roof, and trying to do Image Processing without mathematics is like trying to fly by flapping the arms!

The diversity of people who wish to work and contribute in the general area of Machine Intelligence led me towards writing this book on two levels. One level should be easy to follow with a limited amount of mathematics. This is appropriate for newcomers to the field and undergraduate students. The second level, more sophisticated, going through the mathematical intricacies of the various methods and proofs, is appropriate for the inquisitive student who wishes to know the "why" and the "how" and get at the bottom of things in an uncompromising way. At the lower level, the book can be followed with no reference made at all to the higher level. All material referring to the higher level is presented inside grey boxes, and may be skipped. The book contains numerous examples presented inside frames. Examples that refer to boxed material are clearly marked with a **B** and they may be ignored alongside the advanced material if so desired. The basic mathematical background required by the reader is the knowledge of how to add and subtract matrices. Knowledge of eigenvalue analysis of matrices is also important. However, there are several fully worked examples, so that even if somebody is hardly familiar with the subject, they can easily learn the nuts and bolts of it by working through this book. This approach is also carried to the stochastic methods presented: one can start learning from the basic concept of a random variable and reach the level of understanding and using the concept of ergodicity.

I would like to take this opportunity to thank the numerous MSc students who over the years helped shape this book, sometimes with their penetrating questions, and sometimes with their seemingly naive(!) questions. However, there are no naive questions when one is learning: the naivety is with those who do not ask the questions! My students' questions helped formulate the route to learning and gave me the idea to present the material in the form of questions and answers.

Writing this book was a learning process for Panagiota and me too. We had a lot of fun working through the example images and discovering the secrets of the methods. One thing that struck us as most significant was the divergence between the continuous and the discrete methods. An analytically derived formula appropriate for the continuous domain often has very little to do with the formula one has to program into the computer in order to perform the task. This is very clearly exemplified in Chapter 6 concerned with image restoration. That is the reason we demonstrate all the methods we present using small, manageable discrete images, that allow us to manipulate them "manually" and learn what exactly the computer has to do if a real size image is to be used. When talking about real size images, we would like to thank Constantinos Boukouvalas who helped with the programming of some of the methods presented.

Finally, I would also like to thank my colleagues in the Centre for Vision, Speech and Signal Processing of Surrey University, and in particular the director Josef Kittler for all the opportunities and support he gave me, and our systems manager Graeme Wilford for being always helpful and obliging.

Maria Petrou

List of Figures

Chapter 1

Introduction

Why do we process images?

Image Processing has been developed in response to three major problems concerned with pictures:

- Picture digitization and coding to facilitate transmission, printing and storage of pictures.

- Picture enhancement and restoration in order, for example, to interpret more easily pictures of the surface of other planets taken by various probes.

- Picture segmentation and description as an early stage in Machine Vision.

What is an image?

A monochrome image is a 2-dimensional light intensity function, $f(x, y)$, where x and y are spatial coordinates and the value of f at (x, y) is proportional to the brightness of the image at that point. If we have a multicolour image, f is a vector, each component of which indicates the brightness of the image at point (x, y) at the corresponding colour *band*.

A digital image is an image $f(x, y)$ that has been discretized both in spatial coordinates and in brightness. It is represented by a 2-dimensional integer array, or a series of 2-dimensional arrays, one for each colour band. The digitized brightness value is called the *grey level* value.

Each element of the array is called a *pixel* or a *pel* derived from the term "picture element". Usually, the size of such an array is a few hundred pixels by a few hundred pixels and there are several dozens of possible different grey levels. Thus, a digital image looks like this:

$$f(x, y) = \begin{bmatrix} f(0,0) & f(0,1) & \cdots & f(0, N-1) \\ f(1,0) & f(1,1) & \cdots & f(1, N-1) \\ \vdots & \vdots & & \vdots \\ f(N-1,0) & f(N-1,1) & \cdots & f(N-1, N-1) \end{bmatrix}$$

with $0 \le f(x, y) \le G - 1$ where usually N and G are expressed as integer powers of 2 ($N = 2^n$, $G = 2^m$).

What is the brightness of an image at a pixel position?

Each pixel of an image corresponds to a part of a physical object in the 3D world. This physical object is illuminated by some light which is partly reflected and partly absorbed by it. Part of the reflected light reaches the sensor used to image the scene and is responsible for the value recorded for the specific pixel. The recorded value of course, depends on the type of sensor used to image the scene, and the way this sensor responds to the spectrum of the reflected light. However, as a whole scene is imaged by the same sensor, we usually ignore these details. What is important to remember is that the brightness values of different pixels have significance only relative to each other and they are meaningless in absolute terms. So, pixel values between different images should only be compared if either care has been taken for the physical processes used to form the two images to be identical, or the brightness values of the two images have somehow been normalized so that the effects of the different physical processes have been removed.

Why are images often quoted as being 512 × 512, 256 × 256, 128 × 128 etc?

Many image calculations with images are simplified when the size of the image is a power of 2.

How many bits do we need to store an image?

The number of bits, b, we need to store an image of size $N \times N$ with 2^m different grey levels is:

$$b = N \times N \times m \qquad (1.1)$$

So, for a typical 512 × 512 image with 256 grey levels ($m = 8$) we need 2,097,152 bits or 262,144 8-bit bytes. That is why we often try to reduce m and N, without significant loss in the quality of the picture.

What is meant by image resolution?

The resolution of an image expresses how much detail we can see in it and clearly depends on both N and m.

Keeping m constant and decreasing N results in the *checkerboard effect* (*Figure* 1.1). Keeping N constant and reducing m results in *false contouring* (*Figure* 1.2). Experiments have shown that the more detailed a picture is, the less it improves by keeping N constant and increasing m. So, for a detailed picture, like a picture of crowds (*Figure* 1.3), the number of grey levels we use does not matter much.

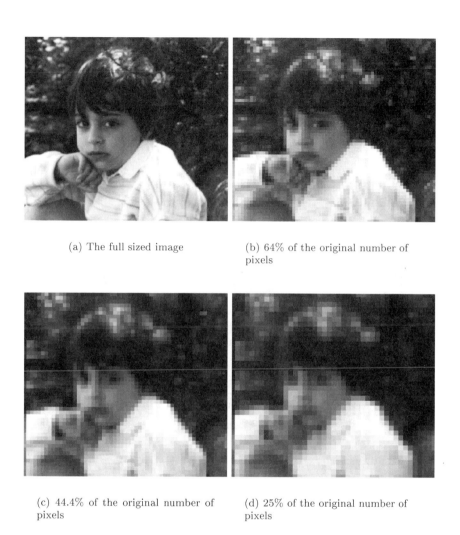

(a) The full sized image

(b) 64% of the original number of pixels

(c) 44.4% of the original number of pixels

(d) 25% of the original number of pixels

Figure 1.1: Keeping m constant and decreasing the size of the image from 338×298 down to a quarter of that produces the checkerboard effect.

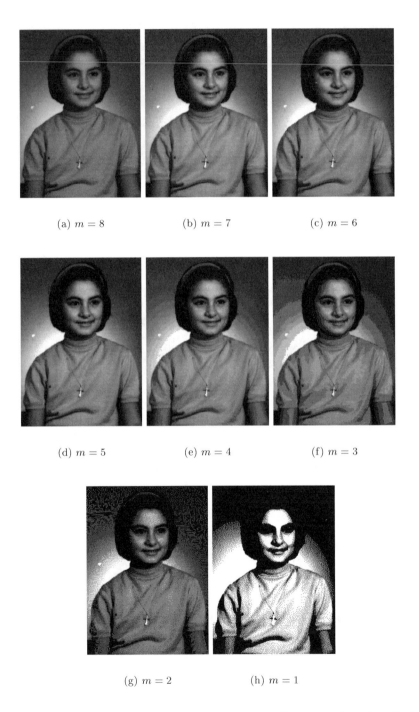

(a) $m = 8$ (b) $m = 7$ (c) $m = 6$

(d) $m = 5$ (e) $m = 4$ (f) $m = 3$

(g) $m = 2$ (h) $m = 1$

Figure 1.2: Keeping the size of the image constant (247×333) and reducing the number of grey levels ($= 2^m$) produces false contouring.

(a) $m = 8$

(b) $m = 6$

(c) $m = 4$

(d) $m = 3$

(e) $m = 2$

(f) $m = 1$

Figure 1.3: When keeping the size of a detailed image constant (1022×677) and reducing the number of grey levels, the result is not affected very much.

How do we do Image Processing?

We perform Image Processing by using Image Transformations. Image Transformations are performed using *Operators*. An Operator takes as input an image and
produces another image. In this book we shall concentrate mainly on a particular
class of operators, called *Linear Operators*.

What is a linear operator?

Consider $\mathcal{O}$ to be an operator which takes images into images. If f is an image, $\mathcal{O}(f)$
is the result of applying $\mathcal{O}$ to f. $\mathcal{O}$ is linear if:

$$\mathcal{O}[af + bg] = a\mathcal{O}[f] + b\mathcal{O}[g] \tag{1.2}$$

for all images f and g and all scalars a and b.

How are operators defined?

Operators are defined in terms of their *point spread functions*. The point spread
function of an operator is what we get out if we apply the operator on a point source:

$$\mathcal{O}[\text{point source}] \quad = \quad \text{point spread function} \tag{1.3}$$

Or:

$$\mathcal{O}[\delta(x - \alpha, y - \beta)] \quad = \quad h(x, \alpha, y, \beta) \tag{1.4}$$

where $\delta(x - \alpha, y - \beta)$ is a point source of brightness 1 centred at point (α, β).

How does an operator transform an image?

If the operator is *linear*, when the point source is a times brighter, the result will be
a times larger:

$$\mathcal{O}[a\delta(x - \alpha, y - \beta)] = ah(x, \alpha, y, \beta) \tag{1.5}$$

An image is a collection of point sources (the *pixels*) each with its own brightness
value. We may say that an image is the sum of these point sources. Then the effect of
an operator characterized by point spread function $h(x, \alpha, y, \beta)$ on an image $f(x, y)$
can be written as:

$$g(\alpha, \beta) = \sum_{x=0}^{N-1} \sum_{y=0}^{N-1} f(x, y) h(x, \alpha, y, \beta) \tag{1.6}$$

where $g(\alpha, \beta)$ is the output "image", $f(x, y)$ is the input image and the size of the
images is $N \times N$.

What is the meaning of the point spread function?

The point spread function $h(x, \alpha, y, \beta)$ expresses how much the input value at position (x, y) influences the output value at position (α, β). If the influence expressed by the point spread function is independent of the actual positions but depends only on the **relative** position of the influencing and the influenced pixels, we have a *shift invariant* point spread function:

$$h(x, \alpha, y, \beta) = h(\alpha - x, \beta - y) \tag{1.7}$$

Then equation (1.6) is a *convolution*:

$$g(\alpha, \beta) = \sum_{x=0}^{N-1} \sum_{y=0}^{N-1} f(x, y) h(\alpha - x, \beta - y) \tag{1.8}$$

If the columns are influenced independently from the rows of the image, then the point spread function is *separable*:

$$h(x, \alpha, y, \beta) \equiv h_c(x, \alpha) h_r(y, \beta) \tag{1.9}$$

where the above expression serves also as the definition of functions $h_c(x, \alpha)$ and $h_r(y, \beta)$. Then equation (1.6) can be written as a cascade of two 1D transformations:

$$g(\alpha, \beta) = \sum_{x=0}^{N-1} h_c(x, \alpha) \sum_{y=0}^{N-1} f(x, y) h_r(y, \beta) \tag{1.10}$$

If the point spread function is both *shift invariant* and *separable*, then equation (1.6) can be written as a cascade of two 1D convolutions:

$$g(\alpha, \beta) = \sum_{x=0}^{N-1} h_c(\alpha - x) \sum_{y=0}^{N-1} f(x, y) h_r(\beta - y) \tag{1.11}$$

B1.1: The formal definition of a point source in the continuous domain

Define an extended source of constant brightness:

$$\delta_n(x, y) \equiv n^2 rect(nx, ny) \tag{1.12}$$

where n is a positive constant and

$$rect(nx, ny) \equiv \begin{cases} 1 \text{ inside a rectangle } |nx| \le \frac{1}{2}, |ny| \le \frac{1}{2} \\ 0 \text{ elsewhere} \end{cases} \tag{1.13}$$

The total brightness of this source is given by

$$\int_{-\infty}^{\infty}\int_{-\infty}^{\infty} \delta_n(x,y)dxdy \;=\; \underbrace{n^2\int_{-\infty}^{\infty}\int_{-\infty}^{\infty} rect(nx,ny)dxdy = 1}_{\text{area of rectangle}} \quad (1.14)$$

and is independent of n.

As $n \to \infty$, we create a sequence, δ_n, of extended square sources which gradually shrink with their brightness remaining constant. At the limit, δ_n becomes Dirac's delta function

$$\delta(x,y) \begin{cases} \neq 0 & \text{for } x = y = 0 \\ = 0 & \text{elsewhere} \end{cases} \quad (1.15)$$

with the property

$$\int_{-\infty}^{\infty}\int_{-\infty}^{\infty} \delta(x,y)dxdy \;=\; 1. \quad (1.16)$$

The integral

$$\int_{-\infty}^{\infty}\int_{-\infty}^{\infty} \delta_n(x,y)g(x,y)dxdy \quad (1.17)$$

is the average of image $g(x,y)$ over a square with sides $\frac{1}{n}$ centred at $(0,0)$. At the limit we have:

$$\int_{-\infty}^{\infty}\int_{-\infty}^{\infty} \delta(x,y)g(x,y)dxdy \;=\; g(0,0) \quad (1.18)$$

which is the value of the image at the origin. Similarly

$$\int_{-\infty}^{\infty}\int_{-\infty}^{\infty} g(x,y)\delta_n(x-a,y-b)dxdy \quad (1.19)$$

is the average value of g over a square $\frac{1}{n} \times \frac{1}{n}$ centred at $x = a$, $y = b$, since:

$$\delta_n(x-a,y-b) \;=\; n^2 rect[n(x-a),n(y-b)]$$
$$= \begin{cases} n^2 & |n(x-a)| \leq \frac{1}{2} \quad |n(y-b)| \leq \frac{1}{2} \\ 0 & \text{elsewhere} \end{cases} \quad (1.20)$$

We can see that this is a square source centred at (a,b) by considering that $|n(x-a)| \leq \frac{1}{2}$ means $-\frac{1}{2} \leq n(x-a) \leq \frac{1}{2}$ i.e. $-\frac{1}{2n} \leq x - a \leq \frac{1}{2n}$ or $a - \frac{1}{2n} \leq x \leq a + \frac{1}{2n}$. Thus we have that $\delta_n(x-a,y-b) = n^2$ in the region $a - \frac{1}{2n} \leq x \leq a + \frac{1}{2n}$, $b - \frac{1}{2n} \leq y \leq b + \frac{1}{2n}$.

At the limit of $n \to \infty$, integral (1.19) is the value of the image g at $x = a$, $y = b$, i.e.

$$\int_{-\infty}^{\infty}\int_{-\infty}^{\infty} g(x,y)\delta_n(x-a,y-b)dxdy \;\; = \;\; g(a,b) \qquad (1.21)$$

This equation is called the *shifting property* of the delta function. This equation also shows that **any** image $g(x,y)$ can be expressed as a superposition of point sources.

How can we express in practice the effect of a linear operator on an image?

This is done with the help of matrices. We can rewrite equation (1.6) as follows:

$$g(\alpha,\beta) =$$
$$f(0,0)h(0,\alpha,0,\beta) + f(1,0)h(1,\alpha,0,\beta) + \ldots + f(N-1,0)h(N-1,\alpha,0,\beta)$$
$$+f(0,1)h(0,\alpha,1,\beta) + f(1,1)h(1,\alpha,1,\beta) + \ldots + f(N-1,1)h(N-1,\alpha,1,\beta)$$
$$+\ldots + f(0,N-1)h(0,\alpha,N-1,\beta) + f(1,N-1)h(1,\alpha,N-1,\beta) + \ldots$$
$$+f(N-1,N-1)h(N-1,\alpha,N-1,\beta) \qquad (1.22)$$

The right hand side of this expression can be thought of as the dot product of vector

$$\mathbf{h}_{\alpha\beta}^T \equiv$$
$$[h(0,\alpha,0,\beta), h(1,\alpha,0,\beta), \ldots, h(N-1,\alpha,0,\beta), h(0,\alpha,1,\beta), h(1,\alpha,1,\beta), \ldots,$$
$$h(N-1,\alpha,1,\beta), \ldots, h(1,\alpha,N-1,\beta), \ldots, h(N-1,\alpha,N-1,\beta)] \qquad (1.23)$$

with vector

$$\mathbf{f}^T \;\; \equiv \;\; [f(0,0), f(1,0), \ldots, f(N-1,0), f(0,1), f(1,1), \ldots, f(N-1,1),$$
$$\ldots, f(0,N-1), f(1,N-1), \ldots, f(N-1,N-1)] \qquad (1.24)$$

This last vector is actually the image $f(x,y)$ written as a vector by stacking its columns one under the other. If we imagine writing $g(\alpha,\beta)$ in the same way, then vectors $\mathbf{h}_{\alpha\beta}^T$ will arrange themselves as the rows of a matrix H, where for $\alpha = 0$, β will run from 0 to $N-1$ to give the first N rows of the matrix, then for $\alpha = 1$, β will run again from 0 to $N-1$ to give the second N rows of the matrix, and so on. Thus, equation (1.6) can be written in a more compact way as:

$$\boxed{\mathbf{g} = H\mathbf{f}} \qquad (1.25)$$

This is the *Fundamental Equation* of linear Image Processing. H here is a square $N^2 \times N^2$ matrix that is made up of $N \times N$ submatrices of size $N \times N$ each, arranged in the following way:

$$H = \begin{pmatrix} \alpha\downarrow\begin{pmatrix} x\to \\ y=0 \\ \beta=0 \end{pmatrix} & \alpha\downarrow\begin{pmatrix} x\to \\ y=1 \\ \beta=0 \end{pmatrix} & \cdots & \alpha\downarrow\begin{pmatrix} x\to \\ y=N-1 \\ \beta=0 \end{pmatrix} \\ \alpha\downarrow\begin{pmatrix} x\to \\ y=0 \\ \beta=1 \end{pmatrix} & \alpha\downarrow\begin{pmatrix} x\to \\ y=1 \\ \beta=1 \end{pmatrix} & \cdots & \alpha\downarrow\begin{pmatrix} x\to \\ y=N-1 \\ \beta=1 \end{pmatrix} \\ \vdots & \vdots & & \vdots \\ \alpha\downarrow\begin{pmatrix} x\to \\ y=0 \\ \beta=N-1 \end{pmatrix} & \alpha\downarrow\begin{pmatrix} x\to \\ y=1 \\ \beta=N-1 \end{pmatrix} & \cdots & \alpha\downarrow\begin{pmatrix} x\to \\ y=N-1 \\ \beta=N-1 \end{pmatrix} \end{pmatrix} \quad (1.26)$$

In this representation each bracketed expression represents an $N \times N$ submatrix made up from function $h(x,\alpha,y,\beta)$ for fixed values of y and β and with variables x and α taking up all their possible values in the directions indicated by the arrows. This schematic structure of matrix H is said to correspond to a *partition* of this matrix into N^2 square submatrices.

B1.2: What is the stacking operator?

The stacking operator allows us to write an $N \times N$ image array as an $N^2 \times 1$ vector, or an $N^2 \times 1$ vector as an $N \times N$ square array.

We define some vectors $\mathbf{V_n}$ and some matrices N_n as:

$$\mathbf{V_n} = \begin{bmatrix} 0 \\ \vdots \\ 0 \\ 1 \\ 0 \\ \vdots \\ 0 \end{bmatrix} \begin{matrix} \left.\right\} \text{rows } 0 \text{ to } n-1 \\ \} \quad \text{row } n \\ \left.\right\} \text{rows } n+1 \text{ to } N \end{matrix}$$

$$N_n = \begin{bmatrix} & \mathbf{0} & \\ 1 & 0 & \cdots & 0 \\ 0 & 1 & \cdots & 0 \\ \vdots & \vdots & & \vdots \\ 0 & 0 & \cdots & 1 \\ & \mathbf{0} & \end{bmatrix}$$

$n-1$ square $N \times N$ matrices on top of each other with all their elements 0

the nth matrix is the unit matrix

$N-n$ square $N \times N$ matrices on the top of each other with all their elements 0

The dimensions of $\mathbf{V_n}$ are $(N \times 1)$ and of N_n $(N^2 \times N)$. Then vector $\mathbf{f}$ which corresponds to the $(N \times N)$ square matrix f is given by:

$$\mathbf{f} = \sum_{n=1}^{N} N_n f \mathbf{V_n} \tag{1.27}$$

It can be shown that if $\mathbf{f}$ is an $N^2 \times 1$ vector, we can write it as an $N \times N$ matrix f the first column of which is made up from the first N elements of $\mathbf{f}$, the second column from the second N elements of $\mathbf{f}$, and so on, by using the following expression:

$$f = \sum_{n=1}^{N} N_n^T \mathbf{f} \mathbf{V_n^T} \tag{1.28}$$

Example 1.1 (B)

You are given a 3×3 image f and you are asked to use the stacking operator to write it in vector form.

Let us say that:

$$f = \begin{pmatrix} f_{11} & f_{12} & f_{13} \\ f_{21} & f_{22} & f_{23} \\ f_{31} & f_{32} & f_{33} \end{pmatrix}$$

We define vectors $\mathbf{V_n}$ and matrices N_n for $n = 1, 2, 3$:

$$\mathbf{V_1} = \begin{pmatrix} 1 \\ 0 \\ 0 \end{pmatrix}, \quad \mathbf{V_2} = \begin{pmatrix} 0 \\ 1 \\ 0 \end{pmatrix}, \quad \mathbf{V_3} = \begin{pmatrix} 0 \\ 0 \\ 1 \end{pmatrix}$$

$$N_1 = \begin{pmatrix} 1 & 0 & 0 \\ 0 & 1 & 0 \\ 0 & 0 & 1 \\ 0 & 0 & 0 \\ 0 & 0 & 0 \\ 0 & 0 & 0 \\ 0 & 0 & 0 \\ 0 & 0 & 0 \\ 0 & 0 & 0 \end{pmatrix}, \quad N_2 = \begin{pmatrix} 0 & 0 & 0 \\ 0 & 0 & 0 \\ 0 & 0 & 0 \\ 1 & 0 & 0 \\ 0 & 1 & 0 \\ 0 & 0 & 1 \\ 0 & 0 & 0 \\ 0 & 0 & 0 \\ 0 & 0 & 0 \end{pmatrix}, \quad N_3 = \begin{pmatrix} 0 & 0 & 0 \\ 0 & 0 & 0 \\ 0 & 0 & 0 \\ 0 & 0 & 0 \\ 0 & 0 & 0 \\ 0 & 0 & 0 \\ 1 & 0 & 0 \\ 0 & 1 & 0 \\ 0 & 0 & 1 \end{pmatrix}.$$

According to equation (1.27):

$$\mathbf{f} = N_1 f \mathbf{V}_1 + N_2 f \mathbf{V}_2 + N_3 f \mathbf{V}_3 \qquad (1.29)$$

We shall calculate each term separately:

$$N_1 f \mathbf{V}_1 = \begin{pmatrix} 1 & 0 & 0 \\ 0 & 1 & 0 \\ 0 & 0 & 1 \\ 0 & 0 & 0 \\ 0 & 0 & 0 \\ 0 & 0 & 0 \\ 0 & 0 & 0 \\ 0 & 0 & 0 \\ 0 & 0 & 0 \end{pmatrix} \begin{pmatrix} f_{11} & f_{12} & f_{13} \\ f_{21} & f_{22} & f_{23} \\ f_{31} & f_{32} & f_{33} \end{pmatrix} \begin{pmatrix} 1 \\ 0 \\ 0 \end{pmatrix} \Rightarrow$$

$$N_1 f \mathbf{V}_1 = \begin{pmatrix} 1 & 0 & 0 \\ 0 & 1 & 0 \\ 0 & 0 & 1 \\ 0 & 0 & 0 \\ 0 & 0 & 0 \\ 0 & 0 & 0 \\ 0 & 0 & 0 \\ 0 & 0 & 0 \\ 0 & 0 & 0 \end{pmatrix} \begin{pmatrix} f_{11} \\ f_{21} \\ f_{31} \end{pmatrix} = \begin{pmatrix} f_{11} \\ f_{21} \\ f_{31} \\ 0 \\ 0 \\ 0 \\ 0 \\ 0 \\ 0 \end{pmatrix}$$

Similarly

$$N_2 f \mathbf{V}_2 = \begin{pmatrix} 0 \\ 0 \\ 0 \\ f_{12} \\ f_{22} \\ f_{32} \\ 0 \\ 0 \\ 0 \end{pmatrix}, \qquad N_3 f \mathbf{V}_3 = \begin{pmatrix} 0 \\ 0 \\ 0 \\ 0 \\ 0 \\ 0 \\ f_{13} \\ f_{23} \\ f_{33} \end{pmatrix} \qquad (1.30)$$

Then by substituting in (1.29) we get vector **f**.

Example 1.2 (B)

You are given a 9×1 vector f. Use the stacking operator to write it as a 3×3 matrix.

Let us say that:

$$\mathbf{f} = \begin{pmatrix} f_{11} \\ f_{21} \\ f_{31} \\ f_{12} \\ f_{22} \\ f_{32} \\ f_{13} \\ f_{23} \\ f_{33} \end{pmatrix}$$

According to equation (1.28)

$$f = N_1^T \mathbf{f} V_1^T + N_2^T \mathbf{f} V_2^T + N_3^T \mathbf{f} V_3^T \tag{1.31}$$

We shall calculate each term separately:

$$N_1^T \mathbf{f} V_1^T = \begin{pmatrix} 1 & 0 & 0 & 0 & 0 & 0 & 0 & 0 & 0 \\ 0 & 1 & 0 & 0 & 0 & 0 & 0 & 0 & 0 \\ 0 & 0 & 1 & 0 & 0 & 0 & 0 & 0 & 0 \end{pmatrix} \begin{pmatrix} f_{11} \\ f_{21} \\ f_{31} \\ f_{12} \\ f_{22} \\ f_{32} \\ f_{13} \\ f_{23} \\ f_{33} \end{pmatrix} \begin{pmatrix} 1 & 0 & 0 \end{pmatrix}$$

$$= \begin{pmatrix} 1 & 0 & 0 & 0 & 0 & 0 & 0 & 0 & 0 \\ 0 & 1 & 0 & 0 & 0 & 0 & 0 & 0 & 0 \\ 0 & 0 & 1 & 0 & 0 & 0 & 0 & 0 & 0 \end{pmatrix} \begin{pmatrix} f_{11} & 0 & 0 \\ f_{21} & 0 & 0 \\ f_{31} & 0 & 0 \\ 0 & 0 & 0 \\ 0 & 0 & 0 \\ 0 & 0 & 0 \\ 0 & 0 & 0 \\ 0 & 0 & 0 \\ 0 & 0 & 0 \end{pmatrix} = \begin{pmatrix} f_{11} & 0 & 0 \\ f_{21} & 0 & 0 \\ f_{31} & 0 & 0 \end{pmatrix}$$

$$\tag{1.32}$$

Similarly

$$N_2^T \mathbf{f} \mathbf{V_2}^T = \begin{pmatrix} 0 & f_{12} & 0 \\ 0 & f_{22} & 0 \\ 0 & f_{32} & 0 \end{pmatrix}, \qquad N_3^T \mathbf{f} \mathbf{V_3}^T = \begin{pmatrix} 0 & 0 & f_{13} \\ 0 & 0 & f_{23} \\ 0 & 0 & f_{33} \end{pmatrix} \qquad (1.33)$$

Then by substituting in (1.31) we get matrix f.

What is the implication of the separability assumption on the structure of matrix H?

According to the separability assumption, we can replace $h(x, \alpha, y, \beta)$ by $h_c(x, \alpha) h_r(y, \beta)$. Then inside each partition of H in equation (1.26), $h_c(x, \alpha)$ remains constant and we may write for H:

$$\begin{pmatrix} h_{c00}\begin{pmatrix} h_{r00} & \cdots & h_{rN-10} \\ h_{r01} & \cdots & h_{rN-11} \\ \vdots & & \vdots \\ h_{r0N-1} & \cdots & h_{rN-1N-1} \end{pmatrix} & \cdots & h_{cN-10}\begin{pmatrix} h_{r00} & \cdots & h_{rN-10} \\ h_{r01} & \cdots & h_{rN-11} \\ \vdots & & \vdots \\ h_{r0N-1} & \cdots & h_{rN-1N-1} \end{pmatrix} \\ h_{c01}\begin{pmatrix} h_{r00} & \cdots & h_{rN-10} \\ h_{r01} & \cdots & h_{rN-11} \\ \vdots & & \vdots \\ h_{r0N-1} & \cdots & h_{rN-1N-1} \end{pmatrix} & \cdots & h_{cN-11}\begin{pmatrix} h_{r00} & \cdots & h_{rN-10} \\ h_{r01} & \cdots & h_{rN-11} \\ \vdots & & \vdots \\ h_{r0N-1} & \cdots & h_{rN-1N-1} \end{pmatrix} \\ \vdots & & \vdots \\ h_{c0N-1}\begin{pmatrix} h_{r00} & \cdots & h_{rN-10} \\ h_{r01} & \cdots & h_{rN-11} \\ \vdots & & \vdots \\ h_{r0N-1} & \cdots & h_{rN-1N-1} \end{pmatrix} & \cdots & h_{cN-1N-1}\begin{pmatrix} h_{r00} & \cdots & h_{rN-10} \\ h_{r01} & \cdots & h_{rN-11} \\ \vdots & & \vdots \\ h_{r0N-1} & \cdots & h_{rN-1N-1} \end{pmatrix} \end{pmatrix}$$

where the arguments of functions $h_c(x, \alpha)$ and $h_r(y, \beta)$ have been written as indices to save space. We say then that matrix H is the *Kronecker product* of matrices h_c^T and h_r^T and we write this as:

$$H = h_c^T \otimes h_r^T \qquad (1.34)$$

How can a separable transform be written in matrix form?

Consider again equation (1.10) which expresses the separable linear transform of an image:

$$g(\alpha, \beta) \quad = \quad \sum_{x=0}^{N-1} h_c(x, \alpha) \sum_{y=0}^{N-1} f(x, y) h_r(y, \beta) \tag{1.35}$$

Notice that factor $\sum_{y=0}^{N-1} f(x, y) h_r(y, \beta)$ actually represents the product of two $N \times N$ matrices, which must be another matrix of the same size. Let us define it as:

$$s(x, \beta) \equiv \sum_{y=0}^{N-1} f(x, y) h_r(y, \beta) = f h_r \tag{1.36}$$

Then (1.35) can be written as:

$$g(\alpha, \beta) = \sum_{x=0}^{N-1} h_c(x, \alpha) s(x, \beta) = h_c^T s \tag{1.37}$$

Thus in matrix form

$$\boxed{g = h_c^T f h_r} \tag{1.38}$$

What is the meaning of the separability assumption?

The separability assumption implies that our operator $\mathcal{O}$ (the point spread function of which, $h(x, \alpha, y, \beta)$, is separable) operates on the rows of the image matrix f independently from the way it operates on its columns. These independent operations are expressed by the two matrices h_r and h_c respectively. That is why we chose subscripts r and c to denote these matrices ($r =$ rows, $c =$ columns).

B1.3 The formal derivation of the separable matrix equation

We can use equations (1.27) and (1.28) with (1.25) as follows: First express the output image g using (1.28) in terms of $\mathbf{g}$:

$$g \quad = \quad \sum_{m=1}^{N} N_m^T \mathbf{g} V_m^T \tag{1.39}$$

Then express $\mathbf{g}$ in terms of H and $\mathbf{f}$ from (1.25) and replace $\mathbf{f}$ in terms of f using (1.27):

$$\mathbf{g} \quad = \quad H \sum_{n=1}^{N} N_n f V_n \tag{1.40}$$

Substitute (1.40) into (1.39) and group factors with the help of brackets to get:

$$g \;=\; \sum_{m=1}^{N}\sum_{n=1}^{N}(N_m^T H N_n)f(\mathbf{V_n V_m^T}) \tag{1.41}$$

H is a $(N^2 \times N^2)$ matrix. We may think of it as partitioned in $N \times N$ submatrices stacked together. Then it can be shown that $N_m^T H N_n$ is the H_{mn} such submatrix. Under the separability assumption, matrix H is the Kronecker product of matrices h_c and h_r:

$$H \;=\; h_c^T \otimes h_r^T \tag{1.42}$$

Then partition H_{mn} is essentially $h_r^T(m,n)h_c^T$. If we substitute this in (1.41) we obtain:

$$g \;=\; \sum_{m=1}^{N}\sum_{n=1}^{N}\underbrace{h_r^T(m,n)}_{\text{a scalar}} h_c^T f(\mathbf{V_n V_m^T}) \Rightarrow$$

$$\Rightarrow g \;=\; h_c^T f \sum_{m=1}^{N}\sum_{n=1}^{N} h_r^T(m,n)\mathbf{V_n V_m^T} \tag{1.43}$$

The product $\mathbf{V_n V_m^T}$ is the product between an $(N \times 1)$ matrix with the only non-zero element at position n, with a $(1 \times N)$ matrix, with the only non-zero element at position m. So, it is an $N \times N$ square matrix with the only non-zero element at position (n,m).

When multiplied by $h_r^T(m,n)$ it places the (m,n) element of the h_r^T matrix in position (n,m) and sets to zero all other elements. The sum over all m's and n's is h_r. So from (1.43) we have:

$$g = h_c^T f h_r \tag{1.44}$$

Example 1.3 (B)

You are given a 9×9 matrix H which is partitioned into nine 3×3 submatrices. Show that $N_2^T H N_3$, where N_2 and N_3 are matrices of the stacking operator, is partition H_{23} of matrix H.

Let us say that:

$$
H = \left(\begin{array}{ccc|ccc|ccc}
h_{11} & h_{12} & h_{13} & h_{14} & h_{15} & h_{16} & h_{17} & h_{18} & h_{19} \\
h_{21} & h_{22} & h_{23} & h_{24} & h_{25} & h_{26} & h_{27} & h_{28} & h_{29} \\
h_{31} & h_{32} & h_{33} & h_{34} & h_{35} & h_{36} & h_{37} & h_{38} & h_{39} \\
\hline
h_{41} & h_{42} & h_{43} & h_{44} & h_{45} & h_{46} & h_{47} & h_{48} & h_{49} \\
h_{51} & h_{52} & h_{53} & h_{54} & h_{55} & h_{56} & h_{57} & h_{58} & h_{59} \\
h_{61} & h_{62} & h_{63} & h_{64} & h_{65} & h_{66} & h_{67} & h_{68} & h_{69} \\
\hline
h_{71} & h_{72} & h_{73} & h_{74} & h_{75} & h_{76} & h_{77} & h_{78} & h_{79} \\
h_{81} & h_{82} & h_{83} & h_{84} & h_{85} & h_{86} & h_{87} & h_{88} & h_{89} \\
h_{91} & h_{92} & h_{93} & h_{94} & h_{95} & h_{96} & h_{97} & h_{98} & h_{99}
\end{array}\right)
$$

The H_{23} submatrix is:

$$
H_{23} = \begin{pmatrix}
h_{47} & h_{48} & h_{49} \\
h_{57} & h_{58} & h_{59} \\
h_{67} & h_{68} & h_{69}
\end{pmatrix}
$$

We shall show that this is given by the following expression:

$$
N_2^T H N_3 = \begin{pmatrix}
0 & 0 & 0 & 1 & 0 & 0 & 0 & 0 & 0 \\
0 & 0 & 0 & 0 & 1 & 0 & 0 & 0 & 0 \\
0 & 0 & 0 & 0 & 0 & 1 & 0 & 0 & 0
\end{pmatrix}
$$

$$
\begin{pmatrix}
h_{11} & h_{12} & h_{13} & h_{14} & h_{15} & h_{16} & h_{17} & h_{18} & h_{19} \\
h_{21} & h_{22} & h_{23} & h_{24} & h_{25} & h_{26} & h_{27} & h_{28} & h_{29} \\
h_{31} & h_{32} & h_{33} & h_{34} & h_{35} & h_{36} & h_{37} & h_{38} & h_{39} \\
h_{41} & h_{42} & h_{43} & h_{44} & h_{45} & h_{46} & h_{47} & h_{48} & h_{49} \\
h_{51} & h_{52} & h_{53} & h_{54} & h_{55} & h_{56} & h_{57} & h_{58} & h_{59} \\
h_{61} & h_{62} & h_{63} & h_{64} & h_{65} & h_{66} & h_{67} & h_{68} & h_{69} \\
h_{71} & h_{72} & h_{73} & h_{74} & h_{75} & h_{76} & h_{77} & h_{78} & h_{79} \\
h_{81} & h_{82} & h_{83} & h_{84} & h_{85} & h_{86} & h_{87} & h_{88} & h_{89} \\
h_{91} & h_{92} & h_{93} & h_{94} & h_{95} & h_{96} & h_{97} & h_{98} & h_{99}
\end{pmatrix}
\begin{pmatrix}
0 & 0 & 0 \\
0 & 0 & 0 \\
0 & 0 & 0 \\
0 & 0 & 0 \\
0 & 0 & 0 \\
0 & 0 & 0 \\
1 & 0 & 0 \\
0 & 1 & 0 \\
0 & 0 & 1
\end{pmatrix}
$$

$$
= \begin{pmatrix}
0 & 0 & 0 & 1 & 0 & 0 & 0 & 0 & 0 \\
0 & 0 & 0 & 0 & 1 & 0 & 0 & 0 & 0 \\
0 & 0 & 0 & 0 & 0 & 1 & 0 & 0 & 0
\end{pmatrix}
\begin{pmatrix}
h_{17} & h_{18} & h_{19} \\
h_{27} & h_{28} & h_{29} \\
h_{37} & h_{38} & h_{39} \\
h_{47} & h_{48} & h_{49} \\
h_{57} & h_{58} & h_{59} \\
h_{67} & h_{68} & h_{69} \\
h_{77} & h_{78} & h_{79} \\
h_{87} & h_{88} & h_{89} \\
h_{97} & h_{98} & h_{99}
\end{pmatrix}
$$

$$
= \begin{pmatrix}
h_{47} & h_{48} & h_{49} \\
h_{57} & h_{58} & h_{59} \\
h_{67} & h_{68} & h_{69}
\end{pmatrix} = H_{23}.
$$

What is the "take home" message of this chapter?

Under the assumption that the operator with which we manipulate an image is *linear* and *separable*, this operation can be expressed by an equation of the form:

$$g = h_c^T f h_r \qquad (1.45)$$

where f and g are the input and output images respectively and h_c and h_r are matrices expressing the point spread function of the operator.

What is the purpose of Image Processing?

The purpose of Image Processing is to solve the following four problems:

- Given an image f *choose* matrices h_c and h_r so that the output image g is "better" than f according to some subjective criteria. This is the problem of *Image Enhancement*.

- Given an image f *choose* matrices h_c and h_r so that g can be represented by fewer bits than f without much loss of detail. This is the problem of *Image Compression*.

- Given an image g **and** an estimate of $h(x, \alpha, y, \beta)$, recover image f. This is the problem of *Image Restoration*.

- Given an image f *choose* matrices h_c and h_r so that output image g salienates certain features of f. This is the problem of *preparation of an image for Automatic Vision*.

Figures 1.4 and 1.5 show examples of these processes.

What is this book about?

This book is about introducing the Mathematical foundations of Image Processing in the context of specific applications in the four main themes of Image Processing as identified above. The themes of Image Enhancement, Image Restoration and preparation for Automatic Vision will be discussed in detail. The theme of Image Compression is only touched upon as this could be the topic of a whole book on its own. The book mainly concentrates on linear methods, but several non-linear techniques relevant to Image Enhancement and preparation for Automatic Vision will also be presented.

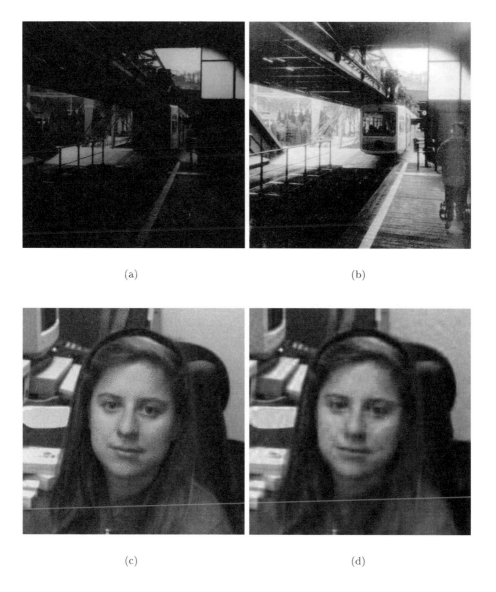

(a)   (b)

(c)   (d)

Figure 1.4: Top row: An original image and its enhanced version. Bottom row: An original image and a version of it represented by 40% of the original number of bits.

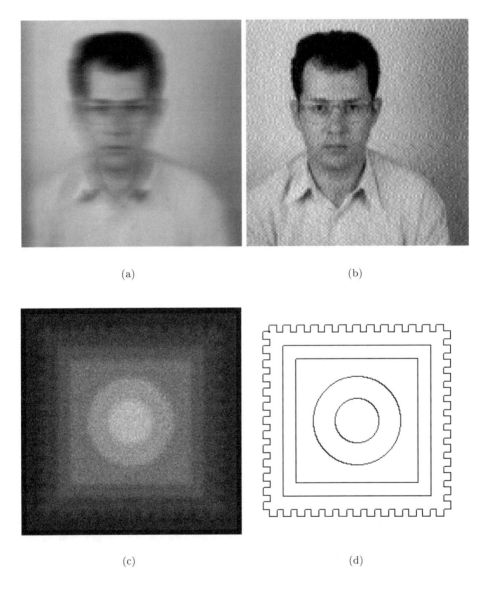

(a)   (b)

(c)   (d)

Figure 1.5: Top row: A blurred original image and its restored version. Bottom row: An original image and its edge map (indicating locations where the brightness of the image changes abruptly).

Chapter 2

Image Transformations

What is this chapter about?

This chapter is concerned with the development of some of the most important tools of linear Image Processing, namely the ways by which we express an image as the linear superposition of some elementary images.

How can we define an elementary image?

We can define an elementary image as the *outer product* of two vectors.

What is the outer product of two vectors?

Consider two vectors $N \times 1$:

$$\mathbf{u_i}^T = (u_{i1}, u_{i2}, \ldots, u_{iN})$$
$$\mathbf{v_j}^T = (v_{j1}, v_{j2}, \ldots, v_{jN})$$

Their outer product is defined as:

$$\mathbf{u_i v_j}^T = \begin{pmatrix} u_{i1} \\ u_{i2} \\ \vdots \\ u_{iN} \end{pmatrix} \begin{pmatrix} v_{j1} & v_{j2} & \ldots v_{jN} \end{pmatrix} = \begin{pmatrix} u_{i1}v_{j1} & u_{i1}v_{j2} & \cdots & u_{i1}v_{jN} \\ u_{i2}v_{j1} & u_{i2}v_{j2} & \cdots & u_{i2}v_{jN} \\ \vdots & \vdots & & \vdots \\ u_{iN}v_{j1} & u_{iN}v_{j2} & \cdots & u_{iN}v_{jN} \end{pmatrix} \tag{2.1}$$

Therefore, the outer product of these two vectors is an $N \times N$ matrix which can be thought of as an image.

How can we expand an image in terms of vector outer products?

We saw in the previous chapter that a general separable linear transformation of an image matrix f can be written as:

$$g = h_c^T f h_r \tag{2.2}$$

where g is the output image and h_c and h_r are the transforming matrices.

We can use the inverse matrices of h_c^T and h_r to solve this expression for f in terms of g as follows: Multiply both sides of the equation with $(h_c^T)^{-1}$ on the left and h_r^{-1} on the right:

$$(h_c^T)^{-1} g h_r^{-1} = (h_c^T)^{-1} h_c^T f h_r h_r^{-1} = f \tag{2.3}$$

Thus we write:

$$f = (h_c^T)^{-1} g h_r^{-1} \tag{2.4}$$

Suppose that we partition matrices $(h_c^T)^{-1}$ and h_r^{-1} in their column and row vectors respectively:

$$[\mathbf{u_1}|\mathbf{u_2}|\ldots|\mathbf{u_N}] , \qquad \begin{pmatrix} \mathbf{v}_1^T \\ -- \\ \mathbf{v}_2^T \\ -- \\ \vdots \\ -- \\ \mathbf{v}_N^T \end{pmatrix} \tag{2.5}$$

Then

$$f = (\, \mathbf{u_1} \quad \mathbf{u_2} \quad \ldots \quad \mathbf{u_N} \,) \, g \begin{pmatrix} \mathbf{v}_1^T \\ \mathbf{v}_2^T \\ \vdots \\ \mathbf{v}_N^T \end{pmatrix} \tag{2.6}$$

We may also write matrix g as a sum of N^2, $N \times N$ matrices, each one having only one non-zero element:

$$g = \begin{pmatrix} g_{11} & 0 & \ldots & 0 \\ 0 & 0 & \ldots & 0 \\ \vdots & \vdots & \ddots & \vdots \\ 0 & 0 & \ldots & 0 \end{pmatrix} + \begin{pmatrix} 0 & g_{12} & \ldots & 0 \\ 0 & 0 & \ldots & 0 \\ \vdots & \vdots & \ddots & \vdots \\ 0 & 0 & \ldots & 0 \end{pmatrix} + \ldots + \begin{pmatrix} 0 & 0 & \ldots & 0 \\ 0 & 0 & \ldots & 0 \\ \vdots & \vdots & \ddots & \vdots \\ 0 & 0 & \ldots & g_{NN} \end{pmatrix} \tag{2.7}$$

Then equation (2.6) can be written as:

$$f = \sum_{i=1}^{N} \sum_{j=1}^{N} g_{ij} \mathbf{u_i} \mathbf{v_j}^T \tag{2.8}$$

This is an expansion of image f in terms of vector outer products. The outer product $\mathbf{u_i} \mathbf{v_j}^T$ may be interpreted as an "image" so that the sum over all combinations of the outer products, appropriately weighted by the g_{ij} coefficients, represents the original image f.

Example 2.1

Derive the term $i = 2$, $j = 1$ in the right hand side of equation (2.8).

If we substitute g from equation (2.7) into equation (2.6), the right hand side of equation (2.6) will consist of N^2 terms of similar form. One such term is:

$$
\begin{pmatrix} \mathbf{u_1} & \mathbf{u_2} & \cdots & \mathbf{u_N} \end{pmatrix}
\begin{pmatrix} 0 & 0 & \cdots & 0 \\ g_{21} & 0 & \cdots & 0 \\ \vdots & \vdots & & \vdots \\ 0 & 0 & \cdots & 0 \end{pmatrix}
\begin{pmatrix} \mathbf{v_1}^T \\ \mathbf{v_2}^T \\ \vdots \\ \mathbf{v_N}^T \end{pmatrix}
$$

$$
= \begin{pmatrix} \mathbf{u_1} & \mathbf{u_2} & \cdots & \mathbf{u_N} \end{pmatrix}
\begin{pmatrix} 0 & 0 & \cdots & 0 \\ g_{21} & 0 & \cdots & 0 \\ \vdots & \vdots & & \vdots \\ 0 & 0 & \cdots & 0 \end{pmatrix}
\begin{pmatrix} v_{11} & v_{12} & \cdots & v_{1N} \\ v_{21} & v_{22} & \cdots & v_{2N} \\ \vdots & \vdots & & \vdots \\ v_{N1} & v_{N2} & \cdots & v_{NN} \end{pmatrix}
$$

$$
= \begin{pmatrix} u_{11} & u_{21} & \cdots & u_{N1} \\ u_{12} & u_{22} & \cdots & u_{N2} \\ \vdots & \vdots & & \vdots \\ u_{1N} & u_{2N} & \cdots & u_{NN} \end{pmatrix}
\begin{pmatrix} 0 & 0 & \cdots & 0 \\ g_{21}v_{11} & g_{21}v_{12} & \cdots & g_{21}v_{1N} \\ \vdots & \vdots & & \vdots \\ 0 & 0 & \cdots & 0 \end{pmatrix}
$$

$$
= \begin{pmatrix} u_{21}g_{21}v_{11} & u_{21}g_{21}v_{12} & \cdots & u_{21}g_{21}v_{1N} \\ u_{22}g_{21}v_{11} & u_{22}g_{21}v_{12} & \cdots & u_{22}g_{21}v_{1N} \\ \cdots & \cdots & & \cdots \\ u_{2N}g_{21}v_{11} & u_{2N}g_{21}v_{12} & \cdots & u_{2N}g_{21}v_{1N} \end{pmatrix}
$$

$$
= g_{21} \begin{pmatrix} u_{21}v_{11} & u_{21}v_{12} & \cdots & u_{21}v_{1N} \\ u_{22}v_{11} & u_{22}v_{12} & \cdots & u_{22}v_{1N} \\ \cdots & \cdots & & \cdots \\ u_{2N}v_{11} & u_{2N}v_{12} & \cdots & u_{2N}v_{1N} \end{pmatrix} = g_{21}\mathbf{u_2}\mathbf{v_1}^T
$$

What is a unitary transform?

If matrices h_c and h_r are chosen to be *unitary*, equation (2.2) represents a *unitary transform* of f, and g is termed the *unitary transform domain* of image f.

What is a unitary matrix?

A matrix U is called *unitary* if its inverse is the complex conjugate of its transpose, i.e.

$$UU^{T*} = I \qquad (2.9)$$

where I is the unit matrix. We often write superscript "H" instead of "$T*$".

If the elements of the matrix are real numbers, we use the term *orthogonal* instead of unitary.

What is the inverse of a unitary transform?

If matrices h_c and h_r in (2.2) are unitary, then the inverse of it is:

$$f = h_c g h_r^H$$

For simplicity, from now on we shall write U instead of h_c and V instead of h_r, so that the expansion of an image f in terms of vector outer products can be written as:

$$f = U g V^H \tag{2.10}$$

How can we construct a unitary matrix?

If we consider equation (2.9) we see that for the matrix U to be unitary the requirement is that the dot product of any two of its columns must be zero while the magnitude of any of its column vectors must be 1. In other words, U is unitary if its columns form a set of *orthonormal* vectors.

How should we choose matrices U and V so that g can be represented by fewer bits than f?

If we wanted to represent image f with fewer than N^2 number of elements, then we could choose matrices U and V so that the transformed image g was a diagonal matrix. Then we could represent image f with the help of equation (2.8) using only the N non-zero elements of g. This can be achieved with a process called *matrix diagonalization*, and it is called *Singular Value Decomposition* (SVD) of the image.

How can we diagonalize a matrix?

It can be shown (see box B2.1) that a matrix g of rank r can be written as:

$$g \quad = \quad U \Lambda^{\frac{1}{2}} V^T \tag{2.11}$$

where U and V are orthogonal matrices of size $N \times r$ and $\Lambda^{\frac{1}{2}}$ is a diagonal $r \times r$ matrix.

Example 2.2

If Λ is a diagonal 2×2 matrix and Λ^m is defined by putting all non-zero elements of Λ to the power of m, show that:

$$\Lambda^{-\frac{1}{2}}\Lambda\Lambda^{-\frac{1}{2}} = I \quad \text{and} \quad \Lambda^{-\frac{1}{2}}\Lambda^{\frac{1}{2}} = I.$$

Indeed,

$$\Lambda^{-\frac{1}{2}}\Lambda\Lambda^{-\frac{1}{2}} = \begin{pmatrix} \lambda_1^{-\frac{1}{2}} & 0 \\ 0 & \lambda_2^{-\frac{1}{2}} \end{pmatrix} \begin{pmatrix} \lambda_1 & 0 \\ 0 & \lambda_2 \end{pmatrix} \begin{pmatrix} \lambda_1^{-\frac{1}{2}} & 0 \\ 0 & \lambda_2^{-\frac{1}{2}} \end{pmatrix}$$

$$= \begin{pmatrix} \lambda_1^{-\frac{1}{2}} & 0 \\ 0 & \lambda_2^{-\frac{1}{2}} \end{pmatrix} \begin{pmatrix} \lambda_1^{\frac{1}{2}} & 0 \\ 0 & \lambda_2^{\frac{1}{2}} \end{pmatrix}$$

$$= \begin{pmatrix} 1 & 0 \\ 0 & 1 \end{pmatrix}$$

This also shows that $\Lambda^{-\frac{1}{2}}\Lambda^{\frac{1}{2}} = I$.

Example 2.3 (B)

Assume that H is a 3×3 matrix and partition it in a 2×3 submatrix H_1 and a 1×3 submatrix H_2. Show that:

$$H^T H = H_1^T H_1 + H_2^T H_2.$$

Let us say that

$$H_1 = \begin{pmatrix} h_{11} & h_{12} & h_{13} \\ h_{21} & h_{22} & h_{23} \end{pmatrix} \quad \text{and} \quad H_2 = \begin{pmatrix} \tilde{h}_{31} & \tilde{h}_{32} & \tilde{h}_{33} \end{pmatrix}$$

$$H = \begin{pmatrix} h_{11} & h_{12} & h_{13} \\ h_{21} & h_{22} & h_{23} \\ \tilde{h}_{31} & \tilde{h}_{32} & \tilde{h}_{33} \end{pmatrix} \quad \text{and} \quad H^T = \begin{pmatrix} h_{11} & h_{21} & \tilde{h}_{31} \\ h_{12} & h_{22} & \tilde{h}_{32} \\ h_{13} & h_{23} & \tilde{h}_{33} \end{pmatrix} = \begin{pmatrix} H_1^T & | & H_2^T \end{pmatrix}$$

Then:

$$H^T H = \begin{pmatrix} h_{11} & h_{21} & \tilde{h}_{31} \\ h_{12} & h_{22} & \tilde{h}_{32} \\ h_{13} & h_{23} & \tilde{h}_{33} \end{pmatrix} \begin{pmatrix} h_{11} & h_{12} & h_{13} \\ h_{21} & h_{22} & h_{23} \\ \tilde{h}_{31} & \tilde{h}_{32} & \tilde{h}_{33} \end{pmatrix} =$$

$$\begin{pmatrix} h_{11}^2 + h_{21}^2 + \tilde{h}_{31}^2 & h_{11}h_{12} + h_{21}h_{22} + \tilde{h}_{31}\tilde{h}_{32} & h_{11}h_{13} + h_{21}h_{23} + \tilde{h}_{31}\tilde{h}_{33} \\ h_{12}h_{11} + h_{22}h_{21} + \tilde{h}_{32}\tilde{h}_{31} & h_{12}^2 + h_{22}^2 + \tilde{h}_{32}^2 & h_{12}h_{13} + h_{22}h_{23} + \tilde{h}_{32}\tilde{h}_{33} \\ h_{13}h_{11} + h_{23}h_{21} + \tilde{h}_{33}\tilde{h}_{31} & h_{13}h_{12} + h_{23}h_{22} + \tilde{h}_{33}\tilde{h}_{32} & h_{13}^2 + h_{23}^2 + \tilde{h}_{33}^2 \end{pmatrix}$$

We shall show that this is equal to $H_1^T H_1 + H_2^T H_2$ by calculating this sum explicitly:

$$H_1^T H_1 = \begin{pmatrix} h_{11} & h_{21} \\ h_{12} & h_{22} \\ h_{13} & h_{23} \end{pmatrix} \begin{pmatrix} h_{11} & h_{12} & h_{13} \\ h_{21} & h_{22} & h_{23} \end{pmatrix}$$

$$= \begin{pmatrix} h_{11}^2 + h_{21}^2 & h_{11}h_{12} + h_{21}h_{22} & h_{11}h_{13} + h_{21}h_{23} \\ h_{12}h_{11} + h_{22}h_{21} & h_{12}^2 + h_{22}^2 & h_{13}h_{12} + h_{22}h_{23} \\ h_{13}h_{11} + h_{21}h_{23} & h_{13}h_{12} + h_{23}h_{22} & h_{13}^2 + h_{23}^2 \end{pmatrix}$$

$$H_2^T H_2 = \begin{pmatrix} \tilde{h}_{31} \\ \tilde{h}_{32} \\ \tilde{h}_{33} \end{pmatrix} \begin{pmatrix} \tilde{h}_{31} & \tilde{h}_{32} & \tilde{h}_{33} \end{pmatrix} = \begin{pmatrix} \tilde{h}_{31}^2 & \tilde{h}_{31}\tilde{h}_{32} & \tilde{h}_{31}\tilde{h}_{33} \\ \tilde{h}_{32}\tilde{h}_{31} & \tilde{h}_{32}^2 & \tilde{h}_{32}\tilde{h}_{33} \\ \tilde{h}_{33}\tilde{h}_{31} & \tilde{h}_{33}\tilde{h}_{32} & \tilde{h}_{33}^2 \end{pmatrix}$$

Adding $H_1^T H_1$ and $H_2^T H_2$ we obtain the same answer as before.

Example 2.4

You are given an image which is represented by a matrix g. Show that matrix gg^T is symmetric.

A matrix is symmetric when it is equal to its transpose. Consider the transpose of gg^T:

$$\left(gg^T\right)^T = \left(g^T\right)^T g^T = gg^T$$

Example 2.5 (B)

Show that if we partition an $N \times N$ matrix S into an $r \times N$ submatrix S_1 and an $(N - r) \times N$ submatrix S_2, the following holds:

$$SAS^T = \left(\begin{array}{c|c} S_1 A S_1^T & S_1 A S_2^T \\ \hline S_2 A S_1^T & S_2 A S_2^T \end{array} \right)$$

where A is an $N \times N$ matrix.

$$SAS^T = \left(\begin{array}{c} S_1 \\ -- \\ S_2 \end{array} \right) A \, (\, S_1^T \quad | \quad S_2^T \,)$$

Consider the multiplication of A with $(S_1^T \quad | \quad S_2^T)$. The rows of A will be multiplied with the columns of $(S_1^T \quad | \quad S_2^T)$. Schematically:

$$\left(\begin{array}{c} \cdots \cdots \\ \cdots \cdots \end{array} \right) \left(\begin{array}{cccccc} : & : & : & | & : & : \\ : & : & : & | & : & : \\ : & : & : & | & : & : \end{array} \right)$$

Then it becomes clear that the result will be $(AS_1^T \quad | \quad AS_2^T)$. Next we consider the multiplication of $\left(\begin{array}{c} S_1 \\ --- \\ S_2 \end{array} \right)$ with $(AS_1^T \quad | \quad AS_2^T)$. The rows of $\left(\begin{array}{c} S_1 \\ --- \\ S_2 \end{array} \right)$ will multiply the columns of $(AS_1^T \quad | \quad AS_2^T)$. Schematically:

$$\left(\begin{array}{c} \cdots \cdots \\ \cdots \cdots \\ ----- \\ \cdots \cdots \\ \cdots \cdots \end{array} \right) \left(\begin{array}{cccccc} : & : & : & | & : & : \\ : & : & : & | & : & : \\ : & : & : & | & : & : \\ : & : & : & | & : & : \\ : & : & : & | & : & : \\ : & : & : & | & : & : \end{array} \right)$$

Then the result is obvious.

Example 2.6 (B)

Show that if $Agg^T A^T = 0$ then $Ag = 0$, where A and g are an $r \times N$ and an $N \times N$ real matrix respectively.

We may write

$$Agg^T A^T = Ag(Ag)^T = 0$$

Ag is an $N \times N$ matrix. Let us call it B. We have therefore, $BB^T = 0$:

$$
\begin{pmatrix} b_{11} & b_{12} & \cdots & b_{1N} \\ b_{21} & b_{22} & \cdots & b_{2N} \\ \cdots & \cdots & & \cdots \\ b_{N1} & b_{N2} & \cdots & b_{NN} \end{pmatrix} \begin{pmatrix} b_{11} & b_{21} & \cdots & b_{N1} \\ b_{12} & b_{22} & \cdots & b_{N2} \\ \cdots & \cdots & & \cdots \\ b_{1N} & b_{2N} & \cdots & b_{NN} \end{pmatrix} = \begin{pmatrix} 0 & 0 & \cdots & 0 \\ 0 & 0 & \cdots & 0 \\ \cdots & \cdots & & \cdots \\ 0 & 0 & \cdots & 0 \end{pmatrix} \Rightarrow
$$

$$
\begin{pmatrix} b_{11}^2 + b_{12}^2 + \ldots + b_{1N}^2 & \cdots & & \cdots \\ & b_{21}^2 + b_{22}^2 + \ldots + b_{2N}^2 & \cdots & \cdots \\ \cdots & \cdots & \cdots & \cdots \\ \cdots & \cdots & \cdots & b_{N1}^2 + b_{N2}^2 + \ldots + b_{NN}^2 \end{pmatrix}
$$

$$
= \begin{pmatrix} 0 & 0 & \cdots & 0 \\ 0 & 0 & \cdots & 0 \\ \cdots & \cdots & & \cdots \\ 0 & 0 & \cdots & 0 \end{pmatrix}
$$

Equating the corresponding elements, we obtain, for example:

$$
b_{11}^2 + b_{12}^2 + \ldots + b_{1N}^2 = 0
$$

The only way that the sum of the squares of N real numbers can be 0 is if each one of them is 0. Similarly for all other diagonal elements of BB^T. Then the result follows.

B2.1: Can we diagonalize ANY image?

Yes. Consider an image g and its transpose g^T. gg^T is a real and symmetric matrix (see Example 2.4) and let us say that it has r non-zero eigenvectors. Let λ_i be its ith eigenvalue. Then it is known that there exists an orthogonal matrix S (made up from the eigenvectors of gg^T) such that:

$$
Sgg^T S^T = \left(\begin{array}{cccc|cccc} \lambda_1 & 0 & \cdots & 0 & 0 & 0 & \cdots & 0 \\ 0 & \lambda_2 & \cdots & 0 & 0 & 0 & \cdots & 0 \\ \vdots & \vdots & & \vdots & \vdots & \vdots & & \vdots \\ 0 & 0 & \cdots & \lambda_r & 0 & 0 & \cdots & 0 \\ \hline 0 & 0 & \cdots & 0 & 0 & 0 & \cdots & 0 \\ 0 & 0 & \cdots & 0 & 0 & 0 & \cdots & 0 \\ \vdots & \vdots & & \vdots & \vdots & \vdots & & \vdots \\ 0 & 0 & \cdots & 0 & 0 & 0 & \cdots & 0 \end{array} \right)
$$

$$
= \left(\begin{array}{c|c} \Lambda & 0 \\ \hline 0 & 0 \end{array} \right) \tag{2.12}
$$

where Λ and 0 represent the partitions of the diagonal matrix above. Similarly we can partition matrix S to an $r \times N$ matrix S_1 and an $(N - r) \times N$ matrix S_2:

$$
S = \begin{pmatrix} S_1 \\ - \\ S_2 \end{pmatrix}
$$

Because S is orthogonal, and using the result of Example 2.3, we have:

$$
S^T S = I \Rightarrow S_1^T S_1 + S_2^T S_2 = I \Rightarrow
$$
$$
S_1^T S_1 = I - S_2^T S_2 \Rightarrow S_1^T S_1 g = g - S_2^T S_2 g \tag{2.13}
$$

From (2.12) and Examples 2.5 and 2.6 we clearly have:

$$
S_1 g g^T S_1^T = \Lambda \tag{2.14}
$$
$$
S_2 g g^T S_2^T = 0 \Rightarrow S_2 g = 0 \tag{2.15}
$$

Using (2.15) in (2.13) we have:

$$
S_1^T S_1 g = g \tag{2.16}
$$

This means that $S_1^T S_1 = I$, i.e. S_1 is an orthogonal matrix. We multiply both sides of equation (2.14) from left and right by $\Lambda^{-\frac{1}{2}}$ to get:

$$
\Lambda^{-\frac{1}{2}} S_1 g g^T S_1^T \Lambda^{-\frac{1}{2}} = \Lambda^{-\frac{1}{2}} \Lambda \Lambda^{-\frac{1}{2}} = I \tag{2.17}
$$

Since $\Lambda^{-\frac{1}{2}}$ is diagonal, $\Lambda^{-\frac{1}{2}} = \left(\Lambda^{-\frac{1}{2}}\right)^T$. So the above equation can be rewritten as:

$$
\Lambda^{-\frac{1}{2}} S_1 g \left(\Lambda^{-\frac{1}{2}} S_1 g\right)^T = I \tag{2.18}
$$

Therefore, there exists a matrix $q \equiv \Lambda^{-\frac{1}{2}} S_1 g$ whose inverse is its transpose (i.e. it is orthogonal). We can express matrix $S_1 g$ as $\Lambda^{\frac{1}{2}} q$ and substitute in (2.16) to get:

$$
S_1^T \Lambda^{\frac{1}{2}} q = g \quad \text{or} \quad g = S_1^T \Lambda^{\frac{1}{2}} q \tag{2.19}
$$

In other words, g is expressed as a diagonal matrix $\Lambda^{\frac{1}{2}}$ made up from the square roots of the non-zero eigenvalues of gg^T, multiplied from left and right by the two orthogonal matrices S_1 and q. This result expresses the diagonalization of image g.

How can we compute matrices U, V and $\Lambda^{\frac{1}{2}}$ needed for the image diagonalization?

If we take the transpose of (2.11) we have:

$$g^T = V\Lambda^{\frac{1}{2}}U^T \tag{2.20}$$

Multiply (2.11) by (2.20) to obtain:

$$gg^T = U\Lambda^{\frac{1}{2}}V^TV\Lambda^{\frac{1}{2}}U^T = U\Lambda^{\frac{1}{2}}\Lambda^{\frac{1}{2}}U^T \Rightarrow gg^T = U\Lambda U^T \tag{2.21}$$

This shows that matrix Λ consists of the r non-zero eigenvalues of matrix gg^T while U is made up from the eigenvectors of the same matrix.

Similarly, if we multiply (2.20) by (2.11) we get:

$$g^Tg = V\Lambda V^T \tag{2.22}$$

This shows that matrix V is made up from the eigenvectors of matrix g^Tg.

B2.2 What happens if the eigenvalues of matrix gg^T are negative?

We shall show that the eigenvalues of gg^T are always non-negative numbers. Let us assume that λ is an eigenvalue of matrix gg^T and $\mathbf{u}$ is the corresponding eigenvector. We have then:

$$gg^T\mathbf{u} = \lambda\mathbf{u}$$

Multiply both sides with $\mathbf{u}^T$ from the left:

$$\mathbf{u}^Tgg^T\mathbf{u} = \mathbf{u}^T\lambda\mathbf{u}$$

λ is a scalar and can change position on the right hand side of the equation. Also, because of the associativity of matrix multiplication, we can write:

$$(\mathbf{u}^Tg)(g^T\mathbf{u}) = \lambda\mathbf{u}^T\mathbf{u}$$

Since $\mathbf{u}$ is an eigenvector, $\mathbf{u}^T\mathbf{u} = 1$. Therefore:

$$\left(g^T\mathbf{u}\right)^T(g^T\mathbf{u}) = \lambda$$

$g^T\mathbf{u}$ is some vector $\mathbf{y}$. Then we have: $\lambda = \mathbf{y}^T\mathbf{y} \geq 0$ since $\mathbf{y}^T\mathbf{y}$ is the square magnitude of vector $\mathbf{y}$.

Example 2.7

If λ_i are the eigenvalues of gg^T and $\mathbf{u_i}$ the corresponding eigenvectors, show that $g^T g$ has the same eigenvalues with the corresponding eigenvectors given by: $\mathbf{v_i} = g^T \mathbf{u_i}$.

By definition

$$gg^T \mathbf{u_i} = \lambda_i \mathbf{u_i}$$

Multiply both sides on the left by g^T:

$$g^T gg^T \mathbf{u_i} = g^T \lambda_i \mathbf{u_i}$$

λ_i is a scalar and can change position. Also, by the associativity of matrix multiplication:

$$g^T g(g^T \mathbf{u_i}) = \lambda_i (g^T \mathbf{u_i})$$

This identifies $g^T \mathbf{u_i}$ as an eigenvector of $g^T g$ with λ_i the corresponding eigenvalue.

Example 2.8

You are given an image: $g = \begin{pmatrix} 1 & 0 & 0 \\ 2 & 1 & 1 \\ 0 & 0 & 1 \end{pmatrix}$. Compute the eigenvectors $\mathbf{u_i}$ of gg^T and $\mathbf{v_i}$ of $g^T g$.

$$g^T = \begin{pmatrix} 1 & 2 & 0 \\ 0 & 1 & 0 \\ 0 & 1 & 1 \end{pmatrix}$$

Then

$$gg^T = \begin{pmatrix} 1 & 0 & 0 \\ 2 & 1 & 1 \\ 0 & 0 & 1 \end{pmatrix} \begin{pmatrix} 1 & 2 & 0 \\ 0 & 1 & 0 \\ 0 & 1 & 1 \end{pmatrix} = \begin{pmatrix} 1 & 2 & 0 \\ 2 & 6 & 1 \\ 0 & 1 & 1 \end{pmatrix}$$

The eigenvalues of gg^T will be computed from the characteristic equation:

$$\begin{vmatrix} 1-\lambda & 2 & 0 \\ 2 & 6-\lambda & 1 \\ 0 & 1 & 1-\lambda \end{vmatrix} = 0 \Rightarrow (1-\lambda)[(6-\lambda)(1-\lambda) - 1] - 2[2(1-\lambda)] = 0$$

$$\Rightarrow (1-\lambda)[(6-\lambda)(1-\lambda) - 1 - 4] = 0$$

One eigenvalue is $\lambda = 1$. *The other two are the roots of:*

$$6 - 6\lambda - \lambda + \lambda^2 - 5 = 0 \Rightarrow \lambda^2 - 7\lambda + 1 = 0 \Rightarrow \lambda = \frac{7 \pm \sqrt{49 - 4}}{2} = \frac{7 \pm 6.7}{2}$$

$$\lambda = 6.85 \ or \ \lambda = 0.146$$

In descending order the eigenvalues are:

$$\lambda_1 = 6.85, \ \lambda_2 = 1, \ \lambda_3 = 0.146$$

Let $\mathbf{u_i} = \begin{pmatrix} x_1 \\ x_2 \\ x_3 \end{pmatrix}$ *be the eigenvector which corresponds to eigenvalue* λ_i. *Then:*

$$\begin{pmatrix} 1 & 2 & 0 \\ 2 & 6 & 1 \\ 0 & 1 & 1 \end{pmatrix} \begin{pmatrix} x_1 \\ x_2 \\ x_3 \end{pmatrix} = \lambda_i \begin{pmatrix} x_1 \\ x_2 \\ x_3 \end{pmatrix} \Rightarrow \begin{array}{l} x_1 + 2x_2 = \lambda_i x_1 \\ 2x_1 + 6x_2 + x_3 = \lambda_i x_2 \\ x_2 + x_3 = \lambda_i x_3 \end{array}$$

For $\lambda_i = 6.85$

$$2x_2 - 5.85x_1 = 0 \tag{2.23}$$
$$2x_1 - 0.85x_2 + x_3 = 0 \tag{2.24}$$
$$x_2 - 5.85x_3 = 0 \tag{2.25}$$

Multiply (2.24) by 5.85 and add equation (2.25) to get:

$$11.7x_1 - 4x_2 = 0 \tag{2.26}$$

Equation (2.26) is the same as (2.23). So we have really only two independent equations for the three unknowns. We choose the value of x_1 *to be 1. Then*

$$x_2 = 2.927 \ and \ from \ (2.24) \ x_3 = -2 + 0.85 \times 2.925 = -2 + 2.5 = 0.5$$

Thus the first eigenvector is

$$\begin{pmatrix} 1 \\ 2.927 \\ 0.5 \end{pmatrix}$$

and after normalization, i.e. division by $\sqrt{1^2 + 2.927^2 + 0.5^2} = 3.133$, *we get:*

$$\mathbf{u_1} = \begin{pmatrix} 0.319 \\ 0.934 \\ 0.160 \end{pmatrix}$$

For $\lambda_i = 1$, *the system of linear equations we have to solve is:*

$$x_1 + 2x_2 = x_1 \Rightarrow x_2 = 0$$
$$2x_1 + x_3 = 0 \Rightarrow x_3 = -2x_1$$

Choose $x_1 = 1$. Then $x_3 = -2$. Since $x_2 = 0$, we must divide all components by $\sqrt{1^2 + 2^2} = \sqrt{5}$ for the eigenvector to have unit length:

$$\mathbf{u_2} = \begin{pmatrix} 0.447 \\ 0 \\ -0.894 \end{pmatrix}$$

For $\lambda_i = 0.146$, the system of linear equations we have to solve is:

$$0.854x_1 + 2x_2 = 0$$
$$2x_1 + 5.854x_2 + x_3 = 0$$
$$x_2 + 0.854x_3 = 0$$

Choose $x_1 = 1$. Then $x_2 = -\frac{0.854}{2} = -0.427$ and $x_3 = -\frac{0.427}{0.854} = 0.5$. Therefore, the third eigenvector is:

$$\begin{pmatrix} 1 \\ -0.427 \\ 0.5 \end{pmatrix},$$

and after division by $\sqrt{1 + 0.427^2 + 0.5^2} = 1.197$ we get:

$$\mathbf{u_3} = \begin{pmatrix} 0.835 \\ -0.357 \\ 0.418 \end{pmatrix}$$

The corresponding eigenvectors of $g^T g$ are given by $g^T \mathbf{u_i}$; i.e. the first one is:

$$\begin{pmatrix} 1 & 2 & 0 \\ 0 & 1 & 0 \\ 0 & 1 & 1 \end{pmatrix} \begin{pmatrix} 0.319 \\ 0.934 \\ 0.160 \end{pmatrix} = \begin{pmatrix} 2.187 \\ 0.934 \\ 1.094 \end{pmatrix}$$

We normalize it by dividing by $\sqrt{2.187^2 + 0.934^2 + 1.094^2} = 2.618$, to get:

$$\mathbf{v_1} = \begin{pmatrix} 0.835 \\ 0.357 \\ 0.418 \end{pmatrix}$$

Similarly

$$\mathbf{v_2} = \begin{pmatrix} 1 & 2 & 0 \\ 0 & 1 & 0 \\ 0 & 1 & 1 \end{pmatrix} \begin{pmatrix} 0.447 \\ 0 \\ -0.894 \end{pmatrix} = \begin{pmatrix} 0.447 \\ 0 \\ -0.894 \end{pmatrix},$$

while the third eigenvector is

$$
\begin{pmatrix} 1 & 2 & 0 \\ 0 & 1 & 0 \\ 0 & 1 & 1 \end{pmatrix} \begin{pmatrix} 0.835 \\ -0.357 \\ 0.418 \end{pmatrix} = \begin{pmatrix} 0.121 \\ -0.357 \\ 0.061 \end{pmatrix}
$$

which after normalization becomes:

$$
\mathbf{v}_3 = \begin{pmatrix} 0.319 \\ -0.934 \\ 0.160 \end{pmatrix}
$$

What is the singular value decomposition of an image?

The Singular Value Decomposition (SVD) of an image g is its expansion in vector outer products where the vectors used are the eigenvectors of gg^T and g^Tg and the coefficients of the expansion are the eigenvalues of these matrices. In that case, equation (2.8), applied for image g instead of image f, can be written as:

$$
g = \sum_{i=1}^{r} \lambda_i^{\frac{1}{2}} \mathbf{u_i} \mathbf{v_i}^T , \tag{2.27}
$$

since the only non-zero terms are those with $i = j$.

How can we approximate an image using SVD?

If in equation (2.27) we decide to keep only $k < r$ terms, we shall reproduce an approximated version of the image as:

$$
g_k = \sum_{i=1}^{k} \lambda_i^{\frac{1}{2}} \mathbf{u_i} \mathbf{v_i}^T \tag{2.28}
$$

Example 2.9

A 256×256 grey image with 256 grey levels is to be transmitted. How many terms can be kept in its SVD before the transmission of the trans-formed image becomes too inefficient in comparison with the transmission of the original image? (Assume that real numbers require 32 bits each.)

Assume that $\lambda_i^{\frac{1}{2}}$ is incorporated into one of the vectors. When we transmit one term of the SVD expansion of the image we must transmit:

$$2 \times 32 \times 256 \; bits.$$

This is because we have to transmit two vectors, and each vector has 256 components in the case of a 256×256 image, and each component requires 32 bits since it is a real number. If we want to transmit the full image, we shall have to transmit $256 \times 256 \times 8$ bits (since each pixel requires 8 bits). Then the maximum number of terms transmitted before the SVD becomes uneconomical is:

$$k = \frac{256 \times 256 \times 8}{2 \times 32 \times 256} = \frac{256}{8} = 32$$

What is the error of the approximation of an image by SVD?

The difference between the original and the approximated image is:

$$D \equiv g - g_k = \sum_{i=k+1}^{r} \lambda_i^{\frac{1}{2}} \mathbf{u_i} \mathbf{v_i}^T \tag{2.29}$$

We can calculate how big this error is by calculating the norm of matrix D, i.e. the sum of the squares of its elements. From (2.29) it is obvious that the mn element of D is:

$$d_{mn} = \sum_{i=k+1}^{r} \lambda_i^{\frac{1}{2}} u_{im} v_{in} \Rightarrow$$

$$d_{mn}^2 = \left(\sum_{i=k+1}^{r} \lambda_i^{\frac{1}{2}} u_{im} v_{in} \right)^2$$

$$= \sum_{i=k+1}^{r} \lambda_i u_{im}^2 v_{in}^2 + 2 \sum_{i=k+1}^{r} \sum_{j=k+1, j \neq i}^{r} \lambda_i^{\frac{1}{2}} \lambda_j^{\frac{1}{2}} u_{im} v_{in} u_{jm} v_{jn}$$

$$\Rightarrow \|D\| = \sum_m \sum_n d_{mn}^2$$

$$= \sum_m \sum_n \sum_{i=k+1}^{r} \lambda_i u_{im}^2 v_{in}^2 + 2 \sum_m \sum_n \sum_{i=k+1}^{r} \sum_{j=k+1, j \neq i}^{r} \lambda_i^{\frac{1}{2}} \lambda_j^{\frac{1}{2}} u_{im} v_{in} u_{jm} v_{jn}$$

$$= \sum_{i=k+1}^{r} \lambda_i \sum_m u_{im}^2 \sum_n v_{in}^2 + 2 \sum_{i=k+1}^{r} \sum_{j=k+1, j \neq i}^{r} \lambda_i^{\frac{1}{2}} \lambda_j^{\frac{1}{2}} \sum_m u_{im} u_{jm} \sum_n v_{in} v_{jn}$$

However, $\mathbf{u_i}$, $\mathbf{v_i}$ are eigenvectors and therefore they form an orthonormal set. So:

$$\sum_m u_{im}^2 = 1, \quad \sum_n v_{in}^2 = 1$$

and $\sum_{n} v_{in}v_{jn} = 0,$ and $\sum_{m} u_{im}u_{jm} = 0$ for $i \neq j$

since $\mathbf{u_i u_j^T} = 0$ and $\mathbf{v_i v_j^T} = 0$ for $i \neq j$. Then

$$||D|| = \sum_{i=k+1}^{r} \lambda_i \qquad (2.30)$$

Therefore, the error of the approximate reconstruction of the image using equation (2.28) is equal to the sum of the omitted eigenvalues.

Example 2.10

For a 3×3 matrix D show that its norm, defined as the trace of $D^T D$, is equal to the sum of the squares of its elements.

Let us assume that:

$$D \equiv \begin{pmatrix} d_{11} & d_{12} & d_{13} \\ d_{21} & d_{22} & d_{23} \\ d_{31} & d_{32} & d_{33} \end{pmatrix}$$

Then

$$D^T D = \begin{pmatrix} d_{11} & d_{21} & d_{31} \\ d_{12} & d_{22} & d_{32} \\ d_{13} & d_{23} & d_{33} \end{pmatrix} \begin{pmatrix} d_{11} & d_{12} & d_{13} \\ d_{21} & d_{22} & d_{23} \\ d_{31} & d_{32} & d_{33} \end{pmatrix} =$$

$$\begin{pmatrix} d_{11}^2 + d_{21}^2 + d_{31}^2 & d_{11}d_{12} + d_{21}d_{22} + d_{31}d_{32} & d_{31}d_{13} + d_{32}d_{23} + d_{31}d_{33} \\ d_{12}d_{11} + d_{22}d_{21} + d_{32}d_{31} & d_{12}^2 + d_{22}^2 + d_{32}^2 & d_{12}d_{13} + d_{22}d_{23} + d_{32}d_{33} \\ d_{13}d_{11} + d_{23}d_{21} + d_{33}d_{31} & d_{13}d_{12} + d_{23}d_{22} + d_{33}d_{32} & d_{13}^2 + d_{23}^2 + d_{33}^2 \end{pmatrix}$$

$$\begin{aligned} trace[D^T D] &= (d_{11}^2 + d_{21}^2 + d_{31}^2) + (d_{12}^2 + d_{22}^2 + d_{32}^2) + (d_{13}^2 + d_{23}^2 + d_{33}^2) \\ &= sum \; of \; all \; terms \; squared. \end{aligned}$$

How can we minimize the error of the reconstruction?

If we arrange the eigenvalues λ_i in decreasing order and truncate the expansion at some integer $k < r$, we approximate the image g by g_k which is the least square error approximation. This is because the sum of the squares of the elements of the difference matrix is minimum, since it is equal to the sum of the unused eigenvalues which have been chosen to be the smallest ones.

Notice that this singular value decomposition of the image is optimal in the *least square error* sense but the base images (eigenimages), with respect to which we expanded the image, are determined by the image itself. (They are determined by the eigenvectors of $g^T g$ and $g g^T$.)

Example 2.11

In the singular value decomposition of the image of Example 2.8 only the first term is kept while the others are set to zero. Verify that the square error of the reconstructed image is equal to the sum of the omitted eigenvalues.

If we keep only the first eigenvalue, the image is approximated by the first eigenimage only, weighted by the square root of the corresponding eigenvalue:

$$g_1 = \sqrt{\lambda_1} \mathbf{u}_1 \mathbf{v}_1^T = \sqrt{6.85} \begin{pmatrix} 0.319 \\ 0.934 \\ 0.160 \end{pmatrix} (0.835 \quad 0.357 \quad 0.418)$$

$$= \begin{pmatrix} 0.835 \\ 2.444 \\ 0.419 \end{pmatrix} (0.835 \quad 0.357 \quad 0.418) = \begin{pmatrix} 0.697 & 0.298 & 0.349 \\ 2.041 & 0.873 & 1.022 \\ 0.350 & 0.150 & 0.175 \end{pmatrix}$$

The error of the reconstruction is given by the difference between g_1 and the original image:

$$g - g_1 = \begin{pmatrix} 0.303 & -0.298 & -0.349 \\ -0.041 & 0.127 & -0.022 \\ -0.350 & -0.150 & 0.825 \end{pmatrix}$$

The sum of the squares of the errors is:

$$0.303^2 + 0.298^2 + 0.349^2 + 0.041^2 + 0.127^2 + 0.022^2 + 0.350^2 + 0.150^2 + 0.825^2$$
$$= 1.146$$

This is exactly equal to the sum of the two omitted eigenvalues λ_2 and λ_3.

What are the elementary images in terms of which SVD expands an image?

There is no specific answer to this because these elementary images are intrinsic to each image; they are its *eigenimages*.

Example 2.12

Perform the singular value decomposition (SVD) of the following image:

$$g = \begin{pmatrix} 1 & 0 & 1 \\ 0 & 1 & 0 \\ 1 & 0 & 1 \end{pmatrix}$$

Thus identify the eigenimages of the above image.

$$gg^T = \begin{pmatrix} 1 & 0 & 1 \\ 0 & 1 & 0 \\ 1 & 0 & 1 \end{pmatrix} \begin{pmatrix} 1 & 0 & 1 \\ 0 & 1 & 0 \\ 1 & 0 & 1 \end{pmatrix} = \begin{pmatrix} 2 & 0 & 2 \\ 0 & 1 & 0 \\ 2 & 0 & 2 \end{pmatrix}$$

The eigenvalues of gg^T are the solutions of:

$$\begin{vmatrix} 2-\lambda & 0 & 2 \\ 0 & 1-\lambda & 0 \\ 2 & 0 & 2-\lambda \end{vmatrix} = 0 \Rightarrow (2-\lambda)^2(1-\lambda) - 4(1-\lambda) = 0 \Rightarrow$$

$$(1-\lambda)(\lambda - 4)\lambda = 0$$

The eigenvalues are: $\lambda_1 = 4$, $\lambda_2 = 1$, $\lambda_3 = 0$. The corresponding eigenvectors are:

$$\begin{pmatrix} 2 & 0 & 2 \\ 0 & 1 & 0 \\ 2 & 0 & 2 \end{pmatrix} \begin{pmatrix} x_1 \\ x_2 \\ x_3 \end{pmatrix} = 4 \begin{pmatrix} x_1 \\ x_2 \\ x_3 \end{pmatrix} \Rightarrow \left. \begin{matrix} 2x_1 + 2x_3 = 4x_1 \\ x_2 = 4x_2 \\ 2x_1 + 2x_3 = 4x_3 \end{matrix} \right| \Rightarrow \begin{matrix} x_1 = x_3 \\ x_2 = 0 \end{matrix}$$

We choose $x_1 = x_3 = \frac{1}{\sqrt{2}}$ so that the eigenvector has unit length. Thus $\mathbf{u}_1^T = \left(\frac{1}{\sqrt{2}} \quad 0 \quad \frac{1}{\sqrt{2}} \right)$.

$$\begin{pmatrix} 2 & 0 & 2 \\ 0 & 1 & 0 \\ 2 & 0 & 2 \end{pmatrix} \begin{pmatrix} x_1 \\ x_2 \\ x_3 \end{pmatrix} = \begin{pmatrix} x_1 \\ x_2 \\ x_3 \end{pmatrix} \Rightarrow \left. \begin{matrix} 2x_1 + 2x_3 = x_1 \\ x_2 = x_2 \\ 2x_1 + 2x_3 = x_3 \end{matrix} \right| \Rightarrow \begin{matrix} x_1 = -2x_3 \\ x_2 = x_2 \end{matrix}$$

We have an extra constraint as this eigenvector must be orthogonal to the first one, that is:

$$\mathbf{u}_1^T \mathbf{u}_2 = 0 \Rightarrow \frac{1}{\sqrt{2}}x_1 + \frac{1}{\sqrt{2}}x_3 = 0 \Rightarrow x_1 = -x_3$$

These equations are satisfied if $x_1 = x_3 = 0$ and x_2 is anything. x_2 is chosen to be 1 so that $\mathbf{u}_2$ has also unit length. Thus: $\mathbf{u}_2^T = (0 \quad 1 \quad 0)$.

Because g is symmetric, $gg^T = g^T g$ and the eigenvectors of gg^T are the same as the eigenvectors of $g^T g$. Then the SVD of g is:

$$g = \sqrt{\lambda_1}\mathbf{u}_1\mathbf{u}_1^T + \sqrt{\lambda_2}\mathbf{u}_2\mathbf{u}_2^T = 2 \begin{pmatrix} \frac{1}{\sqrt{2}} \\ 0 \\ \frac{1}{\sqrt{2}} \end{pmatrix} \begin{pmatrix} \frac{1}{\sqrt{2}} & 0 & \frac{1}{\sqrt{2}} \end{pmatrix} + \begin{pmatrix} 0 \\ 1 \\ 0 \end{pmatrix} \begin{pmatrix} 0 & 1 & 0 \end{pmatrix}$$

$$= \begin{pmatrix} 1 & 0 & 1 \\ 0 & 0 & 0 \\ 1 & 0 & 1 \end{pmatrix} + \begin{pmatrix} 0 & 0 & 0 \\ 0 & 1 & 0 \\ 0 & 0 & 0 \end{pmatrix}$$

These two matrices are the eigenimages of g.

Example 2.13

Perform the singular value decomposition of the following image and identify its eigenimages:

$$g = \begin{pmatrix} 0 & 1 & 0 \\ 1 & 0 & 1 \\ 0 & 1 & 0 \end{pmatrix}$$

$$gg^T = \begin{pmatrix} 0 & 1 & 0 \\ 1 & 0 & 1 \\ 0 & 1 & 0 \end{pmatrix} \begin{pmatrix} 0 & 1 & 0 \\ 1 & 0 & 1 \\ 0 & 1 & 0 \end{pmatrix} = \begin{pmatrix} 1 & 0 & 1 \\ 0 & 2 & 0 \\ 1 & 0 & 1 \end{pmatrix}$$

The eigenvalues of this matrix are given by:

$$\begin{vmatrix} 1-\lambda & 0 & 1 \\ 0 & 2-\lambda & 0 \\ 1 & 0 & 1-\lambda \end{vmatrix} = 0 \Rightarrow (1-\lambda)^2(2-\lambda) - (2-\lambda) = 0$$

$$\Rightarrow (2-\lambda)\left[(1-\lambda)^2 - 1\right] = 0 \Rightarrow (2-\lambda)^2(1-\lambda-1)(1-\lambda+1) = 0$$

where $\lambda_1 = 2$, $\lambda_2 = 2$, $\lambda_3 = 0$.

The first eigenvector is:

$$\begin{vmatrix} 1 & 0 & 1 \\ 0 & 2 & 0 \\ 1 & 0 & 1 \end{vmatrix} \begin{pmatrix} x_1 \\ x_2 \\ x_3 \end{pmatrix} = 2 \begin{pmatrix} x_1 \\ x_2 \\ x_3 \end{pmatrix} \Rightarrow \begin{matrix} x_1 + x_3 = 2x_1 \\ 2x_2 = 2x_2 \\ x_1 + x_3 = 2x_3 \end{matrix} \Rightarrow \begin{matrix} x_1 = x_3 \\ x_2 \ any \ value \end{matrix}$$

Choose $x_1 = x_3 = \frac{1}{\sqrt{2}}$ and $x_2 = 0$, so $\mathbf{u_1} = (\frac{1}{\sqrt{2}}, 0, \frac{1}{\sqrt{2}})^T$.

The second eigenvector must satisfy the same constraints and must be orthogonal to $\mathbf{u_1}$. Therefore:

$$\mathbf{u_2} = (0, 1, 0)^T$$

Calculate the corresponding eigenvectors of $g^T g$ using (see Example 2.7):

$$\mathbf{v_i} = g^T \mathbf{u_i}$$

Therefore:

$$\mathbf{v_1} = \begin{pmatrix} 0 & 1 & 0 \\ 1 & 0 & 1 \\ 0 & 1 & 0 \end{pmatrix} \begin{pmatrix} \frac{1}{\sqrt{2}} \\ 0 \\ \frac{1}{\sqrt{2}} \end{pmatrix} = \begin{pmatrix} 0 \\ \frac{1}{\sqrt{2}} + \frac{1}{\sqrt{2}} \\ 0 \end{pmatrix} = \begin{pmatrix} 0 \\ \sqrt{2} \\ 0 \end{pmatrix}$$

Normalize it so that $|\mathbf{v_1}| = 1$; i.e. set $\mathbf{v_1} = (0, 1, 0)^T$.

$$\mathbf{v_2} = \begin{pmatrix} 0 & 1 & 0 \\ 1 & 0 & 1 \\ 0 & 1 & 0 \end{pmatrix} \begin{pmatrix} 0 \\ 1 \\ 0 \end{pmatrix} = \begin{pmatrix} 1 \\ 0 \\ 1 \end{pmatrix}$$

Normalize it so that $|\mathbf{v_2}| = 1$; i.e. set $\mathbf{v_2} = (\frac{1}{\sqrt{2}}, 0, \frac{1}{\sqrt{2}})^T$. Then the SVD of g is:

$$\begin{aligned} g &= \sqrt{\lambda_1} \mathbf{u_1} \mathbf{v_1}^T + \sqrt{\lambda_2} \mathbf{u_2} \mathbf{v_2}^T \\ &= \sqrt{2} \begin{pmatrix} \frac{1}{\sqrt{2}} \\ 0 \\ \frac{1}{\sqrt{2}} \end{pmatrix} (0 \ \ 1 \ \ 0) + \sqrt{2} \begin{pmatrix} 0 \\ 1 \\ 0 \end{pmatrix} (\frac{1}{\sqrt{2}} \ \ 0 \ \ \frac{1}{\sqrt{2}}) \\ &= \sqrt{2} \begin{pmatrix} 0 & \frac{1}{\sqrt{2}} & 0 \\ 0 & 0 & 0 \\ 0 & \frac{1}{\sqrt{2}} & 0 \end{pmatrix} + \sqrt{2} \begin{pmatrix} 0 & 0 & 0 \\ \frac{1}{\sqrt{2}} & 0 & \frac{1}{\sqrt{2}} \\ 0 & 0 & 0 \end{pmatrix} \\ &= \begin{pmatrix} 0 & 1 & 0 \\ 0 & 0 & 0 \\ 0 & 1 & 0 \end{pmatrix} + \begin{pmatrix} 0 & 0 & 0 \\ 1 & 0 & 1 \\ 0 & 0 & 0 \end{pmatrix} \end{aligned}$$

These two matrices are the eigenimages of g.

Example 2.14

Show the different stages of the SVD of the following image:

$$g = \begin{pmatrix} 255 & 255 & 255 & 255 & 255 & 255 & 255 & 255 \\ 255 & 255 & 255 & 100 & 100 & 100 & 255 & 255 \\ 255 & 255 & 100 & 150 & 150 & 150 & 100 & 255 \\ 255 & 255 & 100 & 150 & 200 & 150 & 100 & 255 \\ 255 & 255 & 100 & 150 & 150 & 150 & 100 & 255 \\ 255 & 255 & 255 & 100 & 100 & 100 & 255 & 255 \\ 255 & 255 & 255 & 255 & 50 & 255 & 255 & 255 \\ 50 & 50 & 50 & 50 & 255 & 255 & 255 & 255 \end{pmatrix}$$

The gg^T matrix is:

$$gg^T = \begin{pmatrix} 520200 & 401625 & 360825 & 373575 & 360825 & 401625 & 467925 & 311100 \\ 401625 & 355125 & 291075 & 296075 & 291075 & 355125 & 381125 & 224300 \\ 360825 & 291075 & 282575 & 290075 & 282575 & 291075 & 330075 & 205025 \\ 373575 & 296075 & 290075 & 300075 & 290075 & 296075 & 332575 & 217775 \\ 360825 & 291075 & 282575 & 290075 & 282575 & 291075 & 330075 & 205025 \\ 401625 & 355125 & 291075 & 296075 & 291075 & 355125 & 381125 & 224300 \\ 467925 & 381125 & 330075 & 332575 & 330075 & 381125 & 457675 & 258825 \\ 311100 & 224300 & 205025 & 217775 & 205025 & 224300 & 258825 & 270100 \end{pmatrix}$$

Its eigenvalues sorted in decreasing order are:

$$2593416.500 \quad 111621.508 \quad 71738.313 \quad 34790.875$$
$$11882.712 \quad 0.009 \quad 0.001 \quad 0.000$$

The last three eigenvalues are practically 0, so we compute only the eigenvectors that correspond to the first five eigenvalues. These eigenvectors are the columns of the following matrix:

$$\begin{pmatrix} -0.441 & 0.167 & 0.080 & -0.388 & 0.764 \\ -0.359 & -0.252 & 0.328 & 0.446 & 0.040 \\ -0.321 & -0.086 & -0.440 & 0.034 & -0.201 \\ -0.329 & -0.003 & -0.503 & 0.093 & 0.107 \\ -0.321 & -0.086 & -0.440 & 0.035 & -0.202 \\ -0.359 & -0.252 & 0.328 & 0.446 & 0.040 \\ -0.407 & -0.173 & 0.341 & -0.630 & -0.504 \\ -0.261 & 0.895 & 0.150 & 0.209 & -0.256 \end{pmatrix}$$

The $g^T g$ matrix is:

$$g^T g = \begin{pmatrix} 457675 & 457675 & 339100 & 298300 & 269025 & 308550 & 349350 & 467925 \\ 457675 & 457675 & 339100 & 298300 & 269025 & 308550 & 349350 & 467925 \\ 339100 & 339100 & 292600 & 228550 & 191525 & 238800 & 302850 & 349350 \\ 298300 & 298300 & 228550 & 220050 & 185525 & 230300 & 238800 & 308550 \\ 269025 & 269025 & 191525 & 185525 & 237550 & 237800 & 243800 & 321300 \\ 308550 & 308550 & 238800 & 230300 & 237800 & 282575 & 291075 & 360825 \\ 349350 & 349350 & 302850 & 238800 & 243800 & 291075 & 355125 & 401625 \\ 467925 & 467925 & 349350 & 308550 & 321300 & 360825 & 401625 & 520200 \end{pmatrix}$$

Its eigenvectors, computed independently, turn out to be the columns of the following matrix:

$$\begin{pmatrix} -0.410 & -0.389 & 0.264 & 0.106 & 0.012 \\ -0.410 & -0.389 & 0.264 & 0.106 & 0.012 \\ -0.316 & -0.308 & -0.537 & -0.029 & -0.408 \\ -0.277 & -0.100 & 0.101 & -0.727 & -0.158 \\ -0.269 & 0.555 & 0.341 & 0.220 & -0.675 \\ -0.311 & 0.449 & -0.014 & -0.497 & 0.323 \\ -0.349 & 0.241 & -0.651 & 0.200 & 0.074 \\ -0.443 & 0.160 & 0.149 & 0.336 & 0.493 \end{pmatrix}$$

In Figure *2.1 the original image and its five eigenimages are shown. Each eigenimage has been scaled so that its grey values vary between 0 and 255. These eigenimages have to be weighted by the square root of the appropriate eigenvalue and added to produce the original image. The five images shown in* Figure *2.2 are the reconstructed images when one, two,..., five eigenvalues were used for the reconstruction.*

    Then we calculate the sum of the squared errors for each reconstructed image according to the formula

$$\sum_{i=1}^{64} (reconstructed\ pixel - original\ pixel)^2$$

We obtain:

Error for image (a):	230033.32	$(\lambda_2 + \lambda_3 + \lambda_4 + \lambda_5 = 230033.41)$
Error for image (b):	118412.02	$(\lambda_3 + \lambda_4 + \lambda_5 = 118411.90)$
Error for image (c):	46673.53	$(\lambda_4 + \lambda_5 = 46673.59)$
Error for image (d):	11882.65	$(\lambda_5 = 11882.71)$
Error for image (e):	0	

We see that the sum of the omitted eigenvalues agrees very well with the error in the reconstructed image.

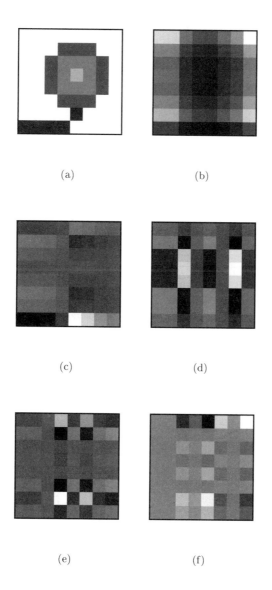

(a)   (b)

(c)   (d)

(e)   (f)

Figure 2.1: The original image and its five eigenimages, each scaled to have values from 0 to 255.

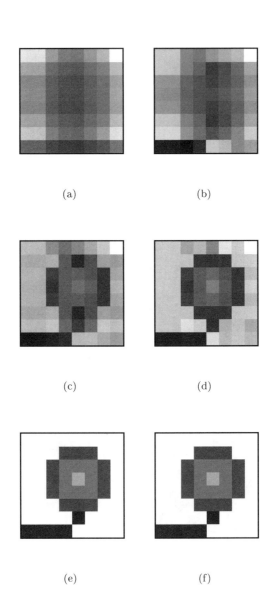

Figure 2.2: Image reconstruction using one, two,..., five eigenimages from top right to bottom left sequentially, with the original image shown in (f).

Are there any sets of elementary images in terms of which ANY image can be expanded?

Yes. They are defined in terms of sets of complete and orthonormal sets of discrete valued discrete functions.

What is a complete and orthonormal set of functions?

A set of functions $S_n(t)$, where n is an integer, is said to be *orthogonal* over an interval $[0, T]$ with weight function $w(t)$ if:

$$\int_0^T w(t)S_n(t)S_m(t)dt = \begin{cases} k \text{ if } n = m \\ 0 \text{ if } n \neq m \end{cases} \tag{2.31}$$

The set is called *orthonormal* if $k = 1$. It is called *complete* if we cannot find any other function which is orthogonal to the set and does not belong to the set. An example of a complete and orthogonal set is the set of functions e^{jnt} which are used as the basis functions of the Fourier transform.

Example 2.15

Show that the columns of an orthogonal matrix form a set of orthonormal vectors.

Let us say that A is an $N \times N$ orthogonal matrix (i.e. $A^T = A^{-1}$), and let us consider its column vectors $\mathbf{u}_1, \mathbf{u}_2, \ldots, \mathbf{u}_N$. We obviously have:

$$A^{-1}A = I \Rightarrow A^T A = I \Rightarrow \begin{pmatrix} \mathbf{u}_1^T \\ \mathbf{u}_2^T \\ \vdots \\ \mathbf{u}_N^T \end{pmatrix} (\mathbf{u}_1 \quad \mathbf{u}_2 \quad \ldots \quad \mathbf{u}_N) = I \Rightarrow$$

$$\begin{pmatrix} \mathbf{u}_1^T\mathbf{u}_1 & \mathbf{u}_1^T\mathbf{u}_2 & \ldots & \mathbf{u}_1^T\mathbf{u}_N \\ \mathbf{u}_2^T\mathbf{u}_1 & \mathbf{u}_2^T\mathbf{u}_2 & \ldots & \mathbf{u}_2^T\mathbf{u}_N \\ \vdots & \vdots & & \vdots \\ \mathbf{u}_N^T\mathbf{u}_1 & \mathbf{u}_N^T\mathbf{u}_2 & \ldots & \mathbf{u}_N^T\mathbf{u}_N \end{pmatrix} = \begin{pmatrix} 1 & 0 & \ldots & 0 \\ 0 & 1 & \ldots & 0 \\ \vdots & \vdots & & \vdots \\ 0 & 0 & \ldots & 1 \end{pmatrix}$$

This proves that the columns of A form an orthonormal set of vectors.

Example 2.16

Show that the inverse of an orthogonal matrix is also orthogonal.

An orthogonal matrix is defined as:

$$A^T = A^{-1} \tag{2.32}$$

To prove that A^{-1} is also orthogonal, it is enough to prove that $(A^{-1})^T = (A^{-1})^{-1}$. This is equivalent to $(A^{-1})^T = A$, which is readily derived if we take the transpose of equation (2.32).

Example 2.17

Show that the rows of an orthogonal matrix also form a set of orthonormal vectors.

Since A is an orthogonal matrix, so is A^{-1} (see Example 2.16). The columns of an orthogonal matrix form a set of orthonormal vectors (see Example 2.15). Therefore, the columns of A^{-1}, which are the rows of A, form a set of orthonormal vectors.

Are there any complete sets of orthonormal discrete valued functions?

Yes. There is, for example, the set of Haar functions which take values from the set $\{0, \pm 1, \pm\sqrt{2^p}, \text{ for } p = 1, 2, 3, \ldots\}$ and the set of Walsh functions which take values in the set $\{+1, -1\}$.

How are the Haar functions defined?

$$
\begin{aligned}
H_0(t) &= 1 \text{ for } 0 \le t \le 1 \\
H_1(t) &= \begin{cases} 1 & \text{if } 0 \le t < \frac{1}{2} \\ -1 & \text{if } \frac{1}{2} \le t < 1 \end{cases} \\
H_{2^p + n}(t) &= \begin{cases} \sqrt{2^p} & \text{for } \frac{n}{2^p} \le t < \frac{(n+0.5)}{2^p} \\ -\sqrt{2^p} & \text{for } \frac{(n+0.5)}{2^p} \le t < \frac{(n+1)}{2^p} \\ 0 & \text{elsewhere} \end{cases}
\end{aligned}
\tag{2.33}
$$

where $p = 1, 2, 3, \ldots$ and $n = 0, 1, \ldots, 2^p - 1$.

How are the Walsh functions defined?

They are defined in various ways all of which can be shown to be equivalent. We use here the definition from the difference equation:

$$W_{2j+q}(t) = (-1)^{[\frac{j}{2}]+q}\{W_j(2t) + (-1)^{j+q}W_j(2t - 1)\} \tag{2.34}$$

where $[\frac{j}{2}]$ means the largest integer which is smaller or equal to $\frac{j}{2}$, $q = 0$ or 1, $j = 0, 1, 2, \ldots$ and

$$W_0(t) = \begin{cases} 1 & \text{for } 0 \leq t < 1 \\ 0 & \text{elsewhere} \end{cases}$$

The above equations define these functions in what is called *natural order*. Other equivalent definitions order these functions in terms of increasing number of zero crossings of the functions, and that is called *sequency order*.

How can we create the image transformation matrices from the Haar and Walsh functions?

We first scale the independent variable t by the size of the matrix we want to create. Then we consider only its integer values i. Then $H_k(i)$ can be written in a matrix form for $k = 0, 1, 2, \ldots, N - 1$ and $i = 0, 1, \ldots, N - 1$ and be used for the transformation of a discrete 2-dimensional image function. We work similarly for $W_k(i)$.

Note that the Haar/Walsh functions defined this way are not orthonormal. Each has to be normalized by being multiplied with $\frac{1}{\sqrt{T}}$ in the continuous case, or by $\frac{1}{\sqrt{N}}$ in the discrete case, if t takes up N equally spaced discrete values.

Example 2.18

Derive the matrix which can be used to calculate the Haar transform of a 4×4 image.

First, by using equation (2.33), we shall calculate and plot the Haar functions of the continuous variable t which are needed for the calculation of the transformation matrix.

$H(0, t) = 1$ *for* $0 \leq t < 1$

$$H(1,t) = \begin{cases} 1 & for\ 0 \le t < \frac{1}{2} \\ -1 & for\ \frac{1}{2} \le t < 1 \end{cases}$$

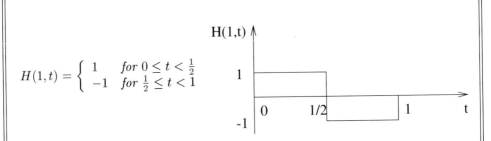

In the definition of the Haar functions, when $p = 1$, n takes the values 0 and 1:

Case $p = 1$, $n = 0$:

$$H(2,t) = \begin{cases} \sqrt{2} & for\ 0 \le t < \frac{1}{4} \\ -\sqrt{2} & for\ \frac{1}{4} \le t < \frac{1}{2} \\ 0 & for\ \frac{1}{2} \le t < 1 \end{cases}$$

Case $p = 1$, $n = 1$:

$$H(3,t) = \begin{cases} 0 & for\ 0 \le t < \frac{1}{2} \\ \sqrt{2} & for\ \frac{1}{2} \le t < \frac{3}{4} \\ -\sqrt{2} & for\ \frac{3}{4} \le t < 1 \end{cases}$$

To transform a 4 × 4 image we need a 4 × 4 matrix. If we scale the t axis by multiplying it by 4 and take only the integer values of t (i.e. $t = 0, 1, 2, 3$) we can construct the transformation matrix. The plots of the scaled functions look like this:

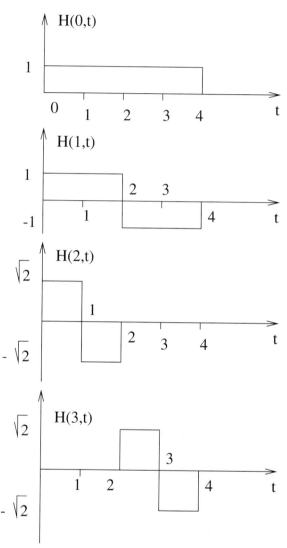

The entries of the transformation matrix are the values of $H(s,t)$ where s and t take the values $0, 1, 2, 3$. Obviously then the transformation matrix is:

$$H = \frac{1}{2} \begin{pmatrix} 1 & 1 & 1 & 1 \\ 1 & 1 & -1 & -1 \\ \sqrt{2} & -\sqrt{2} & 0 & 0 \\ 0 & 0 & \sqrt{2} & -\sqrt{2} \end{pmatrix}$$

The factor $\frac{1}{2}$ appears so that $HH^T = I$, the unit matrix.

Example 2.19

Calculate the Haar transform of the image:

$$
g = \begin{pmatrix} 0 & 1 & 1 & 0 \\ 1 & 0 & 0 & 1 \\ 1 & 0 & 0 & 1 \\ 0 & 1 & 1 & 0 \end{pmatrix}
$$

The Haar transform of the image g is $A = HgH^T$. We shall use matrix H derived in Example 2.18:

$$
A = \frac{1}{4} \begin{pmatrix} 1 & 1 & 1 & 1 \\ 1 & 1 & -1 & -1 \\ \sqrt{2} & -\sqrt{2} & 0 & 0 \\ 0 & 0 & \sqrt{2} & -\sqrt{2} \end{pmatrix} \begin{pmatrix} 0 & 1 & 1 & 0 \\ 1 & 0 & 0 & 1 \\ 1 & 0 & 0 & 1 \\ 0 & 1 & 1 & 0 \end{pmatrix} \begin{pmatrix} 1 & 1 & \sqrt{2} & 0 \\ 1 & 1 & -\sqrt{2} & 0 \\ 1 & -1 & 0 & \sqrt{2} \\ 1 & -1 & 0 & -\sqrt{2} \end{pmatrix}
$$

$$
= \frac{1}{4} \begin{pmatrix} 1 & 1 & 1 & 1 \\ 1 & 1 & -1 & -1 \\ \sqrt{2} & -\sqrt{2} & 0 & 0 \\ 0 & 0 & \sqrt{2} & -\sqrt{2} \end{pmatrix} \begin{pmatrix} 2 & 0 & -\sqrt{2} & \sqrt{2} \\ 2 & 0 & \sqrt{2} & -\sqrt{2} \\ 2 & 0 & \sqrt{2} & -\sqrt{2} \\ 2 & 0 & -\sqrt{2} & \sqrt{2} \end{pmatrix}
$$

$$
= \frac{1}{4} \begin{pmatrix} 8 & 0 & 0 & 0 \\ 0 & 0 & 0 & 0 \\ 0 & 0 & -4 & 4 \\ 0 & 0 & 4 & -4 \end{pmatrix}
$$

$$
= \begin{pmatrix} 2 & 0 & 0 & 0 \\ 0 & 0 & 0 & 0 \\ 0 & 0 & -1 & 1 \\ 0 & 0 & 1 & -1 \end{pmatrix}
$$

Example 2.20

Reconstruct the image of Example 2.19 using an approximation of its Haar transform by setting its bottom right element be equal to 0.

The approximate transformation matrix becomes:

$$\tilde{A} = \begin{pmatrix} 2 & 0 & 0 & 0 \\ 0 & 0 & 0 & 0 \\ 0 & 0 & -1 & 1 \\ 0 & 0 & 1 & 0 \end{pmatrix}$$

The reconstructed image is given by $\tilde{g} = H^T \tilde{A} H$:

$$\tilde{g} = \frac{1}{4} \begin{pmatrix} 1 & 1 & \sqrt{2} & 0 \\ 1 & 1 & -\sqrt{2} & 0 \\ 1 & -1 & 0 & \sqrt{2} \\ 1 & -1 & 0 & -\sqrt{2} \end{pmatrix} \begin{pmatrix} 2 & 0 & 0 & 0 \\ 0 & 0 & 0 & 0 \\ 0 & 0 & -1 & 1 \\ 0 & 0 & 1 & 0 \end{pmatrix} \begin{pmatrix} 1 & 1 & 1 & 1 \\ 1 & 1 & -1 & -1 \\ \sqrt{2} & -\sqrt{2} & 0 & 0 \\ 0 & 0 & \sqrt{2} & -\sqrt{2} \end{pmatrix}$$

$$= \frac{1}{4} \begin{pmatrix} 1 & 1 & \sqrt{2} & 0 \\ 1 & 1 & -\sqrt{2} & 0 \\ 1 & -1 & 0 & \sqrt{2} \\ 1 & -1 & 0 & -\sqrt{2} \end{pmatrix} \begin{pmatrix} 2 & 2 & 2 & 2 \\ 0 & 0 & 0 & 0 \\ -\sqrt{2} & \sqrt{2} & \sqrt{2} & -\sqrt{2} \\ \sqrt{2} & -\sqrt{2} & 0 & 0 \end{pmatrix}$$

$$= \frac{1}{4} \begin{pmatrix} 0 & 4 & 4 & 0 \\ 4 & 0 & 0 & 4 \\ 4 & 0 & 2 & 2 \\ 0 & 4 & 0 & 0 \end{pmatrix} = \begin{pmatrix} 0 & 1 & 1 & 0 \\ 1 & 0 & 0 & 1 \\ 1 & 0 & 0.5 & 0.5 \\ 0 & 1 & 0 & 0 \end{pmatrix}$$

The square error is equal to:

$$0.5^2 + 0.5^2 + 1^2 = 1.5$$

What do the elementary images of the Haar transform look like?

Figure 2.3 shows the basis images for the expansion of an 8×8 image in terms of the Haar functions. Each of these images has been produced by taking the outer product of a discretized Haar function either with itself or with another one. The numbers along the left and the bottom indicate the order of the function used along each row or column respectively. The discrete values of each image have been scaled in the range $[0, 255]$ for display purposes.

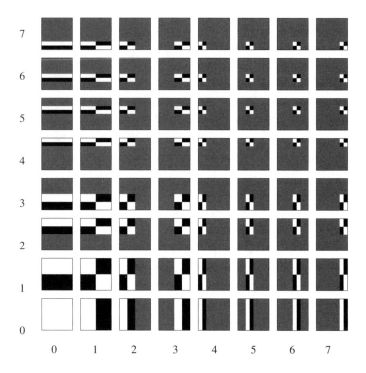

Figure 2.3: Haar transform basis images. In each image, grey means 0, black means a negative and white means a positive number. Note that each image has been scaled separately: black and white indicate different numbers from one image to the next.

Example 2.21

Derive the matrix which can be used to calculate the Walsh transform of a 4×4 image.

First, by using equation (2.34), calculate and plot the Walsh functions of the continuous variable t which are needed for the calculation of the transformation matrix.

$$W(0,t) = \begin{cases} 1 & \text{for } 0 \leq t < 1 \\ 0 & \text{elsewhere} \end{cases}$$

Case $j = 0$, $q = 1$, $\left[\frac{j}{2}\right] = 0$:

$$W(1,t) = -\left\{W(0,2t) - W\left(0, 2(t - \frac{1}{2})\right)\right\}$$

For $0 \le t < \frac{1}{2}$:

$$0 \le 2t < 1 \Rightarrow W(0, 2t) = 1$$

$$-\frac{1}{2} \le t - \frac{1}{2} < 0 \Rightarrow -1 \le 2(t - \frac{1}{2}) < 0 \Rightarrow W\left(0, 2(t - \frac{1}{2})\right) = 0$$

Therefore:

$$W(1,t) = -1 \quad for \quad 0 \le t < \frac{1}{2}$$

For $\frac{1}{2} \le t < 1$:

$$1 \le 2t < 2 \Rightarrow W(0, 2t) = 0$$

$$0 \le t - \frac{1}{2} < \frac{1}{2} \Rightarrow 0 \le 2(t - \frac{1}{2}) \le 1 \Rightarrow W\left(0, 2(t - \frac{1}{2})\right) = 1$$

Therefore:

$$W(1,t) = -(-1) = 1 \quad for \quad \frac{1}{2} \le t < 1$$

$$W(1,t) = \begin{cases} -1 & for\ 0 \le t < \frac{1}{2} \\ 1 & for\ \frac{1}{2} \le t < 1 \end{cases}$$

Case $j = 1$, $q = 0$, $\left[\frac{j}{2}\right] = 0$:

$$W(2,t) = W(1,2t) - W\left(1, 2(t - \frac{1}{2})\right)$$

For $0 \le t < \frac{1}{4}$:

$$0 \leq 2t < \frac{1}{2} \Rightarrow W(1, 2t) = -1$$

$$-\frac{1}{2} \leq t - \frac{1}{2} < -\frac{1}{4} \Rightarrow -1 \leq 2(t - \frac{1}{2}) < -\frac{1}{2} \Rightarrow W(1, 2(t - \frac{1}{2})) = 0$$

Therefore:

$$W(2, t) = -1 \quad for \quad 0 \leq t < \frac{1}{4}$$

For $\frac{1}{4} \leq t < \frac{1}{2}$:

$$\frac{1}{2} \leq 2t < 1 \Rightarrow W(1, 2t) = 1$$

$$-\frac{1}{4} \leq t - \frac{1}{2} < 0 \Rightarrow -\frac{1}{2} \leq 2(t - \frac{1}{2}) < 0 \Rightarrow W\left(1, 2(t - \frac{1}{2})\right) = 0$$

Therefore:

$$W(2, t) = 1 \quad for \quad \frac{1}{4} \leq t < \frac{1}{2}$$

For $\frac{1}{2} \leq t < \frac{3}{4}$:

$$1 \leq 2t < \frac{3}{2} \Rightarrow W(1, 2t) = 0$$

$$0 \leq t - \frac{1}{2} < \frac{1}{4} \Rightarrow 0 \leq 2(t - \frac{1}{2}) < \frac{1}{2} \Rightarrow W\left(1, 2(t - \frac{1}{2})\right) = -1$$

Therefore:

$$W(2, t) = 1 \quad for \quad \frac{1}{2} \leq t < \frac{3}{4}$$

For $\frac{3}{4} \leq t < 1$:

$$\frac{3}{2} \leq 2t < 2 \Rightarrow W(1, 2t) = 0$$

$$\frac{1}{4} \leq t - \frac{1}{2} < \frac{1}{2} \Rightarrow \frac{1}{2} \leq 2(t - \frac{1}{2}) < 1 \Rightarrow W\left(1, 2(t - \frac{1}{2})\right) = 1$$

Therefore:

$$W(2, t) = -1 \quad for \quad \frac{3}{4} \leq t < 1$$

$$W(2,t) = \begin{cases} -1 & \text{for } 0 \le t < \frac{1}{4} \\ 1 & \text{for } \frac{1}{4} \le t < \frac{3}{4} \\ -1 & \text{for } \frac{3}{4} \le t < 1 \end{cases}$$

Case $j = 1$, $q = 1$, $\left[\frac{j}{2}\right] = 0$:

$$W(3,t) = -\left\{W(1,2t) + W\left(1, 2(t - \frac{1}{2})\right)\right\}$$

For $0 \le t < \frac{1}{4}$:

$$W(1,2t) = -1, \qquad W\left(1, 2(t - \frac{1}{2})\right) = 0$$

Therefore:

$$W(3,t) = 1 \quad for \quad 0 \le t < \frac{1}{4}$$

For $\frac{1}{4} \le t < \frac{1}{2}$:

$$W(1,2t) = 1, \qquad W\left(1, 2(t - \frac{1}{2})\right) = 0$$

Therefore:

$$W(3,t) = -1 \quad for \quad \frac{1}{4} \le t < \frac{1}{2}$$

For $\frac{1}{2} \le t < \frac{3}{4}$:

$$W(1,2t) = 0, \qquad W\left(1, 2(t - \frac{1}{2})\right) = -1$$

Therefore:

$$W(3,t) = 1 \quad for \quad \frac{1}{2} \le t < \frac{3}{4}$$

For $\frac{3}{4} \le t < 1$:

$$W(1, 2t) = 0, \qquad W\left(1, 2(t - \frac{1}{2})\right) = 1$$

Therefore:

$$W(3, t) = -1 \quad \text{for} \quad \frac{3}{4} \le t < 1$$

$$W(3, t) = \begin{cases} 1 & \text{for } 0 \le t < \frac{1}{4} \\ -1 & \text{for } \frac{1}{4} \le t < \frac{1}{2} \\ 1 & \text{for } \frac{1}{2} \le t < \frac{3}{4} \\ -1 & \text{for } \frac{3}{4} \le t < 1 \end{cases}$$

To create a 4×4 matrix, we multiply t by 4 and consider only its integer values i.e. $0, 1, 2, 3$. The first row of the matrix will be formed from $W(0, t)$. The second from $W(1, t)$, the third from $W(2, t)$ and so on:

$$W = \frac{1}{2} \begin{pmatrix} 1 & 1 & 1 & 1 \\ -1 & -1 & 1 & 1 \\ -1 & 1 & 1 & -1 \\ 1 & -1 & 1 & -1 \end{pmatrix}$$

This matrix has been normalized by multiplying it by $\frac{1}{2}$ so that $W^T W = I$, where I is the unit matrix.

Example 2.22

Calculate the Walsh transform of the image:

$$g = \begin{pmatrix} 0 & 1 & 1 & 0 \\ 1 & 0 & 0 & 1 \\ 1 & 0 & 0 & 1 \\ 0 & 1 & 1 & 0 \end{pmatrix}$$

In the general formula of a separable linear transform $A = UgV^T$, use $U = V = W$ as derived in Example 2.21:

$$A = \frac{1}{4}\begin{pmatrix} 1 & 1 & 1 & 1 \\ -1 & -1 & 1 & 1 \\ -1 & 1 & 1 & -1 \\ 1 & -1 & 1 & -1 \end{pmatrix}\begin{pmatrix} 0 & 1 & 1 & 0 \\ 1 & 0 & 0 & 1 \\ 1 & 0 & 0 & 1 \\ 0 & 1 & 1 & 0 \end{pmatrix}\begin{pmatrix} 1 & -1 & -1 & 1 \\ 1 & -1 & 1 & -1 \\ 1 & 1 & 1 & 1 \\ 1 & 1 & -1 & -1 \end{pmatrix}$$

$$= \frac{1}{4}\begin{pmatrix} 1 & 1 & 1 & 1 \\ -1 & -1 & 1 & 1 \\ -1 & 1 & 1 & -1 \\ 1 & -1 & 1 & -1 \end{pmatrix}\begin{pmatrix} 2 & 0 & 2 & 0 \\ 2 & 0 & -2 & 0 \\ 2 & 0 & -2 & 0 \\ 2 & 0 & 2 & 0 \end{pmatrix}$$

$$= \frac{1}{4}\begin{pmatrix} 8 & 0 & 0 & 0 \\ 0 & 0 & 0 & 0 \\ 0 & 0 & -8 & 0 \\ 0 & 0 & 0 & 0 \end{pmatrix} = \begin{pmatrix} 2 & 0 & 0 & 0 \\ 0 & 0 & 0 & 0 \\ 0 & 0 & -2 & 0 \\ 0 & 0 & 0 & 0 \end{pmatrix}$$

Can we define an orthogonal matrix with entries only $+1$ or -1?

Yes. There are the Hadamard matrices named after the mathematician who studied them in 1893. The Hadamard matrices are defined only for sizes that are powers of 2, in a recursive way, as follows:

$$H_1 = \begin{pmatrix} 1 & 1 \\ 1 & -1 \end{pmatrix} \quad \text{and} \quad H_{2N} = \begin{pmatrix} H_N & H_N \\ H_N & -H_N \end{pmatrix}$$

For a general size these matrices have been shown to exist only for sizes up to 200×200. The rows of such matrices can be shown to be equal to the discretized form of the Walsh functions of the corresponding order. So the Walsh functions can be calculated in terms of these matrices for $N = 2^n$. The Walsh functions generated in this way are said to be in *Kronecker* or *lexicographic* ordering, i.e. they are not in the same order as those produced by using equation (2.34).

What do the basis images of the Hadamard/Walsh transform look like?

Figure 2.4 shows the basis images for the expansion of an 8×8 image in terms of Walsh functions in Kronecker ordering. These images are binary.

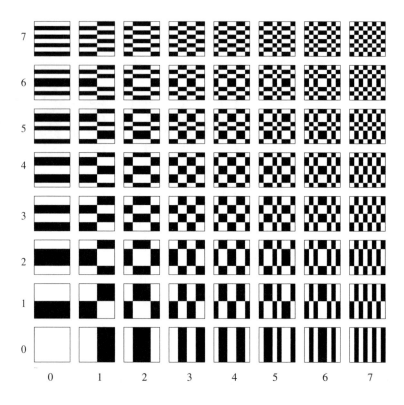

Figure 2.4: Hadamard/Walsh transform basis images.

Example 2.23

Show the different stages of the Haar transform of the image of Example 2.14:

The eight images shown in Figure *2.5 are the reconstructed images when for the reconstruction the basis images created by one, two,..., eight Haar functions are used. For example,* Figure *2.5b is the reconstructed image when only the coefficients that multiply the four basis images at the bottom left corner of* Figure *2.3 are retained. These four basis images are created from the first two Haar functions, $H(0,t)$ and $H(1,t)$. Image 2.5g is reconstructed when all the coefficients that multiply the basis images along the top row and the right column in* Figure *2.3 are set to 0. The basis images used were created from the first seven Haar functions, i.e. $H(0,t), H(1,t), \ldots, H(6,t)$.*

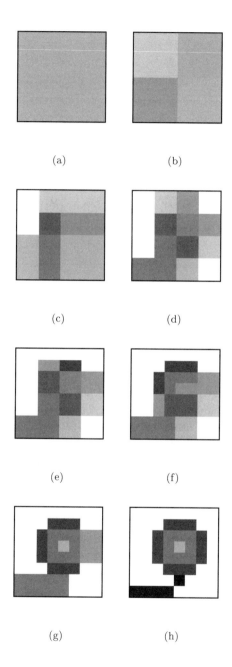

(a) (b)

(c) (d)

(e) (f)

(g) (h)

Figure 2.5: Reconstructed images when the basis images used are those created from the first one, two, three,..., eight Haar functions, from top left to bottom right respectively.

The sum of the squared errors for each reconstructed image is:

$$
\begin{array}{ll}
\textit{Error for image (a):} & 366394 \\
\textit{Error for image (b):} & 356192 \\
\textit{Error for image (c):} & 291740 \\
\textit{Error for image (d):} & 222550 \\
\textit{Error for image (e):} & 192518 \\
\textit{Error for image (f):} & 174625 \\
\textit{Error for image (g):} & 141100 \\
\textit{Error for image (h):} & 0
\end{array}
$$

Example 2.24

Show the different stages of the Walsh/Hadamard transform of the image of Example 2.14.

The eight images shown in Figure *2.6 are the reconstructed images when for the reconstruction the basis images created from the first one, two,..., eight Walsh functions are used. For example,* Figure *2.6f has been reconstructed from the inverse Walsh/Hadamard transform, by setting to 0 all elements of the transformation matrix that multiply the basis images in the top two rows and the two rightmost columns in* Figure *2.4. These omitted basis images are those that are created from functions* $W(6,t)$ *and* $W(7,t)$*.*

The sum of the squared errors for each reconstructed image is:

$$
\begin{array}{ll}
\textit{Error for image (a):} & 366394 \\
\textit{Error for image (b):} & 356190 \\
\textit{Error for image (c):} & 262206 \\
\textit{Error for image (d):} & 222550 \\
\textit{Error for image (e):} & 148029 \\
\textit{Error for image (f):} & 92078 \\
\textit{Error for image (g):} & 55905 \\
\textit{Error for image (h):} & 0
\end{array}
$$

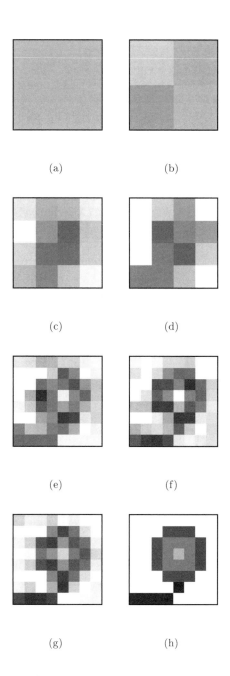

(a) (b)

(c) (d)

(e) (f)

(g) (h)

Figure 2.6: Reconstructed images when the basis images used are those created from the first one, two, three,..., eight Walsh functions, from top left to bottom right respectively.

What are the advantages and disadvantages of the Walsh and the Haar transforms?

From *Figure* 2.3 notice that the higher order Haar basis images use the same basic pattern that scans the whole image, as if every basis image attempts to capture more accurately the local characteristics of the image focusing every time at one place only. For example, the 16 basis images in the top right quadrant of *Figure* 2.3 all use a window of 2×2 pixels to reproduce detail in various parts of the image. If we are not interested in that level of detail, we can set the corresponding 16 coefficients of the transform to zero. So, when we truncate a Haar reconstruction we must keep basis images that are created from the first $2^0, 2^1, 2^2, \ldots$ Haar functions instead of retaining an arbitrary number of them. In this way the reconstruction error will be uniformly distributed over the whole image. Alternatively, we may wish to retain those basis images (i.e. retain their corresponding coefficients) that help reconstruct most accurately those parts of the image that contain most detail. In other words, the Haar basis functions allow us to reconstruct with different levels of detail different parts of an image.

In contrast, higher order Walsh basis images try to approximate the image as a whole, with uniformly distributed detail structure. This is because Walsh functions cannot take the 0 value. Notice how this difference between the two bases is reflected in the reconstructed images: both images 2.5g and 2.6g have been reconstructed by retaining the same number of basis images. In *Figure* 2.5g the flower has been almost fully reconstructed apart from some details on the right and at the bottom, because the omitted basis images were those that would describe the image in those locations, and the image happened to have significant detail there. That is why the reconstructed error in this case is higher for the Haar than the Walsh case. Notice that the error in the Walsh reconstruction is uniformly distributed over the whole image.

On the other hand, Walsh transforms have the advantage over Haar transforms that the Walsh functions take up only two values $+1$ or -1, and thus they are easily implemented in a computer as their values correspond to binary logic.

What is the Haar wavelet?

The property of the Haar basis functions to concentrate on one part of the image at a time is a characteristic property of a more general class of functions called *wavelets*. The *Haar wavelets* are all scaled and translated versions of the same function and for an 8×8 image they are shown in the seven top-most and the seven right-most panels of *Figure* 2.3. The function represented by the bottom left panel of *Figure* 2.3, i.e. the average flat image, is called the *scaling function*. The basis images represented by the other panels in the first column and the bottom row of *Figure* 2.3 are produced from combinations of the scaling function and the wavelet. All panels together constitute a complete basis in terms of which any 8×8 image can be expanded.

What is the discrete version of the Fourier transform?

The 1-dimensional discrete Fourier transform of a function $f(k)$ defined at discrete points k, is defined as:

$$F(m) = \frac{1}{\sqrt{N}} \sum_{k=0}^{N-1} f(k) \exp\left[-j\frac{2\pi mk}{N}\right] \tag{2.35}$$

The 2-dimensional discrete Fourier transform (DFT) for an $N \times N$ image is defined as:

$$\alpha_{mn} = \frac{1}{N} \sum_{k=0}^{N-1} \sum_{l=0}^{N-1} g_{kl} e^{-j2\pi \frac{km+nl}{N}} \tag{2.36}$$

Unlike the other transforms that were developed directly in the discrete domain, this transform was initially developed in the continuous domain. To preserve this "historical" consistency, we shall go back into using function arguments rather than indices. Further, because we shall have to associate Fourier transforms of different functions, we shall use the convention of the Fourier transform being denoted by the same letter as the function, but with a hat on the top. Different numbers of hats will be used to distinguish the Fourier transforms that refer to different versions of the same function. The reason for this will become clear when the case arises. For the time being, however, we define the Fourier transform of an $M \times N$ image as follows:

$$\hat{g}(m,n) = \frac{1}{\sqrt{MN}} \sum_{k=0}^{M-1} \sum_{l=0}^{N-1} g(k,l) e^{-2\pi j[\frac{km}{M} + \frac{ln}{N}]} \tag{2.37}$$

We must think of this formula as a "slot machine" where when we slot in a function, out pops its DFT:

$$\underbrace{\cdots}_{DFT} = \frac{1}{\sqrt{MN}} \sum_{k=0}^{M-1} \sum_{l=0}^{N-1} \underbrace{\cdots}_{function} e^{-2\pi j[\frac{km}{M} + \frac{ln}{N}]} \tag{2.38}$$

B2.3: What is the inverse discrete Fourier transform?

Multiply both sides of (2.37) by $e^{2\pi j[\frac{qm}{M} + \frac{pn}{N}]}\frac{1}{\sqrt{MN}}$ and sum over all m and n from 0 to $M-1$, and $N-1$ respectively. We get:

$$\frac{1}{\sqrt{MN}} \sum_{m=0}^{M-1} \sum_{n=0}^{N-1} \hat{g}(m,n) e^{2\pi j[\frac{qm}{M} + \frac{pn}{N}]}$$

$$= \frac{1}{MN} \sum_{k=0}^{M-1} \sum_{l=0}^{N-1} \sum_{m=0}^{M-1} \sum_{n=0}^{N-1} g(k,l) e^{2\pi j[\frac{m(q-k)}{M} + \frac{n(p-l)}{N}]}$$

$$= \frac{1}{MN} \sum_{k=0}^{M-1} \sum_{l=0}^{N-1} g(k,l) \sum_{m=0}^{M-1} e^{2\pi j m \frac{q-k}{M}} \sum_{n=0}^{N-1} e^{2\pi j n \frac{p-l}{N}} \tag{2.39}$$

Let us examine the sum:

$$\sum_{m=0}^{M-1} e^{2\pi j t \frac{m}{M}} \tag{2.40}$$

where t is an integer. This is a geometric progression with M elements, first term 1 ($m = 0$) and ratio $q = e^{2\pi j \frac{t}{M}}$. The sum of the first n terms of such a geometric progression is given by:

$$\sum_{k=0}^{n-1} q^k = \frac{q^n - 1}{q - 1} \quad \text{for} \quad q \neq 1 \tag{2.41}$$

The sum (2.40) is therefore equal to:

$$\sum_{m=0}^{M-1} e^{2\pi j t \frac{m}{M}} = \frac{e^{2\pi j t} - 1}{e^{2\pi j \frac{k}{M}} - 1} = \frac{\cos 2\pi t + j \sin 2\pi t - 1}{e^{2\pi j \frac{k}{M}} - 1} = 0 \tag{2.42}$$

If however $t = 0$, all terms in (2.40) are equal to 1 and we have $\sum_{m=0}^{M-1} 1 = M$. So:

$$\sum_{m=0}^{M-1} e^{2\pi j t \frac{m}{M}} = \begin{cases} M & \text{if } t = 0 \\ 0 & \text{if } t \neq 0 \end{cases} \tag{2.43}$$

Applying this to equation (2.39) once for $t \equiv q - k$ and once for $t \equiv p - l$ we deduce that the right hand side of (2.39) is:

$$\frac{1}{MN} \sum_{k=0}^{M-1} \sum_{l=0}^{N-1} g(k,l) M \delta(q-k) N \delta(p-l)$$

where $\delta(a - b)$ is 0 unless $a = b$. Therefore, the above expression is $g(q, p)$, i.e.

$$g(q,p) = \frac{1}{\sqrt{MN}} \sum_{m=0}^{M-1} \sum_{n=0}^{N-1} \hat{g}(m,n) e^{2\pi j [\frac{qm}{M} + \frac{pn}{N}]} \tag{2.44}$$

This is the inverse 2-dimensional discrete Fourier transform.

How can we write the discrete Fourier transform in matrix form?

We construct matrix U with elements

$$U_{x\alpha} = \frac{1}{\sqrt{N}} e^{-j\frac{2\pi x \alpha}{N}} \tag{2.45}$$

where x takes the values $0, 1, \ldots, N-1$ along each column and α takes the same values along each row. Notice that U constructed this way is symmetric; i.e. $U^T = U$. Then the 2-dimensional discrete Fourier transform of an image g in matrix form is given by:

$$A = UgU \tag{2.46}$$

Example 2.25

Derive the matrix with which the discrete Fourier transform of a 4×4 image can be obtained.

Apply formula (2.45) with $N = 4$, $0 \leq x \leq 3$, $0 \leq \alpha \leq 3$:

$$U = \frac{1}{\sqrt{4}} \begin{pmatrix} e^{-j\frac{2\pi}{4}\times 0} & e^{-j\frac{2\pi}{4}\times 0} & e^{-j\frac{2\pi}{4}\times 0} & e^{-j\frac{2\pi}{4}\times 0} \\ e^{-j\frac{2\pi}{4}\times 0} & e^{-j\frac{2\pi}{4}\times 1} & e^{-j\frac{2\pi}{4}\times 2} & e^{-j\frac{2\pi}{4}\times 3} \\ e^{-j\frac{2\pi}{4}\times 0} & e^{-j\frac{2\pi}{4}\times 2} & e^{-j\frac{2\pi}{4}\times 0} & e^{-j\frac{2\pi}{4}\times 2} \\ e^{-j\frac{2\pi}{4}\times 0} & e^{-j\frac{2\pi}{4}\times 3} & e^{-j\frac{2\pi}{4}\times 2} & e^{-j\frac{2\pi}{4}\times 1} \end{pmatrix}$$

$$= \frac{1}{2} \begin{pmatrix} 1 & 1 & 1 & 1 \\ 1 & e^{-\frac{\pi}{2}j} & e^{-j\pi} & e^{-\frac{3\pi}{2}j} \\ 1 & e^{-j\pi} & 1 & e^{-j\pi} \\ 1 & e^{-\frac{6\pi}{4}j} & e^{-j\pi} & e^{-\frac{2\pi}{4}j} \end{pmatrix}$$

$$e^{-\frac{2\pi}{4}j} = e^{-\frac{\pi}{2}j} = \cos\frac{\pi}{2} - j\sin\frac{\pi}{2} = -j$$

$$e^{-\pi j} = \cos\pi - j\sin\pi = -1$$

$$e^{-\frac{6\pi}{4}j} = \cos\frac{3\pi}{2} - j\sin\frac{3\pi}{2} = j$$

Therefore:

$$U = \frac{1}{2} \begin{pmatrix} 1 & 1 & 1 & 1 \\ 1 & -j & -1 & j \\ 1 & -1 & 1 & -1 \\ 1 & j & -1 & -j \end{pmatrix}$$

Example 2.26

Use matrix U of Example 2.25 to compute the discrete Fourier transform of the following image:

$$g = \begin{pmatrix} 0 & 0 & 1 & 0 \\ 0 & 0 & 1 & 0 \\ 0 & 0 & 1 & 0 \\ 0 & 0 & 1 & 0 \end{pmatrix}$$

Calculate first gU:

$$\begin{pmatrix} 0 & 0 & 1 & 0 \\ 0 & 0 & 1 & 0 \\ 0 & 0 & 1 & 0 \\ 0 & 0 & 1 & 0 \end{pmatrix} \frac{1}{2} \begin{pmatrix} 1 & 1 & 1 & 1 \\ 1 & -j & -1 & j \\ 1 & -1 & 1 & -1 \\ 1 & j & -1 & -j \end{pmatrix} = \frac{1}{2} \begin{pmatrix} 1 & -1 & 1 & -1 \\ 1 & -1 & 1 & -1 \\ 1 & -1 & 1 & -1 \\ 1 & -1 & 1 & -1 \end{pmatrix}$$

Multiply the result by U from the left to get $UgU = A$ (the discrete Fourier transform):

$$\frac{1}{4} \begin{pmatrix} 1 & 1 & 1 & 1 \\ 1 & -j & -1 & j \\ 1 & -1 & 1 & -1 \\ 1 & j & -1 & -j \end{pmatrix} \begin{pmatrix} 1 & -1 & 1 & -1 \\ 1 & -1 & 1 & -1 \\ 1 & -1 & 1 & -1 \\ 1 & -1 & 1 & -1 \end{pmatrix}$$

$$= \frac{1}{4} \begin{pmatrix} 4 & -4 & 4 & -4 \\ 0 & 0 & 0 & 0 \\ 0 & 0 & 0 & 0 \\ 0 & 0 & 0 & 0 \end{pmatrix} = \begin{pmatrix} 1 & -1 & 1 & -1 \\ 0 & 0 & 0 & 0 \\ 0 & 0 & 0 & 0 \\ 0 & 0 & 0 & 0 \end{pmatrix}$$

Is matrix U used for DFT unitary?

We must show that any row of this matrix is orthogonal to any other row. Because U is a complex matrix, when we test for orthogonality we must use the complex conjugate of one of the two rows we multiply. The product of rows obtained for $x = x_1$ and $x = x_2$ is given by:

$$\sum_{\alpha=0}^{N-1} e^{-j\frac{2\pi x_1 \alpha}{N}} e^{j\frac{2\pi x_2 \alpha}{N}} = \sum_{\alpha=0}^{N-1} e^{-j\frac{2\pi (x_1 - x_2)\alpha}{N}} \tag{2.47}$$

By observing that this sum is the sum of the first N terms of a geometric progression with first element 1 and ratio $e^{-j\frac{2\pi (x_1 - x_2)\alpha}{N}}$, we can show that it is zero.

For the matrix to be unitary, we must also show that the magnitude of each row vector is 1. If we set $x_2 = x_1$ in (2.47), we have for the square magnitude:

$$\sum_{\alpha=0}^{N-1} e^{-j\frac{2\pi x_1 \alpha}{N}} e^{j\frac{2\pi x_1 \alpha}{N}} = \sum_{\alpha=0}^{N-1} 1 = N \tag{2.48}$$

This cancels with the normalizing constant that multiplies the matrix.

Example 2.27

Derive the matrix U needed for the calculation of DFT of an 8×8 image.

By applying formula (2.45) with $N = 8$ we obtain:

$$\frac{1}{\sqrt{8}} \begin{pmatrix}
1 & 1 & 1 & 1 & 1 & 1 & 1 & 1 \\
1 & e^{-j\frac{2\pi}{8}\times 1} & e^{-j\frac{2\pi}{8}\times 2} & e^{-j\frac{2\pi}{8}\times 3} & e^{-j\frac{2\pi}{8}\times 4} & e^{-j\frac{2\pi}{8}\times 5} & e^{-j\frac{2\pi}{8}\times 6} & e^{-j\frac{2\pi}{8}\times 7} \\
1 & e^{-j\frac{2\pi}{8}\times 2} & e^{-j\frac{2\pi}{8}\times 4} & e^{-j\frac{2\pi}{8}\times 6} & e^{-j\frac{2\pi}{8}\times 8} & e^{-j\frac{2\pi}{8}\times 10} & e^{-j\frac{2\pi}{8}\times 12} & e^{-j\frac{2\pi}{8}\times 14} \\
1 & e^{-j\frac{2\pi}{8}\times 3} & e^{-j\frac{2\pi}{8}\times 6} & e^{-j\frac{2\pi}{8}\times 9} & e^{-j\frac{2\pi}{8}\times 12} & e^{-j\frac{2\pi}{8}\times 15} & e^{-j\frac{2\pi}{8}\times 18} & e^{-j\frac{2\pi}{8}\times 21} \\
1 & e^{-j\frac{2\pi}{8}\times 4} & e^{-j\frac{2\pi}{8}\times 8} & e^{-j\frac{2\pi}{8}\times 12} & e^{-j\frac{2\pi}{8}\times 16} & e^{-j\frac{2\pi}{8}\times 20} & e^{-j\frac{2\pi}{8}\times 24} & e^{-j\frac{2\pi}{8}\times 28} \\
1 & e^{-j\frac{2\pi}{8}\times 5} & e^{-j\frac{2\pi}{8}\times 10} & e^{-j\frac{2\pi}{8}\times 15} & e^{-j\frac{2\pi}{8}\times 20} & e^{-j\frac{2\pi}{8}\times 25} & e^{-j\frac{2\pi}{8}\times 30} & e^{-j\frac{2\pi}{8}\times 35} \\
1 & e^{-j\frac{2\pi}{8}\times 6} & e^{-j\frac{2\pi}{8}\times 12} & e^{-j\frac{2\pi}{8}\times 18} & e^{-j\frac{2\pi}{8}\times 24} & e^{-j\frac{2\pi}{8}\times 30} & e^{-j\frac{2\pi}{8}\times 36} & e^{-j\frac{2\pi}{8}\times 42} \\
1 & e^{-j\frac{2\pi}{8}\times 7} & e^{-j\frac{2\pi}{8}\times 14} & e^{-j\frac{2\pi}{8}\times 21} & e^{-j\frac{2\pi}{8}\times 28} & e^{-j\frac{2\pi}{8}\times 35} & e^{-j\frac{2\pi}{8}\times 42} & e^{-j\frac{2\pi}{8}\times 49}
\end{pmatrix}$$

Since

$$e^{-j\frac{2\pi}{N}\times x} = e^{-j\frac{2\pi}{N}\times [mod_N(x)]}$$

the above matrix can be simplified to:

$$U = \frac{1}{\sqrt{8}} \begin{pmatrix}
1 & 1 & 1 & 1 & 1 & 1 & 1 & 1 \\
1 & e^{-j\frac{2\pi}{8}\times 1} & e^{-j\frac{2\pi}{8}\times 2} & e^{-j\frac{2\pi}{8}\times 3} & e^{-j\frac{2\pi}{8}\times 4} & e^{-j\frac{2\pi}{8}\times 5} & e^{-j\frac{2\pi}{8}\times 6} & e^{-j\frac{2\pi}{8}\times 7} \\
1 & e^{-j\frac{2\pi}{8}\times 2} & e^{-j\frac{2\pi}{8}\times 4} & e^{-j\frac{2\pi}{8}\times 6} & e^{-j\frac{2\pi}{8}\times 0} & e^{-j\frac{2\pi}{8}\times 2} & e^{-j\frac{2\pi}{8}\times 4} & e^{-j\frac{2\pi}{8}\times 6} \\
1 & e^{-j\frac{2\pi}{8}\times 3} & e^{-j\frac{2\pi}{8}\times 6} & e^{-j\frac{2\pi}{8}\times 1} & e^{-j\frac{2\pi}{8}\times 4} & e^{-j\frac{2\pi}{8}\times 7} & e^{-j\frac{2\pi}{8}\times 2} & e^{-j\frac{2\pi}{8}\times 5} \\
1 & e^{-j\frac{2\pi}{8}\times 4} & e^{-j\frac{2\pi}{8}\times 0} & e^{-j\frac{2\pi}{8}\times 4} & e^{-j\frac{2\pi}{8}\times 0} & e^{-j\frac{2\pi}{8}\times 4} & e^{-j\frac{2\pi}{8}\times 0} & e^{-j\frac{2\pi}{8}\times 4} \\
1 & e^{-j\frac{2\pi}{8}\times 5} & e^{-j\frac{2\pi}{8}\times 2} & e^{-j\frac{2\pi}{8}\times 7} & e^{-j\frac{2\pi}{8}\times 4} & e^{-j\frac{2\pi}{8}\times 1} & e^{-j\frac{2\pi}{8}\times 6} & e^{-j\frac{2\pi}{8}\times 3} \\
1 & e^{-j\frac{2\pi}{8}\times 6} & e^{-j\frac{2\pi}{8}\times 4} & e^{-j\frac{2\pi}{8}\times 2} & e^{-j\frac{2\pi}{8}\times 0} & e^{-j\frac{2\pi}{8}\times 6} & e^{-j\frac{2\pi}{8}\times 4} & e^{-j\frac{2\pi}{8}\times 2} \\
1 & e^{-j\frac{2\pi}{8}\times 7} & e^{-j\frac{2\pi}{8}\times 6} & e^{-j\frac{2\pi}{8}\times 5} & e^{-j\frac{2\pi}{8}\times 4} & e^{-j\frac{2\pi}{8}\times 3} & e^{-j\frac{2\pi}{8}\times 2} & e^{-j\frac{2\pi}{8}\times 1}
\end{pmatrix}$$

Which are the elementary images in terms of which DFT expands an image?

As the kernel of DFT is a complex function, these images are complex. They can be created by taking the outer product between any two rows of matrix U. *Figure* 2.7 shows the real parts of these elementary images and *Figure* 2.8 the imaginary parts, for the U matrix computed in Example 2.27.

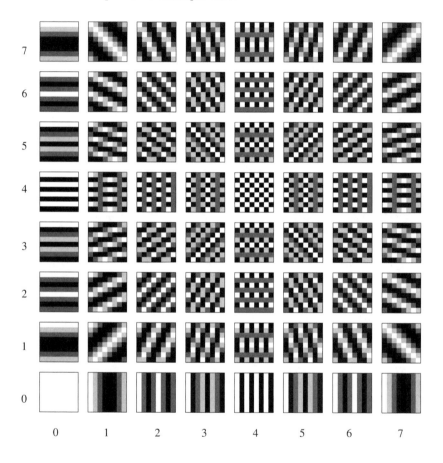

Figure 2.7: Real part of the Fourier transform basis images.

The values of each image have been linearly scaled to vary between 0 (black) and 255 (white). The numbers along the left and the bottom indicate which row of matrix U was multiplied with which row to produce the corresponding image.

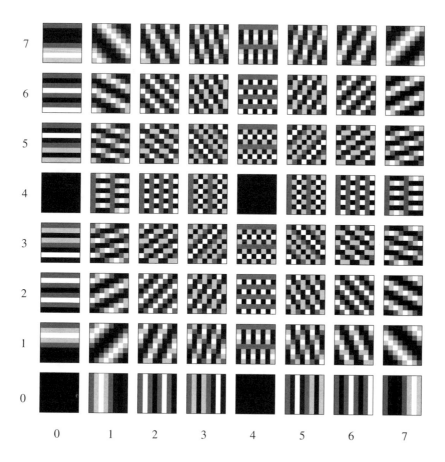

Figure 2.8: Imaginary part of the Fourier transform basis images.

Example 2.28

Compute the real and imaginary parts of the discrete Fourier transform of image:

$$g = \begin{pmatrix} 0 & 0 & 0 & 0 \\ 0 & 1 & 1 & 0 \\ 0 & 1 & 1 & 0 \\ 0 & 0 & 0 & 0 \end{pmatrix}$$

We shall use matrix U of Example 2.25.

$$A = UgU$$

$$gU = \frac{1}{2}\begin{pmatrix} 0 & 0 & 0 & 0 \\ 0 & 1 & 1 & 0 \\ 0 & 1 & 1 & 0 \\ 0 & 0 & 0 & 0 \end{pmatrix}\begin{pmatrix} 1 & 1 & 1 & 1 \\ 1 & -j & -1 & j \\ 1 & -1 & 1 & -1 \\ 1 & j & -1 & -j \end{pmatrix} = \frac{1}{2}\begin{pmatrix} 0 & 0 & 0 & 0 \\ 2 & -1-j & 0 & j-1 \\ 2 & -1-j & 0 & j-1 \\ 0 & 0 & 0 & 0 \end{pmatrix}$$

$$A = \frac{1}{4}\begin{pmatrix} 1 & 1 & 1 & 1 \\ 1 & -j & -1 & j \\ 1 & -1 & 1 & -1 \\ 1 & j & -1 & -j \end{pmatrix}\begin{pmatrix} 0 & 0 & 0 & 0 \\ 2 & -1-j & 0 & j-1 \\ 2 & -1-j & 0 & j-1 \\ 0 & 0 & 0 & 0 \end{pmatrix}$$

$$= \frac{1}{4}\begin{pmatrix} 4 & -2-2j & 0 & 2j-2 \\ -2j-2 & 2j & 0 & 2 \\ 0 & 0 & 0 & 0 \\ 2j-2 & 2 & 0 & -2j \end{pmatrix}$$

$$= \begin{pmatrix} 1 & -\frac{1+j}{2} & 0 & \frac{j-1}{2} \\ -\frac{1+j}{2} & \frac{j}{2} & 0 & \frac{1}{2} \\ 0 & 0 & 0 & 0 \\ \frac{j-1}{2} & \frac{1}{2} & 0 & -\frac{j}{2} \end{pmatrix}$$

$$Re(A) = \begin{pmatrix} 1 & -\frac{1}{2} & 0 & -\frac{1}{2} \\ -\frac{1}{2} & 0 & 0 & \frac{1}{2} \\ 0 & 0 & 0 & 0 \\ -\frac{1}{2} & \frac{1}{2} & 0 & 0 \end{pmatrix} \quad and \quad Im(A) = \begin{pmatrix} 0 & -\frac{1}{2} & 0 & \frac{1}{2} \\ -\frac{1}{2} & \frac{1}{2} & 0 & 0 \\ 0 & 0 & 0 & 0 \\ \frac{1}{2} & 0 & 0 & -\frac{1}{2} \end{pmatrix}$$

Example 2.29

Show the different stages of the Fourier transform of the image of Example 2.14.

The eight images shown in Figure 2.9 *are the reconstructed images when one, two,..., eight lines of the matrix U were used for the reconstruction. The sum of the squared errors for each reconstructed image are:*

$$
\begin{array}{lr}
Error\ for\ image\ a: & 366394 \\
Error\ for\ image\ b: & 285895 \\
Error\ for\ image\ c: & 234539 \\
Error\ for\ image\ d: & 189508 \\
Error\ for\ image\ e: & 141481 \\
Error\ for\ image\ f: & 119612 \\
Error\ for\ image\ g: & 71908 \\
Error\ for\ image\ h: & 0 \\
\end{array}
$$

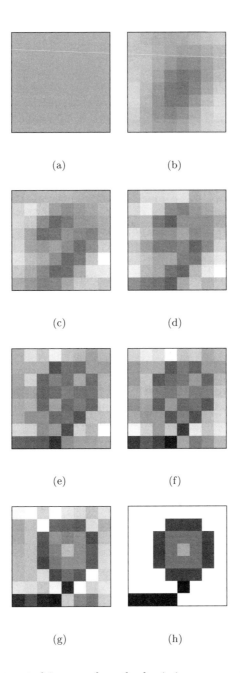

(a) (b)

(c) (d)

(e) (f)

(g) (h)

Figure 2.9: Reconstructed image when the basis images used are those created from the first one, two,..., eight lines of matrix U of Example 2.27, from top left to bottom right respectively.

Why is the discrete Fourier transform more commonly used than the other transforms?

The major advantage of the discrete Fourier transform over the Walsh transform is that it can be defined in such a way that it obeys the convolution theorem. One can define a corresponding theorem for the Walsh functions but the relationship between the Walsh transform and the convolution is not as simple and it cannot be implemented cheaply on a computer. The convolution theorem makes the Fourier transform by far the most attractive in image processing.

Apart from that, the Fourier transform uses very detailed basis functions, so it can approximate an image with smaller error than the other transforms for fixed number of terms retained. This can be judged from the reconstruction errors of Example 2.29, when compared with the reconstruction errors of Examples 2.23 and 2.24. We must compare the errors for reconstructed images (a), (b) and (d) which correspond to keeping the first 2^0, 2^1, and 2^2 basis functions respectively.

Note, however, that when we say we retained n number of basis images, in the case of the Fourier transform we require $2n$ coefficients for the reconstruction, while in the case of Haar and Walsh transforms we require only n coefficients. This is because the Fourier coefficients are complex and both their real and imaginary parts have to be stored or transmitted.

What does the convolution theorem state?

The convolution theorem states that: The Fourier transform of the convolution of two functions is equal to the product of the Fourier transforms of the two functions. If the functions are images defined over a finite space, this theorem is true only if we assume that each image is repeated periodically in all directions.

B2.4: If a function is the convolution of two other functions, what is the relationship of its DFT with the DFTs of the two functions?

Suppose that we convolve two discrete 2-dimensional functions $g(n,m)$ and $w(n,m)$ to produce another function $v(n,m)$:

$$v(n,m) = \sum_{n'=0}^{N-1} \sum_{m'=0}^{M-1} g(n-n', m-m') w(n',m')$$ (2.49)

Let us say that the discrete Fourier transforms of these three functions are $\hat{v}, \hat{g}$ and $\hat{w}$ respectively. To find a relationship between them, we shall try to calculate the DFT of $v(n,m)$. For this purpose, we multiply both sides of equation (2.49) with the kernel

$$\frac{1}{\sqrt{NM}} \exp\left[-j2\pi\left(\frac{pn}{N} + \frac{qm}{M}\right)\right] \tag{2.50}$$

and sum over all m and n. Equation (2.49) then becomes:

$$\frac{1}{\sqrt{NM}} \sum_{n=0}^{N-1}\sum_{m=0}^{M-1} v(n,m) e^{-2\pi j[\frac{pn}{N} + \frac{qm}{M}]}$$

$$= \frac{1}{\sqrt{NM}} \sum_{n'=0}^{N-1}\sum_{m'=0}^{M-1}\sum_{n=0}^{N-1}\sum_{m=0}^{M-1} g(n-n', m-m') w(n', m') e^{-2\pi j[\frac{pn}{N} + \frac{qm}{M}]} \tag{2.51}$$

We recognize the left hand side of this expression as being the discrete Fourier transform of v; i.e.

$$\hat{v}(p,q) = \frac{1}{\sqrt{NM}} \sum_{n'=0}^{N-1}\sum_{m'=0}^{M-1}\sum_{n=0}^{N-1}\sum_{m=0}^{M-1} g(n-n', m-m') w(n', m') e^{-2\pi j\frac{pn}{N}} e^{-2\pi j\frac{qm}{M}}$$

We would like to split this into the product of two double sums. To achieve this, we must have independent indices for g and w. We introduce new indices:

$$n - n' \equiv n'', \qquad m - m' \equiv m''$$

Then $n = n' + n''$, $m = m' + m''$ and we must find the limits of m'' and n''. To do that, we map the area over which we sum in the (n, m) space into the corresponding area in the (n'', m'') space:

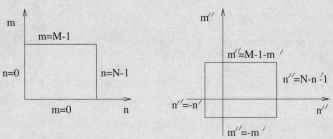

The area over which we sum in the (n, m) space is enclosed by four lines with equations given on the left hand side of the list below. Each of these lines is transformed to a line in the (n'', m'') space, given on the right hand side of the list below. These transformed lines define the new limits of summation.

$$m = 0 \quad \rightarrow \quad m'' = -m'$$
$$m = M - 1 \quad \rightarrow \quad m'' = M - m' - 1$$
$$n = 0 \quad \rightarrow \quad n'' = -n'$$
$$n = N - 1 \quad \rightarrow \quad n'' = N - n' - 1$$

Then the last expression for $\hat{v}(p,q)$ becomes:

$$\hat{v}(p,q) = \frac{1}{\sqrt{NM}} \sum_{n'=0}^{N-1} \sum_{m'=0}^{M-1} w(n',m') e^{-2\pi j [\frac{pn'}{N} + \frac{qm'}{M}]}$$

$$\sum_{m''=m'}^{M-1-m'} \sum_{n''=n'}^{N-1-n'} g(n'',m'') e^{-2\pi j [\frac{pn''}{N} + \frac{qm''}{M}]} \tag{2.52}$$

Let us concentrate on the last two sums of (2.52). Let us call that factor T. We can separate the negative from the positive indices and write:

$$T \equiv \left[\sum_{m''=-m'}^{-1} + \sum_{m''=0}^{M-1-m'} \right] \left[\sum_{n''=-n'}^{-1} + \sum_{n''=0}^{N-1-n'} \right] g(n'',m'') e^{-2\pi j [\frac{pn''}{N} + \frac{qm''}{M}]} \tag{2.53}$$

Clearly the two arrays g and w are not defined for negative indices. We may choose to extend their definition for indices outside the range $[0, N-1]$, $[0, M-1]$ in any which way suits us. Let us examine the factor:

$$\sum_{m''=-m'}^{-1} g(n'',m'') e^{-2\pi j q \frac{m''}{M}}$$

We define a new variable $m''' \equiv M + m''$. Then the above expression becomes:

$$\sum_{m'''=M-m'}^{M-1} g(n'', -M + m''') e^{-2\pi j q \frac{m'''}{M}} e^{-2\pi j q}$$

Now if we **choose** to define: $g(n'', m''' - M) \equiv g(n'', m''')$, the above sum is:

$$\sum_{m'''=M-m'}^{M-1} g(n'', m''') e^{-2\pi j q \frac{m'''}{M}}, \quad \text{since} \quad e^{-2\pi j q} = 1$$

m''' is a dummy index and we can call it anything we like. Say we call it m''. Then the above expression becomes:

$$\sum_{m''=M-m'}^{M-1} g(n'', m'') e^{-2\pi j q \frac{m''}{M}}$$

This term is added to the term

$$\sum_{m''=0}^{M-m'-1} g(n'', m'') e^{-2\pi j q \frac{m''}{M}}$$

and the two together can be written as:

$$\sum_{m''=0}^{M-1} g(n'', m'') e^{-2\pi j q \frac{m''}{M}}$$

We can work in a similar way for the summation over index n'', and assume that g is periodic also in its first index with period N. Then, under the assumption we made about the definition of g outside its real area of definition, the double sum we called T is:

$$T = \sum_{m''=0}^{M-1} \sum_{n''=0}^{N-1} g(n'', m'') e^{-2\pi j [\frac{pn''}{N} + \frac{qm''}{M}]} \tag{2.54}$$

This does not contain indices m', n' and therefore it is a factor that multiplies the double sum over n' and m' in (2.52). Further, it is recognized to be $\sqrt{MN}\hat{g}(p, q)$. Similarly in (2.52) we recognize the discrete Fourier transform of w and thus (2.52) becomes:

$$\boxed{\hat{v}(p, q) = \sqrt{MN}\hat{g}(p, q)\hat{w}(p, q)} \tag{2.55}$$

under the **assumptions** that:

$$
\begin{aligned}
g(n, m) &\equiv g(n - N, m - M) \\
w(n, m) &\equiv w(n - N, m - M) \\
g(n, m) &\equiv g(n, m - M) \\
w(n, m) &\equiv w(n, m - M) \\
g(n, m) &\equiv g(n - N, m) \\
w(n, m) &\equiv w(n - N, m)
\end{aligned}
$$

i.e. we assume that the image arrays g and w are defined in the whole (n, m) space periodically, with period M and N in the two directions respectively. This corresponds to the *time convolution theorem*.

The *frequency convolution theorem* would have exactly the same form. Because of the symmetry between the discrete Fourier transform and its inverse, this implies that the discrete Fourier transforms of these functions are also periodic in the whole (q, p) space, with periods N, M respectively.

Note: The factor $\sqrt{MN}$ in (2.55) appears because we defined the discrete Fourier transform so that the direct and the inverse ones are entirely symmetric.

Example 2.30 (B)

$g(n,m)$ and $w(n,m)$ are two $N \times M$ images. Their DFTs are $\hat{g}(p,q)$ and $\hat{w}(p,q)$ respectively. We create the image $x(n,m)$ by

$$x(n,m) \;=\; g(n,m) \times w(n,m) \qquad\qquad (2.56)$$

Express the DFT of $x(n,m)$, $\hat{x}(p,q)$, in terms of $\hat{g}(p,q)$ and $\hat{w}(p,q)$.

Let us take the DFT of both sides of equation (2.56):

$$\hat{x}(k,l) \;=\; \frac{1}{\sqrt{MN}} \sum_{m=0}^{M-1}\sum_{n=0}^{N-1} g(n,m)w(n,m)e^{-2\pi j\left[\frac{km}{M}+\frac{ln}{N}\right]}$$

Substitute g and w in terms of $\hat{g}$ and $\hat{w}$:

$$\hat{x}(k,l) = \frac{1}{\sqrt{MN}}\frac{1}{MN}\sum_{m=0}^{M-1}\sum_{n=0}^{N-1}\sum_{q=0}^{M-1}\sum_{p=0}^{N-1}\hat{g}(p,q)e^{-2\pi j\left[\frac{pn}{N}+\frac{qm}{M}\right]}$$

$$\sum_{r=0}^{M-1}\sum_{s=0}^{N-1}\hat{w}(s,r)e^{-2\pi j\left[\frac{ns}{N}+\frac{rm}{M}\right]}e^{-2\pi j\left[\frac{km}{M}+\frac{ln}{N}\right]} =$$

$$\frac{1}{\sqrt{MN}}\frac{1}{MN}\sum_{m=0}^{M-1}\sum_{n=0}^{N-1}\sum_{q=0}^{M-1}\sum_{p=0}^{N-1}\sum_{r=0}^{M-1}\sum_{s=0}^{N-1}\hat{g}(p,q)\hat{w}(s,r)e^{2\pi j\left[\frac{n(s+p)}{N}+\frac{m(r+q)}{M}\right]}e^{-2\pi j\left[\frac{km}{M}+\frac{ln}{N}\right]}$$

Let us introduce some new variables u and v instead of s and r:

$$u \equiv r + q \rightarrow limits: q \; to \; M-1+q$$
$$v \equiv p + s \rightarrow limits: p \; to \; N-1+p$$

Therefore

$$\hat{x}(k,l) = \frac{1}{\sqrt{MN}}\frac{1}{MN}\sum_{m=0}^{M-1}\sum_{n=0}^{N-1}\sum_{q=0}^{M-1}\sum_{p=0}^{N-1}\sum_{u=q}^{M-1+q}\sum_{v=p}^{N-1+p}\hat{g}(p,q)$$

$$\hat{w}(v-p,u-q)e^{-2\pi j\left[\frac{nv}{N}+\frac{mu}{M}\right]}e^{-2\pi j\left[\frac{km}{M}+\frac{ln}{N}\right]}$$

$$\Rightarrow \hat{x}(k,l) = \frac{1}{\sqrt{MN}}\frac{1}{MN}\sum_{q=0}^{M-1}\sum_{p=0}^{N-1}\sum_{u=q}^{M-1+q}\sum_{v=p}^{N-1+p}\hat{g}(p,q)$$

$$\hat{w}(v-p,u-q)\sum_{m=0}^{M-1}\sum_{n=0}^{N-1}e^{2\pi j\left[\frac{m(u-k)}{M}+\frac{n(v-l)}{N}\right]}$$

We know that (see equation (2.43)):

$$\sum_{m=0}^{M-1} e^{2\pi j \frac{m}{M} t} \begin{cases} M & \text{for } t = 0 \\ 0 & \text{for } t \neq 0 \end{cases}$$

Therefore:

$$\hat{x}(k, l) = \frac{1}{\sqrt{MN}} \frac{1}{MN} \sum_{q=0}^{M-1} \sum_{p=0}^{N-1} \sum_{u=q}^{M-1+q} \sum_{v=p}^{N-1+p} \hat{g}(p, q)$$

$$\hat{w}(v-p, u-q) M\delta(u-q) N\delta(v-p)$$

$$= \underbrace{\frac{1}{\sqrt{MN}} \sum_{q=0}^{M-1} \sum_{p=0}^{N-1} \hat{g}(p, q) \hat{w}(k-p, l-q)}_{\text{Convolution of } \hat{g} \text{ with } \hat{w}}$$

Example 2.31

Show that if $g(k, l)$ is an $M \times N$ image defined as a periodic function with periods N and M in the whole (k, l) space, its DFT $\hat{g}(m, n)$ is also periodic in the (m, n) space, with the same periods.

We must show that $\hat{g}(m + M, n + N) = \hat{g}(m, n)$. We start from the definition of $\hat{g}(m, n)$:

$$\hat{g}(m, n) = \frac{1}{\sqrt{MN}} \sum_{k=0}^{M-1} \sum_{l=0}^{N-1} g(k, l) e^{-2\pi j \left[\frac{km}{M} + \frac{ln}{N} \right]}$$

Then:

$$\hat{g}(m + M, n + N) = \frac{1}{\sqrt{MN}} \sum_{k=0}^{M-1} \sum_{l=0}^{N-1} g(k, l) e^{-2\pi j \left[\frac{k(m+M)}{M} + \frac{l(n+N)}{N} \right]}$$

$$= \frac{1}{\sqrt{MN}} \sum_{k=0}^{M-1} \sum_{l=0}^{N-1} g(k, l) e^{-2\pi jk} e^{-2\pi jl} = \hat{g}(m, n)$$

Example 2.32 (B)

Show that if $v(n, m)$ is defined by:

$$v(n, m) \equiv \sum_{n'=0}^{N-1} \sum_{m'=0}^{M-1} g(n - n', m - m')w(n', m') \tag{2.57}$$

where $g(n, m)$ and $w(n, m)$ are two periodically defined images with periods N and M in the two variables respectively, $v(n, m)$ is also given by:

$$v(n, m) = \sum_{n'=0}^{N-1} \sum_{m'=0}^{M-1} w(n - n', m - m')g(n', m') \tag{2.58}$$

Define some new variables:

$$n - n' \equiv k \Rightarrow n' = n - k$$
$$m - m' \equiv l \Rightarrow m' = m - l$$

Then equation (2.57) becomes:

$$\Rightarrow v(n, m) = \sum_{k=n}^{n-N+1} \sum_{l=m}^{m-M+1} g(k, l)w(n - k, m - l) \tag{2.59}$$

Consider the sum:

$$\sum_{k=n}^{n-N+1} g(k, l)w(n - k, m - l) = \sum_{k=-N+n+1}^{n} g(k, l)w(n - k, m - l)$$

$$= \underbrace{\sum_{k=-N}^{-1} g(k, l)w(n - k, m - l)}_{\text{Change variable } \tilde{k}=k+N} - \underbrace{\sum_{k=-N}^{-N+n} g(k, l)w(n - k, m - l)}_{\text{Change variable } \tilde{\tilde{k}}=k+N}$$

$$+ \sum_{k=0}^{n} g(k, l)w(n - k, m - l)$$

$$= \sum_{\tilde{k}=0}^{N-1} g(\tilde{k} - N, l)w(n - \tilde{k} + N, m - l) - \sum_{\tilde{\tilde{k}}=0}^{n} g(\tilde{\tilde{k}} - N, l)w(n - \tilde{\tilde{k}} + N, m - l)$$

$$+ \sum_{k=0}^{n} g(k, l)w(n - k, m - l)$$

> $g \text{ periodic} \Rightarrow g(k - N, l) = g(k, l)$
> $w \text{ periodic} \Rightarrow w(s + N, t) = w(s, t)$
>
> *Therefore, the last two sums are identical and cancel each other, and the summation over k in (2.59) is from 0 to $N - 1$. Similarly, we can show that the summation over l in (2.59) is from 0 to $M - 1$, and thus prove equation (2.58).*

How can we display the discrete Fourier transform of an image?

Suppose that the discrete Fourier transform of an image is $\hat{g}(p, q)$. These quantities, $\hat{g}(p, q)$, are the coefficients of the expansion of the image into discrete Fourier functions each of which corresponds to a different pair of spatial frequencies in the 2-dimensional (k, l) plane. As p and q increase, the contribution of these high frequencies to the image becomes less and less important and thus the values of the corresponding coefficients $\hat{g}(p, q)$ become smaller. If we want to display these coefficients we may find it difficult because their values will span a great range. So, for displaying purposes, and only for that, people use instead the logarithmic function:

$$d(p, q) \equiv \log(1 + |\hat{g}(p, q)|).$$

This function is then scaled into a displayable range of grey values and displayed instead of $\hat{g}(p, q)$. Notice that when $\hat{g}(p, q) = 0$, $d(p, q) = 0$ too. This function has the property of reducing the ratio between the high values of $\hat{g}$ and the small ones, so that small and large values can be displayed in the same scale. For example, if $\hat{g}_{max} = 10$ and $\hat{g}_{min} = 0.1$, to draw these numbers on the same graph is rather difficult, as their ratio is 100. However, $log(11) = 1.041$ and $log(1.1) = 0.041$ and their ratio is 25. So both numbers can be drawn on the same scale more easily.

What happens to the discrete Fourier transform of an image if the image is rotated?

We rewrite here the definition of the discrete Fourier transform, i.e. equation (2.37) for a square image (set $M = N$):

$$\hat{g}(m, n) = \frac{1}{N} \sum_{k=0}^{N-1} \sum_{l=0}^{N-1} g(k, l) e^{-2\pi j \frac{km + ln}{N}} \tag{2.60}$$

We can introduce polar coordinates on the planes (k, l) and (m, n) as follows: $k = r \cos\theta$, $l = r \sin\theta$, $m = \omega \cos\phi$, $n = \omega \sin\phi$. We note that $km + ln = r\omega(\cos\theta \cos\phi + \sin\theta \sin\phi) = r\omega \cos(\theta - \phi)$. Then equation (2.60) becomes:

$$\hat{g}(\omega, \phi) = \frac{1}{N} \sum_{k=0}^{N-1} \sum_{l=0}^{N-1} g(r, \theta) e^{-2\pi j \frac{r\omega \cos(\theta - \phi)}{N}} \tag{2.61}$$

Variables k and l over which we sum do not appear in the summand explicitly. However, they are there implicitly, and the summation is supposed to happen over all relevant points. From the values of k and l, we are supposed to find the corresponding values of r and θ.

Suppose now that we rotate $g(r, \theta)$ by an angle θ_0. It becomes $g(r, \theta + \theta_0)$. We want to find the discrete Fourier transform of this rotated function. Formula (2.61) is another "slot machine". We slot in the appropriate place the function the transform of which we require, and out comes its DFT. Therefore, we shall use formula (2.61) to calculate the DFT of $g(r, \theta + \theta_0)$ by simply replacing $g(r, \theta)$ by $g(r, \theta + \theta_0)$. We denote the DFT of $g(r, \theta + \theta_0)$ by $\hat{\hat{g}}(\omega, \phi)$. We get:

$$\hat{\hat{g}}(\omega, \phi) = \frac{1}{N} \underbrace{\sum\sum}_{\text{all points}} g(r, \theta + \theta_0)e^{-2\pi j \frac{r\omega \cos(\theta - \phi)}{N}} \tag{2.62}$$

To find the relationship between $\hat{\hat{g}}(\omega, \phi)$ and $\hat{g}(\omega, \phi)$ we have somehow to make $g(r, \theta)$ appear on the right hand side of this expression. For this purpose, we introduce a new variable, $\tilde{\theta} \equiv \theta + \theta_0$ in (2.62):

$$\hat{\hat{g}}(\omega, \phi) = \frac{1}{N} \underbrace{\sum\sum}_{\text{all points}} g(r, \tilde{\theta})e^{-2\pi j \frac{r\omega \cos(\tilde{\theta} - \theta_0 - \phi)}{N}} \tag{2.63}$$

Then on the right hand side we recognize the DFT of the unrotated image calculated at $\phi + \theta_0$ instead of ϕ: $\hat{g}(\omega, \phi + \theta_0)$. That is, we have:

$$\hat{\hat{g}}(\omega, \phi) = \hat{g}(\omega, \phi + \theta_0) \tag{2.64}$$

The DFT of the image rotated by θ_0 = the DFT of the unrotated image rotated by the same angle θ_0.

Example 2.33

Rotate the image of Example 2.26 anticlockwise by 90^0 and recalculate its discrete Fourier transform. Thus, verify the relationship between the discrete Fourier transform of a 2D image and the discrete Fourier transform of the same image rotated by angle θ_0.

The rotated by 90^0 image is:

$$\begin{pmatrix} 0 & 0 & 0 & 0 \\ 1 & 1 & 1 & 1 \\ 0 & 0 & 0 & 0 \\ 0 & 0 & 0 & 0 \end{pmatrix}$$

To calculate its DFT we multiply it first from the right with matrix U of Example
2.25, and then multiply the result from the left with the same matrix U:

$$\begin{pmatrix} 0 & 0 & 0 & 0 \\ 1 & 1 & 1 & 1 \\ 0 & 0 & 0 & 0 \\ 0 & 0 & 0 & 0 \end{pmatrix} \frac{1}{2} \begin{pmatrix} 1 & 1 & 1 & 1 \\ 1 & -j & -1 & j \\ 1 & -1 & 1 & -1 \\ 1 & j & -1 & -j \end{pmatrix} = \frac{1}{2} \begin{pmatrix} 0 & 0 & 0 & 0 \\ 4 & 0 & 0 & 0 \\ 0 & 0 & 0 & 0 \\ 0 & 0 & 0 & 0 \end{pmatrix}$$

$$\frac{1}{2}\begin{pmatrix} 1 & 1 & 1 & 1 \\ 1 & -j & -1 & j \\ 1 & -1 & 1 & -1 \\ 1 & j & -1 & -j \end{pmatrix} \frac{1}{2}\begin{pmatrix} 0 & 0 & 0 & 0 \\ 4 & 0 & 0 & 0 \\ 0 & 0 & 0 & 0 \\ 0 & 0 & 0 & 0 \end{pmatrix}$$

$$= \frac{1}{4}\begin{pmatrix} 4 & 0 & 0 & 0 \\ -4 & 0 & 0 & 0 \\ 4 & 0 & 0 & 0 \\ -4 & 0 & 0 & 0 \end{pmatrix} = \begin{pmatrix} 1 & 0 & 0 & 0 \\ -1 & 0 & 0 & 0 \\ 1 & 0 & 0 & 0 \\ -1 & 0 & 0 & 0 \end{pmatrix}$$

By comparing the above result with the result of Example 2.26 we see that the
discrete Fourier transform of the rotated image is the discrete Fourier transform
of the unrotated image rotated by $90°$.

**What happens to the discrete Fourier transform of an image if the image
is shifted?**

Suppose that we shift the image to the point (k_0, l_0), so that it becomes $g(k-k_0, l-l_0)$.
To calculate the DFT of the shifted image, we slot this function into formula (2.60).
We denote the DFT of $g(k - k_0, l - l_0)$ by $\hat{\hat{g}}(m, n)$ and obtain:

$$\hat{\hat{g}}(m,n) = \frac{1}{N}\sum_{k=0}^{N-1}\sum_{l=0}^{N-1} g(k - k_0, l - l_0)e^{-2\pi j \frac{km+ln}{N}} \tag{2.65}$$

To find a relationship between $\hat{\hat{g}}(m, n)$ and $\hat{g}(m, n)$, we must somehow make $g(k, l)$
appear on the right hand side of this expression. For this purpose, we define new
variables $k' \equiv k - k_0$ and $l' \equiv l - l_0$. Then:

$$\hat{\hat{g}}(m,n) = \frac{1}{N}\sum_{k'=-k_0}^{N-1-k_0}\sum_{l'=-l_0}^{N-1-l_0} g(k', l')e^{-2\pi j \frac{k'm+l'n}{N}}e^{-2\pi j \frac{k_0 m + l_0 n}{N}} \tag{2.66}$$

Because of the assumed periodic repetition of the image in all directions, where
exactly we perform the summation (i.e. between which indices) does not really matter,
as long as the right sized window is used for the summation. In other words, as long
as summation indices k' and l' are allowed to take N consecutive values each, it does
not matter where they start from. So, we may assume in the expression above that

k' and l' take values from 0 to $N-1$. We also notice that factor $e^{-2\pi j \frac{k_0 m + l_0 n}{N}}$ is independent of k' and l' and therefore can come out of the summation. Then we recognize in (2.66) the DFT of $g(k, l)$ appearing on the right hand side. (Note that k', l' are dummy indices and it makes no difference whether we call them k', l' or k, l.) We have therefore:

$$\hat{\hat{g}}(m, n) = \hat{g}(m, n) e^{-2\pi j \frac{k_0 m + l_0 n}{N}} \tag{2.67}$$

The DFT of the shifted image = the DFT of the unshifted image $\times e^{-2\pi j \frac{k_0 m + l_0 n}{N}}$

Similarly, one can show that:

The shifted DFT of an image = the DFT of $[$ image $\times e^{2\pi j \frac{m_0 k + n_0 l}{N}}]$

or

$$\hat{g}(m - m_0, n - n_0) = \text{DFT of } [\text{ image} \times e^{2\pi j \frac{m_0 k + n_0 l}{N}}]$$

What is the relationship between the average value of a function and its DFT?

The average value of a function is given by:

$$\bar{g} = \frac{1}{N^2} \sum_{k=0}^{N-1} \sum_{l=0}^{N-1} g(k, l) \tag{2.68}$$

If we set $m = n = 0$ in (2.60) we get:

$$\hat{g}(0, 0) = \frac{1}{N} \sum_{k=0}^{N-1} \sum_{l=0}^{N-1} g(k, l) \tag{2.69}$$

Therefore, the mean of an image and the *direct component* (or *dc*) of its DFT are related by:

$$\bar{g} = \frac{1}{N} \hat{g}(0, 0) \tag{2.70}$$

Example 2.34

Confirm the relationship between the average of image

$$g = \begin{pmatrix} 0 & 0 & 0 & 0 \\ 0 & 1 & 1 & 0 \\ 0 & 1 & 1 & 0 \\ 0 & 0 & 0 & 0 \end{pmatrix}$$

and its discrete Fourier transform.

Apply the discrete Fourier transform formula (2.37) for $N = M = 4$ and for $m = n = 0$:

$$\hat{g}(0,0) \equiv \frac{1}{4} \sum_{k=0}^{3} \sum_{l=0}^{3} g(k,l) = \frac{1}{4}(0+0+0+0+0+1+1+0+0+1+1+0$$

$$+0+0+0+0+0) = 1$$

The mean of g is:

$$\bar{g} \equiv \frac{1}{16} \sum_{k=0}^{3} \sum_{l=0}^{3} g(k,l) = \frac{1}{16}(0+0+0+0+0+1+1+0+0+1+1+0$$

$$+0+0+0+0+0) = \frac{4}{16} = \frac{1}{4}$$

Thus $N\bar{g} = 4 \times \frac{1}{4} = 1$ and (2.70) is confirmed.

What happens to the DFT of an image if the image is scaled?

When we take the average of a discretized function over an area over which this function is defined, we implicitly perform the following operation: We divide the area into small elementary areas of size $\Delta x \times \Delta y$ say, take the value of the function at the centre of each of these little tiles and assume that it represents the value of the function over the whole tile. Thus, we sum and divide by the total number of tiles. So, really the average of a function is defined as:

$$\bar{g} = \frac{1}{N^2} \sum_{x=0}^{N-1} \sum_{y=0}^{N-1} g(x,y) \Delta x \Delta y \tag{2.71}$$

We simply omit Δx and Δy because x and y are incremented by 1 at a time, so $\Delta x = 1$, $\Delta y = 1$. We also notice, from the definition of the discrete Fourier transform, that really, the discrete Fourier transform is a weighted average, where the value of $g(k,l)$ is multiplied by a different weight inside each little tile. Seeing the DFT that way, we realize that the correct definition of the discrete Fourier transform should include a factor $\Delta k \times \Delta l$ too, as the area of the little tile over which we assume the value of the function g to be constant. We omit it because $\Delta k = \Delta l = 1$. So, the formula for DFT that explicitly states this is:

$$\hat{g}(m,n) = \frac{1}{N} \sum_{k=0}^{N-1} \sum_{l=0}^{N-1} g(k,l) e^{-2\pi j(\frac{km+ln}{N})} \Delta k \Delta l \tag{2.72}$$

Now suppose that we change the scales in the (k,l) plane and $g(k,l)$ becomes $g(\alpha k, \beta l)$. We denote the discrete Fourier transform of the scaled g by $\hat{\hat{g}}(m,n)$. In order to calculate it, we must slot function $g(\alpha k, \beta l)$ in place of $g(k,l)$ in formula (2.72). We obtain:

$$\hat{\hat{g}}(m,n) = \frac{1}{N} \sum_{k=0}^{N-1} \sum_{l=0}^{N-1} g(\alpha k, \beta l) e^{-2\pi j \frac{km+ln}{N}} \Delta k \Delta l \tag{2.73}$$

We wish to find a relationship between $\hat{\hat{g}}(m,n)$ and $\hat{g}(m,n)$. Therefore, somehow we must make $g(k,l)$ appear on the right hand side of equation (2.73). For this purpose, we define new variables of summation $k' \equiv \alpha k$, $l' \equiv \beta l$. Then:

$$\hat{\hat{g}}(m,n) \;=\; \frac{1}{N} \sum_{k'=0}^{\alpha(N-1)} \sum_{l'=0}^{\beta(N-1)} g(k',l') e^{-2\pi j \frac{k'\frac{m}{\alpha}+l'\frac{n}{\beta}}{N}} \frac{\Delta k'}{\alpha} \frac{\Delta l'}{\beta} \tag{2.74}$$

The summation that appears in this expression spans all points over which function $g(k',l')$ is defined, except that the summation variables k' and l' are not incremented by 1 in each step. We recognize again the DFT of $g(k,l)$ on the right hand side of (2.74), calculated not at point (m,n) but at point $(\frac{m}{\alpha}, \frac{n}{\beta})$. Therefore, we can write:

$$\hat{\hat{g}}(m,n) \;=\; \frac{1}{\alpha\beta} \hat{g}(\frac{m}{\alpha}, \frac{n}{\beta}) \tag{2.75}$$

The DFT of the scaled function $= \dfrac{1}{|\text{product of scaling factors}|} \times$ the DFT of the unscaled function calculated at the same point inversely scaled.

B2.5: What is the Fast Fourier Transform?

All the transforms we have dealt with so far are separable. This means that they can be computed as two 1-dimensional transforms as opposed to one 2-dimensional transform. The discrete Fourier transform in 2 dimensions can be computed as two discrete Fourier transforms in 1 dimension, using special algorithms which are especially designed for speed and efficiency. We shall describe briefly here the Fast Fourier Transform algorithm called *successive doubling*. We shall work in 1 dimension. The discrete Fourier transform is defined by:

$$\hat{f}(u) = \frac{1}{N} \sum_{x=0}^{N-1} f(x) w_N^{ux} \tag{2.76}$$

where $w_N \equiv e^{\frac{-2\pi j}{N}}$. Assume now that $N = 2^n$. Then we can write N as $2M$ and substitute above:

$$\hat{f}(u) = \frac{1}{2M} \sum_{x=0}^{2M-1} f(x) w_{2M}^{ux} \tag{2.77}$$

We can separate the odd and even values of the argument of f. Then:

$$\hat{f}(u) = \frac{1}{2} \left\{ \frac{1}{M} \sum_{x=0}^{M-1} f(2x) w_{2M}^{u(2x)} + \frac{1}{M} \sum_{x=0}^{M-1} f(2x+1) w_{2M}^{u(2x+1)} \right\} \tag{2.78}$$

Obviously $w_{2M}^{2ux} = w_M^{ux}$ and $w_{2M}^{2ux+u} = w_M^{ux} w_{2M}^{u}$. Then:

$$\hat{f}(u) = \frac{1}{2} \left\{ \frac{1}{M} \sum_{x=0}^{M-1} f(2x) w_M^{ux} + \frac{1}{M} \sum_{x=0}^{M-1} f(2x+1) w_M^{ux} w_{2M}^{u} \right\} \tag{2.79}$$

We can write:

$$\hat{f}(u) \equiv \frac{1}{2} \left\{ \hat{f}_{even}(u) + \hat{f}_{odd}(u) w_{2M}^{u} \right\} \tag{2.80}$$

where we have defined $\hat{f}_{even}(u)$ to be the DFT of the even samples of function f and $\hat{f}_{odd}$ to be the DFT of the odd samples of function f. Formula (2.80), however, defines $\hat{f}(u)$ only for $u \leq M$. We need to define $\hat{f}(u)$ for $u = 0, 1, \ldots, N$, i.e. for u up to $2M$. For this purpose we apply formula (2.79) for argument of $\hat{f}$, $u + M$:

$$\hat{f}(u+M) = \frac{1}{2} \left\{ \frac{1}{M} \sum_{x=0}^{M-1} f(2x) w_M^{ux+Mx} + \frac{1}{M} \sum_{x=0}^{M-1} f(2x+1) w_M^{ux+Mx} w_{2M}^{u+M} \right\} \tag{2.81}$$

However:

$$w_M^{u+M} = w_M^u w_M^M = w_M^u$$
$$w_{2M}^{u+M} = w_{2M}^u w_{2M}^M = -w_{2M}^u \tag{2.82}$$

So:

$$\hat{f}(u+M) = \frac{1}{2} \left\{ \hat{f}_{even}(u) - \hat{f}_{odd}(u) w_{2M}^{u} \right\} \tag{2.83}$$

where

$$\hat{f}_{even}(u) \equiv \frac{1}{M} \sum_{x=0}^{M-1} f(2x) w_M^{ux}, \qquad \hat{f}_{odd}(u) \equiv \frac{1}{M} \sum_{x=0}^{M-1} f(2x+1) w_M^{ux} \tag{2.84}$$

We note that formulae (2.80) and (2.83) with definitions (2.84) fully define $\hat{f}(u)$. Thus, an N point transform can be computed as two $N/2$ point transforms given by equations (2.84). Then equations (2.80) and (2.83) can be used to calculate the full transform. It can be shown that the number of operations required reduces from being proportional to N^2 to being proportional to Nn; i.e. $N\log_2 N$. This is another reason why images with dimensions powers of 2 are preferred.

What is the discrete cosine transform?

If the rows of matrices h_c and h_r are discrete versions of a certain class of Chebyshev polynomials, we have the *even symmetrical cosine transform*. It is called this because it is equivalent to assuming that the image is reflected about two adjacent edges to form an image of size $2N \times 2N$. Then the Fourier transform of this symmetric image is taken and this is really a cosine transform. The *discrete cosine transform (DCT)* defined in this way has found wide use in JPEG coding according to which each image is divided into blocks of size 8×8. The cosine transform of each block is computed, and the coefficients of the transformation are coded and transmitted.

What is the "take home" message of this chapter?

This chapter presented the linear, unitary and separable transforms we apply to images. These transforms analyse each image into a linear superposition of elementary basis images. Usually these elementary images are arranged in increasing order of structure (detail). This allows us to represent an image with as much detail as we wish, by using only as many of these basis images as we like, starting from the first one. The optimal way to do that is to use as basis images those that are defined by the image itself, the eigenimages of the image. This, however, is not very efficient, as our basis images change from one image to the next. Alternatively, some bases of predefined images can be created with the help of orthonormal sets of functions. These bases try to capture the basic characteristics of all images. Once the basis used has been agreed, images can be communicated between different agents by simply transmitting the weights with which each of the basis images has to be multiplied before all of them are added to create the original image. The first one of these basis images is always a uniform image. The form of the rest in each basis depends on the orthonormal set of functions used to generate them. As these basic images are used to represent a large number of images, more of them are needed to represent a single image than if the eigenimages of the image itself were used for its representation. However, the gains in number of bits used come from the fact that the basis images are pre-agreed and they do not need to be stored or transmitted with each image separately.

The bases constructed with the help of orthonormal sets of discrete functions are more easy to implement in hardware. However, the basis constructed with the help of the orthonormal set of complex exponential functions is by far the most popular. The representation of an image in terms of it is called a discrete Fourier transform. Its popularity stems from the fact that manipulation of the weights with which the basis images are superimposed to form the original image, for the purpose of omitting, for example, certain details in the image, can be achieved by manipulating the image itself with the help of a simple convolution.

Chapter 3

Statistical Description of Images

What is this chapter about?

This chapter provides the necessary background for the statistical description of images from the signal processing point of view.

Why do we need the statistical description of images?

In various applications, we often have to deal with sets of images of a certain type; for example, X-ray images, traffic scene images, etc. Each image in the set may be different from all the others, but at the same time all images may share certain common characteristics. We need the statistical description of images so that we capture these common characteristics and use them in order to represent an image with fewer bits and reconstruct it with the minimum error "on average".

The first idea is then to try to minimize the **mean square error** in the reconstruction of the image, if the same image or a collection of similar images were to be transmitted and reconstructed several times, as opposed to minimizing the **square error** of each image separately. The second idea is that the data with which we would like to represent the image must be **uncorrelated**. Both these ideas lead to the **statistical** description of images.

Is there an image transformation that allows its representation in terms of uncorrelated data that can be used to approximate the image in the least mean square error sense?

Yes. It is called *Karhunen–Loeve* or *Hotelling transform*. It is derived by treating the image as an instantiation of a random field.

What is a random field?

A random field is a spatial function that assigns a random variable at each spatial position.

What is a random variable?

A *random variable* is the value we assign to the outcome of a random experiment.

How do we describe random variables?

Random variables are described in terms of their *distribution functions* which in turn are defined in terms of *the probability* of an *event* happening. An event is a collection of outcomes of the random experiment.

What is the probability of an event?

The *probability* of an event happening is a *non-negative* number which has the following properties:

(A) The probability of the event which includes all possible outcomes of the experiment is 1.

(B) The probability of two events which do not have any common outcomes is the sum of the probabilities of the two events separately.

What is the distribution function of a random variable?

The distribution function of a random variable f is a function which tells us how likely it is for f to be less than the argument of the function:

$$\underbrace{P_f(z)}_{\substack{\text{Distribution} \\ \text{function of } f}} = \underbrace{P}_{\text{probability}} \{ \underbrace{f}_{\substack{\text{random} \\ \text{variable}}} \leq \underbrace{z}_{\text{a number}} \} \qquad (3.1)$$

Clearly, $P_f(-\infty) = 0$ and $P_f(+\infty) = 1$.

Example 3.1

If $z_1 \leq z_2$, show that $P_f(z_1) \leq P_f(z_2)$.

Suppose that A is the event (i.e. the set of outcomes) which makes $f \leq z_1$ and B is the event which makes $f \leq z_2$. Since $z_1 \leq z_2$, $A \subseteq B \Rightarrow B = (B - A) \cup A$; i.e. the events $(B - A)$ and A do not have common outcomes (see the figure on the next page).

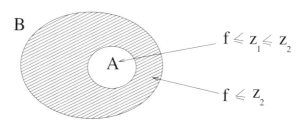

Then by property (B) in the definition of the probability of an event:

$$\mathcal{P}(B) \;=\; \mathcal{P}(B-A) + \mathcal{P}(A) \Rightarrow$$
$$P_f(z_2) \;=\; \underbrace{\mathcal{P}(B-A)}_{\substack{non\text{-}negative \\ number}} \;+\; P_f(z_1) \Rightarrow$$

$$P_f(z_2) \geq P_f(z_1) \tag{3.2}$$

Example 3.2

Show that:

$$\mathcal{P}(z_1 \leq f \leq z_2) = P_f(z_2) - P_f(z_1)$$

According to the notation of Example 3.1, $z_1 \leq f \leq z_2$ when the outcome of the random experiment belongs to $B - A$ (the shaded area in the above figure); i.e. $\mathcal{P}(z_1 \leq f \leq z_2) = P_f(B-A)$. Since $B = (B-A) \cup A$, $P_f(B-A) = P_f(B) - P_f(A)$ and the result follows.

What is the probability of a random variable taking a specific value?

If the random variable takes values from the set of real numbers, it has zero probability of taking a specific value. (This can be seen if in the result of example 3.2 we set $f = z_1 = z_2$.) However, it may have non-zero probability of taking a value within an infinitesimally small range of values. This is expressed by its *probability density function*.

What is the probability density function of a random variable?

The derivative of the distribution function of a random variable is called the *probability density function* of the random variable:

$$p_f(z) \equiv \frac{dP_f(z)}{dz} \tag{3.3}$$

The *expected* or *mean value* of the random variable f is defined by:

$$\mu_f \equiv E\{f\} \equiv \int_{-\infty}^{\infty} z p_f(z) dz \tag{3.4}$$

and the *variance* by:

$$\sigma_f^2 \equiv E\{(f - \mu_f)^2\} \equiv \int_{-\infty}^{+\infty} (z - \mu_f)^2 p_f(z) dz \tag{3.5}$$

The *standard deviation* is the positive square root of the variance, i.e. σ_f.

How do we describe many random variables?

If we have n random variables we can define their *joint distribution function*:

$$P_{f_1 f_2 \ldots f_n}(z_1, z_2, \ldots, z_n) \equiv P\{f_1 \leq z_1, f_2 \leq z_2, \ldots, f_n \leq z_n\} \tag{3.6}$$

We can also define their *joint probability density function*:

$$p_{f_1 f_2 \ldots f_n}(z_1, z_2, \ldots, z_n) \equiv \frac{\partial^n P_{f_1 f_2 \ldots f_n}(z_1, z_2, \ldots, z_n)}{\partial z_1 \partial z_2 \ldots \partial z_n} \tag{3.7}$$

What relationships may n random variables have with each other?

If the distribution of n random variables can be written as:

$$P_{f_1 f_2 \ldots f_n}(z_1, z_2, \ldots, z_n) = P_{f_1}(z_1) P_{f_2}(z_2) \ldots P_{f_n}(z_n) \tag{3.8}$$

then these random variables are called *independent*. They are called *uncorrelated* if:

$$E\{f_i f_j\} = E\{f_i\} E\{f_j\}, \; \forall i, j, \; i \neq j \tag{3.9}$$

Any two random variables are *orthogonal* to each other if:

$$E\{f_i f_j\} = 0 \tag{3.10}$$

The *covariance* of any two random variables is defined as:

$$c_{ij} \equiv E\{(f_i - \mu_{f_i})(f_j - \mu_{f_j})\} \tag{3.11}$$

Example 3.3

Show that if the covariance c_{ij} of two random variables is zero, the two
variables are uncorrelated.

Expanding the right hand side of the definition of the covariance we get:

$$\begin{aligned}
c_{ij} &= E\{f_i f_j - \mu_{f_i} f_j - \mu_{f_j} f_i + \mu_{f_i}\mu_{f_j}\} \\
&= E\{f_i f_j\} - \mu_{f_i} E\{f_j\} - \mu_{f_j} E\{f_i\} + \mu_{f_i}\mu_{f_j} \\
&= E\{f_i f_j\} - \mu_{f_i}\mu_{f_j} - \mu_{f_j}\mu_{f_i} + \mu_{f_i}\mu_{f_j} \\
&= E\{f_i f_j\} - \mu_{f_i}\mu_{f_j} \quad\quad\quad\quad\quad\quad\quad\quad (3.12)
\end{aligned}$$

*Notice that the operation of taking the expectation value of a fixed number has no
effect on it; i.e.* $E\{\mu_{fi}\} = \mu_{fi}$. *If* $c_{ij} = 0$, *we get:*

$$E\{f_i f_j\} = \mu_{f_i}\mu_{f_j} = E\{f_i\}E\{f_j\} \quad\quad\quad\quad (3.13)$$

which shows that f_i *and* f_j *are uncorrelated.*

How do we then define a random field?

If we define a random variable at every point in a 2-dimensional space we say that
we have a 2-dimensional *random field.* The position of the space where a random
variable is defined is like a parameter of the random field:

$$f(\mathbf{r}; \omega_i) \quad\quad\quad\quad\quad\quad\quad (3.14)$$

This function for fixed $\mathbf{r}$ is a random variable but for fixed ω_i (outcome) is a 2-
dimensional function in the plane, an image, say. As ω_i scans all possible outcomes of
the underlying statistical experiment, the random field represents a series of images.
On the other hand, for a given outcome, (fixed ω_i), the random field gives the grey
level values at the various positions in an image.

Example 3.4

Using an unloaded die, we conducted a series of experiments. Each
experiment consisted of throwing the die four times. The outcomes
$\{\omega_1, \omega_2, \omega_3, \omega_4\}$ of sixteen experiments are given below:

$\{1,2,1,6\},\{3,5,2,4\},\{3,4,6,6\},\{1,1,3,2\},\{3,4,4,4\},\{2,6,4,2\},$
$\{1,5,3,6\},\{1,2,6,4\},\{6,5,2,4\},\{3,2,5,6\},\{1,2,4,5\},\{5,1,1,6\},$
$\{2,5,3,1\},\{3,1,5,6\},\{1,2,1,5\},\{3,2,5,4\}$

If r is a 2−dimensional vector taking values:

$\{(1,1),(1,2),(1,3),(1,4),(2,1),(2,2),(2,3),(2,4),$
$(3,1),(3,2),(3,3),(3,4),(4,1),(4,2),(4,3),(4,4)\}$

give the series of images defined by the random field $f(\mathbf{r};\omega_i)$.

The first image is formed by placing the first outcome of each experiment in the corresponding position, the second by using the second outcome of each experiment, and so on. The ensemble of images we obtain is:

$$
\begin{pmatrix} 1 & 3 & 3 & 1 \\ 3 & 2 & 1 & 1 \\ 6 & 3 & 1 & 5 \\ 2 & 3 & 1 & 3 \end{pmatrix}
\begin{pmatrix} 2 & 5 & 4 & 1 \\ 4 & 6 & 5 & 2 \\ 5 & 2 & 2 & 1 \\ 5 & 1 & 2 & 2 \end{pmatrix}
\begin{pmatrix} 1 & 2 & 6 & 3 \\ 4 & 4 & 3 & 6 \\ 2 & 5 & 4 & 1 \\ 3 & 5 & 1 & 5 \end{pmatrix}
\begin{pmatrix} 6 & 4 & 6 & 2 \\ 4 & 2 & 6 & 4 \\ 4 & 6 & 5 & 6 \\ 1 & 6 & 5 & 4 \end{pmatrix}
\qquad (3.15)
$$

How can we relate two random variables that appear in the same random field?

For fixed $\mathbf{r}$ a random field becomes a random variable with an expectation value which depends on $\mathbf{r}$:

$$\mu_f(\mathbf{r}) = E\{f(\mathbf{r};\omega_i)\} = \int_{-\infty}^{+\infty} z p_f(z;\mathbf{r})dz \qquad (3.16)$$

Since for different values of $\mathbf{r}$ we have different random variables, $f(\mathbf{r_1};\omega_i)$ and $f(\mathbf{r_2};\omega_i)$, we can define their correlation, called *autocorrelation* (we use "auto" because the two variables come from the same random field) as:

$$R_{ff}(\mathbf{r_1},\mathbf{r_2}) = E\{f(\mathbf{r_1};\omega_i)f(\mathbf{r_2};\omega_i)\} = \int_{-\infty}^{+\infty}\int_{-\infty}^{+\infty} z_1 z_2 p_f(z_1,z_2;\mathbf{r_1},\mathbf{r_2})dz_1 dz_2 \quad (3.17)$$

The *autocovariance* $C(\mathbf{r_1},\mathbf{r_2})$ is defined by:

$$C_{ff}(\mathbf{r_1},\mathbf{r_2}) = E\{[f(\mathbf{r_1};\omega_i) - \mu_f(\mathbf{r_1})][f(\mathbf{r_2};\omega_i) - \mu_f(\mathbf{r_2})]\} \qquad (3.18)$$

Example 3.5

Show that:

$$C_{ff}(\mathbf{r_1}, \mathbf{r_2}) = R_{ff}(\mathbf{r_1}, \mathbf{r_2}) - \mu_f(\mathbf{r_1})\mu_f(\mathbf{r_2})$$

Starting from equation (3.18):

$$
\begin{aligned}
C_{ff}(\mathbf{r_1}, \mathbf{r_2}) &= E\left\{[f(\mathbf{r_1}; \omega_i) - \mu_f(\mathbf{r_1})][f(\mathbf{r_2}; \omega_i) - \mu_f(\mathbf{r_2})]\right\} \\
&= E\left\{f(\mathbf{r_1}; \omega_i)f(\mathbf{r_2}; \omega_i) - f(\mathbf{r_1}; \omega_i)\mu_f(\mathbf{r_2}) - \mu_f(\mathbf{r_1})f(\mathbf{r_2}; \omega_i)\right. \\
&\quad \left.+\mu_f(\mathbf{r_1})\mu_f(\mathbf{r_2})\right\} \\
&= E\{f(\mathbf{r_1}; \omega_i)f(\mathbf{r_2}; \omega_i)\} - E\{f(\mathbf{r_1}; \omega_i)\}\mu_f(\mathbf{r_2}) - \mu_f(\mathbf{r_1})E\{f(\mathbf{r_2}; \omega_i)\} \\
&\quad +\mu_f(\mathbf{r_1})\mu_f(\mathbf{r_2}) \\
&= R_{ff}(\mathbf{r_1}, \mathbf{r_2}) - \mu_f(\mathbf{r_1})\mu_f(\mathbf{r_2}) - \mu_f(\mathbf{r_1})\mu_f(\mathbf{r_2}) + \mu_f(\mathbf{r_1})\mu_f(\mathbf{r_2}) \\
&= R_{ff}(\mathbf{r_1}, \mathbf{r_2}) - \mu_f(\mathbf{r_1})\mu_f(\mathbf{r_2}) \quad\quad\quad\quad (3.19)
\end{aligned}
$$

How can we relate two random variables that belong to two different random fields?

If we have two random fields, i.e. two series of images generated by two different underlying random experiments, represented by f and g, we can define their *cross correlation*:

$$R_{fg}(\mathbf{r_1}, \mathbf{r_2}) = E\{f(\mathbf{r_1}; \omega_i)g(\mathbf{r_2}; \omega_j)\} \quad\quad\quad\quad (3.20)$$

and their *cross covariance*:

$$
\begin{aligned}
C_{fg}(\mathbf{r_1}, \mathbf{r_2}) &= E\{[f(\mathbf{r_1}; \omega_i) - \mu_f(\mathbf{r_1})][g(\mathbf{r_2}; \omega_j) - \mu_g(\mathbf{r_2})]\} \\
&= R_{fg}(\mathbf{r_1}, \mathbf{r_2}) - \mu_f(\mathbf{r_1})\mu_g(\mathbf{r_2}) \quad\quad\quad\quad (3.21)
\end{aligned}
$$

Two random fields are called *uncorrelated* if for any $\mathbf{r_1}$ and $\mathbf{r_2}$:

$$C_{fg}(\mathbf{r_1}, \mathbf{r_2}) = 0 \quad\quad\quad\quad (3.22)$$

This is equivalent to:

$$E\{f(\mathbf{r_1}; \omega_i)g(\mathbf{r_2}; \omega_j)\} = E\{f(\mathbf{r_1}; \omega_i)\}E\{g(\mathbf{r_2}; \omega_j)\} \quad\quad\quad\quad (3.23)$$

Example 3.6

Show that for two uncorrelated random fields we have:

$$E\left\{f(\mathbf{r_1};\omega_i)g(\mathbf{r_2};\omega_j)\right\} = E\left\{f(\mathbf{r_1};\omega_i)\right\}E\left\{g(\mathbf{r_2};\omega_j)\right\}$$

It follows trivially from the definition of uncorrelated random fields $(C_{fg}(\mathbf{r_1},\mathbf{r_2}) = 0)$ *and the expression:*

$$C_{fg}(\mathbf{r_1},\mathbf{r_2}) = E\{f(\mathbf{r_1};\omega_i)g(\mathbf{r_2};\omega_j)\} - \mu_f(\mathbf{r_1})\mu_g(\mathbf{r_2}) \qquad (3.24)$$

which can be proven in a similar way as Example 3.5.

Since we always have just one version of an image how do we calculate the expectation values that appear in all previous definitions?

We make the assumption that the image we have is a *homogeneous* random field and *ergodic*. The theorem of ergodicity which we then invoke allows us to replace the ensemble statistics with the spatial statistics of an image.

When is a random field homogeneous?

If the expectation value of a random field does not depend on $\mathbf{r}$, and if its autocorrelation function is translation invariant, then the field is called *homogeneous*.

   A translation invariant autocorrelation function depends on only one argument, the relative shifting of the positions at which we calculate the values of the random field:

$$R_{ff}(\mathbf{r_0}) = E\{f(\mathbf{r};\omega_i)f(\mathbf{r}+\mathbf{r_0};\omega_i)\} \qquad (3.25)$$

Example 3.7

Show that the autocorrelation function $R(\mathbf{r_1},\mathbf{r_2})$ **of a homogeneous random field depends only on the difference vector** $\mathbf{r_1} - \mathbf{r_2}$.

The autocorrelation function of a homogeneous random field is translation invariant. Therefore, for any translation vector $\mathbf{r_0}$ *we can write:*

$$\begin{aligned} R_{ff}(\mathbf{r_1},\mathbf{r_2}) &= E\{f(\mathbf{r_1};\omega_i)f(\mathbf{r_2};\omega_i)\} &= E\{f(\mathbf{r_1}+\mathbf{r_0};\omega_i)f(\mathbf{r_2}+\mathbf{r_0};\omega_i)\} \\ &= R_{ff}(\mathbf{r_1}+\mathbf{r_0},\mathbf{r_2}+\mathbf{r_0}) & \forall \mathbf{r_0} \qquad (3.26) \end{aligned}$$

Choosing $\mathbf{r_0} = -\mathbf{r_2}$ we see that for a homogeneous random field:

$$R_{ff}(\mathbf{r_1}, \mathbf{r_2}) = R_{ff}(\mathbf{r_1} - \mathbf{r_2}, 0) = R_{ff}(\mathbf{r_1} - \mathbf{r_2}) \tag{3.27}$$

How can we calculate the spatial statistics of a random field?

Given a random field we can define its spatial average as:

$$\mu(\omega_i) \equiv \lim_{S \to \infty} \frac{1}{S} \int_S f(\mathbf{r}; \omega_i) dx dy \tag{3.28}$$

where $\int_S$ is the integral over the whole space S with area S and $\mathbf{r} = (x, y)$. The result $\mu(\omega_i)$ is clearly a function of the outcome on which f depends; i.e. $\mu(\omega_i)$ is a random variable.

The spatial autocorrelation function of the random field is defined as:

$$R(\mathbf{r_0}; \omega_i) \equiv \lim_{S \to \infty} \frac{1}{S} \int_S f(\mathbf{r}; \omega_i) f(\mathbf{r} + \mathbf{r_0}; \omega_i) dx dy \tag{3.29}$$

This is another random variable.

When is a random field ergodic?

A random field is ergodic when it is ergodic with respect to the mean and with respect to the autocorrelation function.

When is a random field ergodic with respect to the mean?

A random field is said to be ergodic with respect to the mean, if it is homogeneous and its spatial average, defined by (3.28), is independent of the outcome on which f depends; i.e. it is a constant and is equal to the ensemble average defined by equation (3.16):

$$E\{f(\mathbf{r}; \omega_i)\} = \lim_{S \to \infty} \frac{1}{S} \int_S f(\mathbf{r}; \omega_i) dx dy = \mu = \text{a constant} \tag{3.30}$$

When is a random field ergodic with respect to the autocorrelation function?

A random field is said to be ergodic with respect to the autocorrelation function if it is homogeneous and its spatial autocorrelation function, defined by (3.29), is independent of the outcome of the experiment on which f depends, and depends only on the displacement $\mathbf{r_0}$, and it is equal to the ensemble autocorrelation function defined by equation (3.25):

$$E\{f(\mathbf{r};\omega_i)f(\mathbf{r} + \mathbf{r_0};\omega_i)\} = \lim_{S \to \infty} \frac{1}{S} \int_S f(\mathbf{r};\omega_i)f(\mathbf{r} + \mathbf{r_0};\omega_i)dxdy = R(\mathbf{r_0}) \quad (3.31)$$

Example 3.8

Assuming ergodicity, compute the autocorrelation matrix of the following image:

$$\begin{pmatrix} 1 & 2 & 1 \\ 1 & 2 & 1 \\ 1 & 2 & 1 \end{pmatrix}$$

A 3×3 image has the form:

$$\begin{pmatrix} g_{11} & g_{12} & g_{13} \\ g_{21} & g_{22} & g_{23} \\ g_{31} & g_{32} & g_{33} \end{pmatrix} \quad (3.32)$$

To compute its autocorrelation function we write it as a column vector by stacking its columns one under the other:

$$\mathbf{g} = \begin{pmatrix} g_{11} & g_{21} & g_{31} & g_{12} & g_{22} & g_{32} & g_{13} & g_{23} & g_{33} \end{pmatrix}^T \quad (3.33)$$

The autocorrelation matrix is given by: $C = E\{\mathbf{g}\mathbf{g}^T\}$. Instead of averaging over all possible versions of the image, we average over all pairs of pixels at the same relative position in the image since ergodicity is assumed. Thus, the autocorrelation matrix will have the following structure:

	g_{11}	g_{21}	g_{31}	g_{12}	g_{22}	g_{32}	g_{13}	g_{23}	g_{33}
g_{11}	A	B	C	D	E	F	G	H	I
g_{21}	B	A	B	J	D	E	K	G	H
g_{31}	C	B	A	L	J	D	M	K	G
g_{12}	D	J	L	A	B	C	D	E	F
g_{22}	E	D	J	B	A	B	J	D	E
g_{32}	F	E	D	C	B	A	L	J	D
g_{13}	G	K	M	D	J	L	A	B	C
g_{23}	H	G	K	E	D	J	B	A	B
g_{33}	I	H	G	F	E	D	C	B	A

(3.34)

The top row and the left-most column of this matrix show which elements of the image are associated with which in order to produce the corresponding entry in the matrix. A is the average square element:

$$\frac{\sum_{ij} g_{ij}^2}{9} = \frac{6 \times 1^2 + 3 \times 2^2}{9} = \frac{18}{9} = 2 \qquad (3.35)$$

B is the average value of the product of vertical neighbours. We have six such pairs. We must sum the product of their values and divide. The question is whether we must divide by the actual number of pairs of vertical neighbours we have, i.e. 6, or divide by the total number of pixels we have, i.e. 9. This issue is relevant to the calculation of all entries of matrix (3.34) apart from entry A. If we divide by the actual number of pairs, the correlation of the most distant neighbours (for which very few pairs are available) will be exaggerated. Thus, we chose to divide by the total number of pixels in the image knowing that this dilutes the correlation between distant neighbours, although this might be significant. This problem arises because of the finite size of the images. Note that formulae (3.29) and (3.28) really apply for infinite sized images. The problem is more significant in the case of this example which deals with a very small image for which border effects are exaggerated.

$$B = \frac{4 \times 1 + 2 \times 4}{9} = \frac{12}{9} = 1.33$$

C is the average product of vertical neighbours once removed. We have three such pairs:

$$C = \frac{2 \times 1 + 4}{9} = \frac{6}{9} = 0.67$$

D is the average product of horizontal neighbours. There are six such pairs:

$$D = \frac{6 \times 2}{9} = 1.33$$

E is the average product of diagonal neighbours. There are four such pairs:

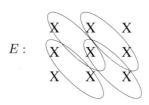

$$E: \qquad E = \frac{4 \times 2}{9} = 0.89$$

$$F: \qquad F = \frac{2 \times 2}{9} = 0.44$$

$$G: \qquad G = \frac{3 \times 1}{9} = 0.33$$

$$H: \qquad H = \frac{2 \times 1}{9} = 0.22$$

$$I: \qquad I = \frac{1 \times 1}{9} = 0.11$$

$$J: \qquad J = \frac{4 \times 2}{9} = 0.89$$

$$K: \qquad K = \frac{2 \times 1}{9} = 0.22$$

$$L: \begin{pmatrix} X & X & X \\ X & X & X \\ X & X & X \end{pmatrix} \qquad L = \frac{2\times 2}{9} = 0.44$$

$$M: \begin{pmatrix} X & X & X \\ X & X & X \\ X & X & X \end{pmatrix} \qquad M = \frac{1}{9} = 0.11$$

So, the autocorrelation matrix is:

$$\begin{pmatrix}
2 & 1.33 & 0.67 & 1.33 & 0.89 & 0.44 & 0.33 & 0.22 & 0.11 \\
1.33 & 2 & 1.33 & 0.89 & 1.33 & 0.89 & 0.22 & 0.33 & 0.22 \\
0.67 & 1.33 & 2 & 0.44 & 0.89 & 1.33 & 0.11 & 0.22 & 0.22 \\
1.33 & 0.89 & 0.44 & 2 & 1.33 & 0.67 & 1.33 & 0.89 & 0.44 \\
0.89 & 1.33 & 0.89 & 1.33 & 2 & 1.33 & 0.89 & 1.33 & 0.89 \\
0.44 & 0.89 & 1.33 & 0.67 & 1.33 & 2 & 0.44 & 0.89 & 1.33 \\
0.33 & 0.22 & 0.11 & 1.33 & 0.89 & 0.44 & 2 & 1.33 & 0.67 \\
0.22 & 0.33 & 0.22 & 0.89 & 1.33 & 0.89 & 1.33 & 2 & 1.33 \\
0.11 & 0.22 & 0.33 & 0.44 & 0.89 & 1.33 & 0.67 & 1.33 & 2
\end{pmatrix}$$

Example 3.9

The following ensemble of images is given:

$$\begin{pmatrix} 5 & 4 & 6 & 2 \\ 5 & 3 & 4 & 3 \\ 6 & 6 & 7 & 1 \\ 5 & 4 & 2 & 3 \end{pmatrix}, \begin{pmatrix} 4 & 2 & 2 & 1 \\ 7 & 2 & 4 & 9 \\ 3 & 5 & 4 & 5 \\ 4 & 6 & 6 & 2 \end{pmatrix}, \begin{pmatrix} 3 & 5 & 2 & 3 \\ 5 & 4 & 4 & 3 \\ 2 & 2 & 6 & 6 \\ 6 & 5 & 4 & 6 \end{pmatrix}, \begin{pmatrix} 6 & 4 & 2 & 8 \\ 3 & 5 & 6 & 4 \\ 4 & 3 & 2 & 2 \\ 5 & 3 & 3 & 4 \end{pmatrix},$$

$$\begin{pmatrix} 4 & 3 & 5 & 4 \\ 6 & 5 & 6 & 2 \\ 4 & 3 & 3 & 4 \\ 3 & 3 & 6 & 5 \end{pmatrix}, \begin{pmatrix} 4 & 5 & 4 & 5 \\ 1 & 6 & 2 & 6 \\ 4 & 8 & 4 & 4 \\ 1 & 3 & 2 & 7 \end{pmatrix}, \begin{pmatrix} 2 & 7 & 6 & 4 \\ 2 & 4 & 2 & 4 \\ 6 & 3 & 4 & 7 \\ 4 & 3 & 6 & 2 \end{pmatrix}, \begin{pmatrix} 5 & 3 & 6 & 6 \\ 4 & 4 & 5 & 2 \\ 4 & 2 & 3 & 4 \\ 5 & 5 & 4 & 4 \end{pmatrix},$$

Is this ensemble of images ergodic with respect to the mean? Is it ergodic with respect to the autocorrelation?

It is ergodic with respect to the mean because the average of each image is 4.125 and the average at each pixel position over all eight images is also 4.125.

It is not ergodic with respect to the autocorrelation function. To prove this let us calculate one element of the autocorrelation matrix, say element $E\{g_{23}g_{34}\}$ which is the average of product values of all pixels at position $(2,3)$ and $(3,4)$ over all images:

$$E\{g_{23}g_{34}\} = \frac{4 \times 1 + 4 \times 5 + 4 \times 6 + 6 \times 2 + 6 \times 4 + 2 \times 4 + 2 \times 7 + 5 \times 4}{8}$$

$$= \frac{4 + 20 + 24 + 12 + 24 + 8 + 14 + 20}{8} = \frac{126}{8} = 15.75$$

This should be equal to the element of the autocorrelation function which expresses the spatial average of pairs of pixels which are diagonal neighbours from top left to bottom right direction. Consider the last image in the ensemble. We have:

$$< g_{ij}g_{i+1,j+1} >$$

$$= \frac{5 \times 4 + 3 \times 5 + 6 \times 2 + 4 \times 2 + 4 \times 3 + 5 \times 4 + 4 \times 5 + 2 \times 4 + 3 \times 4}{16}$$

$$= \frac{20 + 15 + 12 + 8 + 12 + 20 + 20 + 8 + 12}{16} = 7.9375$$

The two numbers are not the same, and therefore the ensemble is not ergodic with respect to the autocorrelation function.

What is the implication of ergodicity?

If an ensemble of images is ergodic, then we can calculate its mean and autocorrelation function by simply calculating spatial averages over *any* image of the ensemble we happen to have.

For example, suppose that we have a collection of M images of similar type $\{g_1(x,y), g_2(x,y), \ldots, g_M(x,y)\}$. The mean and autocorrelation function of this collection can be calculated by taking averages over all images in the collection. On the other hand, if we assume ergodicity, we can pick up only one of these images and calculate the mean and the autocorrelation function from it with the help of spatial averages. This will be correct if the natural variability of all the different images is statistically the same as the natural variability exhibited by the contents of each single image separately.

How can we exploit ergodicity to reduce the number of bits needed for representing an image?

Suppose that we have an ergodic image g which we would like to transmit over a communication channel. We would like the various bits of the image we transmit to be uncorrelated so that we do not duplicate information already transmitted; i.e.

given the number of transmitted bits, we would like to maximize the transmitted information concerning the image.

The autocorrelation function of a random field that has this property is of a special form. After we decide how the image should be transformed so that it consists of uncorrelated pixel values, we can invoke ergodicity to calculate the necessary transformation from the statistics of a single image, rather than from the statistics of a whole ensemble of images.

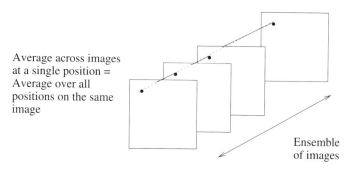

Figure 3.1: Ergodicity in a nutshell

What is the form of the autocorrelation function of a random field with uncorrelated random variables?

The autocorrelation function $R_{ff}(\mathbf{r_1}, \mathbf{r_2})$ of the two random variables defined at positions $\mathbf{r_1}$ and $\mathbf{r_2}$ will be equal to $E\{f(\mathbf{r_1};\omega_i)\}E\{f(\mathbf{r_2};\omega_i)\}$ if these two random variables are uncorrelated (see Example 3.3). If we assume that we are dealing only with random variables with zero mean, (i.e. $E\{f(\mathbf{r_1};\omega_i)\} = E\{f(\mathbf{r_2};\omega_i)\} = 0$), then the autocorrelation function will be zero for all values of its arguments, except for $\mathbf{r_1} = \mathbf{r_2}$, in which case it will be equal to $E\{f(\mathbf{r_1};\omega_i)^2\}$; i.e. equal to the variance of the random variable defined at position $\mathbf{r_1}$.

If an image $\mathbf{g}$ is represented by a column vector, then instead of having vectors $\mathbf{r_1}$ and $\mathbf{r_2}$ to indicate positions of pixels, we have integer indices, i and j, say to indicate components of each column vector. Then the autocorrelation function $R_{\mathbf{gg}}$ becomes a 2-dimensional matrix. For uncorrelated zero mean data this matrix will be diagonal with the non-zero elements along the diagonal equal to the variance at each pixel position. (In the notation used for the autocorrelation matrix of Example 3.8, $A \neq 0$, but all other entries must be 0.)

How can we transform the image so that its autocorrelation matrix is diagonal?

Let us say that the original image is g and its transformed version is $\tilde{g}$. We shall use the vector versions of them, $\mathbf{g}$ and $\tilde{\mathbf{g}}$ respectively; i.e. stack the columns of the two matrices one on top of the other to create two $N^2 \times 1$ vectors. We assume that the transformation we are seeking has the form:

$$\tilde{\mathbf{g}} = A(\mathbf{g} - \mathbf{m}) \tag{3.36}$$

where the transformation matrix A is $N^2 \times N^2$ and the arbitrary vector $\mathbf{m}$ is $N^2 \times 1$. We assume that the image is ergodic. The mean vector of the transformed image is given by:

$$\boldsymbol{\mu}_{\tilde{\mathbf{g}}} = E\{\tilde{\mathbf{g}}\} = E\{A(\mathbf{g} - \mathbf{m})\} = AE\{\mathbf{g}\} - A\mathbf{m} = A(\boldsymbol{\mu}_{\mathbf{g}} - \mathbf{m}) \tag{3.37}$$

where we have used the fact that $\mathbf{m}$ is a non-random vector, and therefore the expectation value operator leaves it unaffected. Notice that although we talk about expectation value and use the same notation as the notation used for ensemble averaging, because of the assumed ergodicity, $E\{\tilde{\mathbf{g}}\}$ means nothing else than finding the average grey value of image $\tilde{g}$ and creating an $N^2 \times 1$ vector all the elements of which are equal to this average grey value. If ergodicity had not been assumed, $E\{\tilde{\mathbf{g}}\}$ would have meant that the averaging would have to be done over all the versions of image $\tilde{g}$.

We can conveniently choose $\mathbf{m} = \boldsymbol{\mu}_{\mathbf{g}} = E\{\mathbf{g}\}$ in (3.37). Then $\boldsymbol{\mu}_{\tilde{\mathbf{g}}} = 0$; i.e. the transformed image will have zero mean.

The autocorrelation function of $\tilde{\mathbf{g}}$ then is the same as its autocovariance function and is defined by:

$$
\begin{aligned}
C_{\tilde{\mathbf{g}}\tilde{\mathbf{g}}} &= E\{\tilde{\mathbf{g}}\tilde{\mathbf{g}}^T\} = E\{A(\mathbf{g} - \boldsymbol{\mu}_{\mathbf{g}})[A(\mathbf{g} - \boldsymbol{\mu}_{\mathbf{g}})]^T\} \\
&= E\{A(\mathbf{g} - \boldsymbol{\mu}_{\mathbf{g}})(\mathbf{g} - \boldsymbol{\mu}_{\mathbf{g}})^T A^T\} \\
&= A \quad \underbrace{E\left\{(\mathbf{g} - \boldsymbol{\mu}_{\mathbf{g}})(\mathbf{g} - \boldsymbol{\mu}_{\mathbf{g}})^T\right\}}_{\text{Definition of autocovariance of the untransformed image}} \quad A^T \tag{3.38}
\end{aligned}
$$

Notice that again the expectation operator refers to spatial averaging, and because matrix A is not a random field, it is not affected by it.

So: $C_{\tilde{\mathbf{g}}\tilde{\mathbf{g}}} = AC_{\mathbf{gg}}A^T$. Then it is obvious that $C_{\tilde{\mathbf{g}}\tilde{\mathbf{g}}}$ is the diagonalized version of the covariance matrix of the untransformed image. Such a diagonalization is achieved if the transformation matrix A is the matrix formed by the eigenvectors of the autocovariance matrix of the image, used as rows, and the diagonal elements of $C_{\tilde{\mathbf{g}}\tilde{\mathbf{g}}}$ are the eigenvalues of the same matrix. The autocovariance matrix of the image can be calculated from the image itself since we assumed ergodicity (no large ensemble of similar images is needed).

Is the assumption of ergodicity realistic?

The assumption of ergodicity is not realistic. It is unrealistic to expect that a single image will be so large and it will include so much variation in its content that all the diversity represented by a collection of images will be captured by it. Only images consisting of pure random noise satisfy this assumption. So, people often divide an image into small patches, which are expected to be uniform, apart from variation due to noise, and apply the ergodicity assumption to each patch separately.

B3.1: How can we calculate the spatial autocorrelation matrix of an image?

To define a general formula for the spatial autocorrelation matrix of an image, we must first establish a correspondence between the index of an element of the vector representation of the image and the two indices that identify the position of a pixel in the image. Since the vector representation of an image is created by placing its columns one under the other, pixel (k_i, l_i) will be the ith element of the vector, where:

$$i = (l_i - 1)N + k_i + 1 \tag{3.39}$$

with N being the number of elements in each column of the image. We can solve the above expression for l_i and k_i in terms of i as follows:

$$\begin{aligned} k_i &= (i-1) \text{ modulo } N \\ l_i &= 1 + \frac{i - 1 - k_i}{N} \end{aligned} \tag{3.40}$$

Element C_{ij} of the autocorrelation matrix can be written as:

$$C_{ij} \equiv \ <g_i g_j> \tag{3.41}$$

where $<>$ means averaging over all pairs of pixels which are in the same relative position, i.e. for which $l_i - l_j$ and $k_i - k_j$ are the same. First we calculate $l_i - l_j$ and $k_i - k_j$:

$$\begin{aligned} k_0 &\equiv k_i - k_j = (i-1) \text{ modulo } N - (j-1) \text{ modulo } N \\ l_0 &\equiv l_i - l_j = \frac{i - j - k_0}{N} \end{aligned} \tag{3.42}$$

Therefore:

$$C_{ij} = \ <g_{kl} g_{k-k_0, l-l_0}> \tag{3.43}$$

The values which $k, l, k - k_0$ and $l - l_0$ take must be in the range of allowable values, that is in the range $[1, N]$ for an $N \times N$ image:

$$\left. \begin{array}{c} 1 \leq k - k_0 \leq N \quad \Rightarrow \quad 1 + k_0 \leq k \leq N + k_0 \\ \text{Also} \quad 1 \leq k \leq N \end{array} \right\} \Rightarrow$$

$$\Rightarrow \max(1, 1 + k_0) \leq k \leq \min(N, N + k_0) \tag{3.44}$$

Similarly

$$\max(1, 1 + l_0) \le l \le \min(N, N + l_0) \qquad (3.45)$$

Thus, finally:

$$C_{ij} = \frac{1}{N^2} \sum_{k=max(1,1+k_0)}^{min(N,N+k_0)} \sum_{l=max(1,1+l_0)}^{min(N,N+l_0)} g_{kl} g_{k-k_0,l-l_0} \qquad (3.46)$$

with k_0, l_0 defined by equations (3.42) in terms of i and j.

Example 3.10 (B)

For the image:

$$\begin{pmatrix} 3 & 5 & 2 & 3 \\ 5 & 4 & 4 & 3 \\ 2 & 2 & 6 & 6 \\ 6 & 5 & 4 & 6 \end{pmatrix}$$

calculate the K-L transform assuming ergodicity.

*First we compute the autocorrelation matrix of the image, using the formula derived in Box **B3.1**. It turns out to be:*

$$\begin{pmatrix}
19.12 & 12.12 & 8.12 & 4.31 & 12.81 & 9.37 & 7.06 & 2.94 \\
12.12 & 19.12 & 12.12 & 8.12 & 9.06 & 12.81 & 9.37 & 7.06 \\
8.12 & 12.12 & 19.12 & 12.12 & 5.50 & 9.06 & 12.81 & 9.37 \\
4.31 & 8.12 & 12.12 & 19.12 & 3.25 & 5.50 & 9.06 & 12.81 \\
12.81 & 9.06 & 5.50 & 3.25 & 19.12 & 12.12 & 8.12 & 4.31 \\
9.37 & 12.81 & 9.06 & 5.50 & 12.12 & 19.12 & 12.12 & 8.12 \\
7.06 & 9.37 & 12.81 & 9.06 & 8.12 & 12.12 & 19.12 & 12.12 \\
2.94 & 7.06 & 9.37 & 12.81 & 4.31 & 8.12 & 12.12 & 19.12 \\
8.19 & 6.37 & 3.06 & 1.69 & 12.81 & 9.06 & 5.50 & 3.25 \\
6.31 & 8.19 & 6.37 & 3.06 & 9.37 & 12.81 & 9.06 & 5.50 \\
5.75 & 6.31 & 8.19 & 6.37 & 7.06 & 9.37 & 12.81 & 9.06 \\
2.62 & 5.75 & 6.31 & 8.19 & 2.94 & 7.06 & 9.37 & 12.81 \\
4.50 & 3.56 & 1.50 & 1.12 & 8.19 & 6.37 & 3.06 & 1.69 \\
3.19 & 4.50 & 3.56 & 1.50 & 6.31 & 8.19 & 6.37 & 3.06 \\
3.00 & 3.19 & 4.50 & 3.56 & 5.75 & 6.31 & 8.19 & 6.37 \\
1.12 & 3.00 & 3.19 & 4.50 & 2.62 & 5.75 & 6.31 & 8.19
\end{pmatrix}$$

$$
\left(
\begin{array}{cccccccc}
8.19 & 6.31 & 5.75 & 2.62 & 4.50 & 3.19 & 3.00 & 1.12 \\
6.37 & 8.19 & 6.31 & 5.75 & 3.56 & 4.50 & 3.19 & 3.00 \\
3.06 & 6.37 & 8.18 & 6.31 & 1.50 & 3.56 & 4.50 & 3.19 \\
1.69 & 3.06 & 6.37 & 8.19 & 1.12 & 1.50 & 3.56 & 4.50 \\
12.81 & 9.37 & 7.06 & 2.94 & 8.19 & 6.31 & 5.75 & 2.62 \\
9.06 & 12.81 & 9.37 & 7.06 & 6.37 & 8.19 & 6.31 & 5.75 \\
5.50 & 9.06 & 12.81 & 9.37 & 3.06 & 6.37 & 8.19 & 6.31 \\
3.25 & 5.50 & 9.06 & 12.81 & 1.69 & 3.06 & 6.37 & 8.19 \\
19.12 & 12.12 & 8.125 & 4.31 & 12.81 & 9.37 & 7.06 & 2.94 \\
12.12 & 19.12 & 12.12 & 8.12 & 9.06 & 12.81 & 9.37 & 7.06 \\
8.12 & 12.12 & 19.12 & 12.12 & 5.50 & 9.06 & 12.81 & 9.37 \\
4.31 & 8.12 & 12.12 & 19.12 & 3.25 & 5.50 & 9.06 & 12.81 \\
12.81 & 9.06 & 5.50 & 3.25 & 19.12 & 12.12 & 8.12 & 4.31 \\
9.37 & 12.81 & 9.06 & 5.50 & 12.12 & 19.12 & 12.12 & 8.12 \\
7.06 & 9.37 & 12.81 & 9.06 & 8.12 & 12.12 & 19.12 & 12.12 \\
2.94 & 7.06 & 9.37 & 12.81 & 4.31 & 8.12 & 12.12 & 19.12
\end{array}
\right)
$$

After we compute the eigenvectors of the autocorrelation matrix, and sort them so that their corresponding eigenvalues are in decreasing order, we use them as rows to form the transformation matrix A with which our image can be transformed:

$$
\left(
\begin{array}{cccccccc}
0.21 & 0.25 & 0.24 & 0.19 & 0.24 & 0.30 & 0.29 & 0.24 \\
-0.065 & 0.11 & 0.30 & 0.39 & -0.22 & -0.03 & 0.21 & 0.36 \\
-0.42 & -0.35 & -0.21 & -0.06 & -0.30 & -0.19 & -0.04 & 0.10 \\
-0.23 & -0.09 & 0.15 & 0.53 & -0.09 & -0.22 & -0.11 & 0.23 \\
-0.26 & 0.08 & 0.39 & -0.06 & -0.35 & 0.09 & 0.24 & -0.26 \\
0.39 & 0.30 & 0.18 & 0.12 & -0.07 & -0.20 & -0.28 & -0.26 \\
-0.22 & 0.38 & -0.13 & -0.05 & -0.21 & 0.39 & -0.27 & 0.08 \\
-0.18 & -0.16 & -0.12 & -0.04 & 0.35 & 0.36 & 0.30 & 0.25 \\
0.08 & 0.13 & 0.09 & -0.50 & -0.26 & -0.02 & 0.37 & 0.06 \\
-0.39 & -0.02 & 0.46 & -0.06 & 0.09 & 0.07 & 0.00 & -0.32 \\
-0.28 & 0.52 & -0.22 & -0.03 & -0.10 & 0.13 & -0.19 & 0.14 \\
-0.01 & -0.10 & -0.23 & 0.36 & 0.00 & 0.39 & -0.06 & -0.37 \\
-0.30 & 0.02 & 0.24 & -0.06 & 0.51 & -0.02 & -0.20 & -0.19 \\
-0.19 & 0.31 & -0.30 & 0.19 & 0.04 & -0.17 & 0.34 & -0.28 \\
0.18 & -0.25 & -0.04 & 0.18 & -0.32 & 0.37 & 0.16 & -0.30 \\
0.07 & -0.19 & 0.25 & -0.13 & -0.14 & 0.36 & -0.42 & 0.21
\end{array}
\right)
$$

$$\begin{pmatrix}
0.24 & 0.29 & 0.30 & 0.24 & 0.19 & 0.24 & 0.25 & 0.21 \\
-0.36 & -0.21 & 0.03 & 0.22 & -0.39 & -0.30 & -0.11 & 0.06 \\
-0.10 & 0.04 & 0.19 & 0.30 & 0.06 & 0.21 & 0.35 & 0.42 \\
0.23 & -0.11 & -0.22 & -0.09 & 0.53 & 0.15 & -0.09 & -0.23 \\
-0.26 & 0.24 & 0.09 & -0.35 & -0.06 & 0.39 & 0.08 & -0.26 \\
-0.26 & -0.28 & -0.20 & -0.07 & 0.12 & 0.18 & 0.30 & 0.39 \\
-0.08 & 0.27 & -0.39 & 0.21 & 0.05 & 0.13 & -0.38 & 0.22 \\
-0.25 & -0.30 & -0.36 & -0.35 & 0.04 & 0.12 & 0.16 & 0.18 \\
-0.06 & -0.37 & 0.02 & 0.26 & 0.50 & -0.09 & -0.13 & -0.08 \\
0.32 & -0.00 & -0.07 & -0.09 & 0.06 & -0.46 & 0.02 & 0.39 \\
0.14 & -0.19 & 0.13 & -0.10 & -0.03 & -0.22 & 0.52 & -0.28 \\
-0.37 & -0.06 & 0.39 & 0.00 & 0.36 & -0.23 & -0.10 & -0.01 \\
-0.19 & -0.20 & -0.02 & 0.51 & -0.06 & 0.24 & 0.02 & -0.30 \\
0.28 & -0.34 & 0.17 & -0.04 & -0.19 & 0.30 & -0.31 & 0.19 \\
0.30 & -0.16 & -0.37 & 0.32 & -0.18 & 0.04 & 0.25 & -0.18 \\
0.21 & -0.42 & 0.36 & -0.14 & -0.13 & 0.25 & -0.19 & 0.07
\end{pmatrix}$$

The corresponding eigenvalues are:

$$\begin{array}{cccccccc}
130.86 & 43.63 & 38.58 & 15.01 & 13.99 & 12.07 & 11.42 & 8.51 \\
6.29 & 5.22 & 4.97 & 4.29 & 3.28 & 3.22 & 2.50 & 2.15
\end{array}$$

The mean value of the image is $\frac{66}{16} = 4.125$. We subtract this from all the elements of the image and stack the columns of the matrix that results one below the other to form vector $\{\mathbf{g} - \boldsymbol{\mu_g}\}$:

$$(\mathbf{g} - \boldsymbol{\mu_g})^T = (\,-1.125 \quad 0.875 \quad -2.125 \quad -1.125 \quad 0.875 \quad -0.125 \quad -0.125$$
$$-1.125 \quad -2.125 \quad -2.125 \quad 1.875 \quad 1.875 \quad 1.875 \quad 0.875 \quad -0.125 \quad 1.875\,)$$

We then multiply this vector with matrix A to derive the Karhunen–Loeve transform of the image. In matrix form this is given by:

$$\tilde{g} = \begin{pmatrix}
-0.04531 & -0.72899 & 2.45651 & -1.20855 \\
-1.15508 & 1.30250 & 0.59357 & 0.67163 \\
2.21824 & -0.60028 & 0.35528 & 2.55183 \\
1.90206 & 1.92720 & -1.62346 & -0.02844
\end{pmatrix}$$

To check that this image is made up from uncorrelated data, we calculate its autocorrelation matrix $C_{\tilde{g}\tilde{g}}$:

$$C_{\tilde{g}\tilde{g}} =$$

$$
\begin{pmatrix}
130.862 & 1\times10^{-5} & 3\times10^{-5} & -6\times10^{-6} & 3\times10^{-5} & 1\times10^{-5} \\
1\times10^{-5} & 43.6336 & -2\times10^{-5} & -1\times10^{-5} & 1\times10^{-6} & 1\times10^{-5} \\
3\times10^{-5} & -2\times10^{-5} & 38.5786 & -1\times10^{-6} & -5\times10^{-6} & -3\times10^{-6} \\
-6\times10^{-6} & -1\times10^{-5} & -1\times10^{-6} & 15.0068 & -1\times10^{-7} & 1\times10^{-5} \\
3\times10^{-5} & 1\times10^{-6} & -5\times10^{-6} & -1\times10^{-7} & 13.9865 & 1\times10^{-5} \\
1\times10^{-5} & 1\times10^{-5} & -3\times10^{-6} & 1\times10^{-5} & 1\times10^{-5} & 12.0737 \\
-2\times10^{-6} & -1\times10^{-5} & 3\times10^{-6} & 4\times10^{-6} & 8\times10^{-6} & -2\times10^{-6} \\
-3\times10^{-5} & -1\times10^{-6} & 1\times10^{-5} & 2\times10^{-6} & -5\times10^{-6} & -7\times10^{-7} \\
6\times10^{-7} & -2\times10^{-5} & 1\times10^{-5} & 3\times10^{-6} & -3\times10^{-6} & 7\times10^{-7} \\
-4\times10^{-5} & -5\times10^{-6} & -1\times10^{-6} & -5\times10^{-6} & -2\times10^{-6} & 6\times10^{-6} \\
-1\times10^{-5} & -5\times10^{-6} & 2\times10^{-5} & 2\times10^{-6} & 4\times10^{-6} & 1\times10^{-6} \\
-2\times10^{-5} & -4\times10^{-6} & -2\times10^{-5} & -4\times10^{-6} & 9\times10^{-7} & -2\times10^{-6} \\
-2\times10^{-5} & -1\times10^{-5} & -3\times10^{-6} & -1\times10^{-6} & 3\times10^{-6} & -4\times10^{-7} \\
2\times10^{-5} & 4\times10^{-6} & 1\times10^{-5} & -2\times10^{-7} & 1\times10^{-6} & -1\times10^{-6} \\
-1\times10^{-7} & -1\times10^{-5} & 1\times10^{-5} & -4\times10^{-7} & 1\times10^{-6} & -1\times10^{-6} \\
3\times10^{-5} & -4\times10^{-6} & -2\times10^{-5} & 5\times10^{-6} & 1\times10^{-6} & -3\times10^{-6} \\
\end{pmatrix}
$$

$$
\begin{array}{cccccc}
-2\times10^{-6} & -3\times10^{-5} & 6\times10^{-7} & -4\times10^{-5} & -1\times10^{-5} & -2\times10^{-5} \\
-1\times10^{-5} & -1\times10^{-6} & -2\times10^{-5} & -5\times10^{-6} & -5\times10^{-6} & -4\times10^{-6} \\
3\times10^{-6} & 1\times10^{-5} & 1\times10^{-5} & -1\times10^{-6} & 2\times10^{-5} & -2\times10^{-5} \\
4\times10^{-6} & 2\times10^{-6} & 3\times10^{-6} & -5\times10^{-6} & 2\times10^{-6} & -4\times10^{-6} \\
8\times10^{-6} & -5\times10^{-6} & -3\times10^{-6} & -2\times10^{-6} & 4\times10^{-6} & 9\times10^{-7} \\
-2\times10^{-6} & -7\times10^{-7} & 7\times10^{-7} & 6\times10^{-6} & 1\times10^{-6} & -2\times10^{-6} \\
11.4241 & 5\times10^{-6} & -2\times10^{-6} & 6\times10^{-7} & -3\times10^{-6} & 9\times10^{-7} \\
5\times10^{-6} & 8.51163 & -5\times10^{-6} & -4\times10^{-6} & 1\times10^{-6} & 1\times10^{-6} \\
-2\times10^{-6} & -5\times10^{-6} & 6.28745 & 1\times10^{-6} & 2\times10^{-6} & 3\times10^{-7} \\
6\times10^{-7} & -4\times10^{-6} & 1\times10^{-6} & 5.21938 & 7\times10^{-7} & -1\times10^{-6} \\
-3\times10^{-6} & 1\times10^{-6} & 2\times10^{-6} & 7\times10^{-7} & 4.97084 & 2\times10^{-6} \\
9\times10^{-7} & 1\times10^{-6} & 3\times10^{-7} & -1\times10^{-6} & 2\times10^{-6} & 4.29183 \\
-1\times10^{-6} & -1\times10^{-6} & -1\times10^{-6} & -1\times10^{-6} & 1\times10^{-6} & -7\times10^{-8} \\
9\times10^{-7} & 6\times10^{-7} & 1\times10^{-7} & -1\times10^{-6} & 1\times10^{-6} & -1\times10^{-6} \\
4\times10^{-6} & 1\times10^{-6} & 1\times10^{-6} & -5\times10^{-7} & -1\times10^{-7} & 1\times10^{-6} \\
2\times10^{-6} & 2\times10^{-6} & 3\times10^{-7} & -2\times10^{-6} & -2\times10^{-6} & 1\times10^{-6} \\
\end{array}
$$

Continued

$$
\begin{pmatrix}
-2 \times 10^{-5} & 2 \times 10^{-5} & -1 \times 10^{-7} & 3 \times 10^{-5} \\
-1 \times 10^{-5} & 4 \times 10^{-6} & -1 \times 10^{-5} & -4 \times 10^{-6} \\
-3 \times 10^{-6} & 1 \times 10^{-5} & 1 \times 10^{-5} & -2 \times 10^{-5} \\
-1 \times 10^{-6} & -2 \times 10^{-7} & -4 \times 10^{-7} & 5 \times 10^{-6} \\
3 \times 10^{-6} & 1 \times 10^{-6} & 1 \times 10^{-6} & 1 \times 10^{-6} \\
-4 \times 10^{-7} & -1 \times 10^{-6} & -1 \times 10^{-6} & -3 \times 10^{-6} \\
-1 \times 10^{-6} & 9 \times 10^{-7} & 4 \times 10^{-6} & 2 \times 10^{-6} \\
-1 \times 10^{-6} & 6 \times 10^{-7} & 1 \times 10^{-6} & 2 \times 10^{-6} \\
-1 \times 10^{-6} & 1 \times 10^{-7} & 1 \times 10^{-6} & 3 \times 10^{-7} \\
-1 \times 10^{-6} & -1 \times 10^{-6} & -5 \times 10^{-7} & -2 \times 10^{-6} \\
1 \times 10^{-6} & 1 \times 10^{-6} & -1 \times 10^{-7} & -2 \times 10^{-6} \\
-7 \times 10^{-8} & -1 \times 10^{-6} & 1 \times 10^{-6} & 1 \times 10^{-6} \\
3.28074 & -2 \times 10^{-6} & -3 \times 10^{-7} & -8 \times 10^{-7} \\
-2 \times 10^{-6} & 3.21776 & 9 \times 10^{-7} & -4 \times 10^{-8} \\
-3 \times 10^{-7} & 9 \times 10^{-7} & 2.50252 & 1 \times 10^{-6} \\
-8 \times 10^{-7} & -4 \times 10^{-8} & 1 \times 10^{-6} & 2.15221
\end{pmatrix}
$$

We can see that to high accuracy this is a diagonal matrix with the eigenvalues of the original image along its diagonal.

How can we approximate an image using its K-L transform?

The K-L transform of an image is given by:

$$\tilde{\mathbf{g}} = A(\mathbf{g} - \boldsymbol{\mu_g})$$

where $\boldsymbol{\mu_g}$ is an $N^2 \times 1$ vector with elements equal to the average grey value of the image, and A is a matrix made up from the eigenvectors of the autocorrelation matrix of image $\mathbf{g}$, used as rows and arranged in decreasing order of the corresponding eigenvalues. The inverse transform is:

$$\mathbf{g} = A^T \tilde{\mathbf{g}} + \boldsymbol{\mu_g}$$

If we set equal to 0 the last few eigenvalues of the autocorrelation matrix of $\mathbf{g}$, matrix A will have its corresponding rows replaced by zeroes, and so will the transformed image $\tilde{\mathbf{g}}$.

What is the error with which we approximate an image when we truncate its K-L expansion?

It can be shown (see Box **B3.2**) that the approximation error in this case is equal to the sum of the omitted eigenvalues of the autocovariance matrix of the image.

What are the basis images in terms of which the Karhunen–Loeve transform expands an image?

Since $\tilde{\mathbf{g}} = A(\mathbf{g} - \boldsymbol{\mu_g})$ and A is an orthogonal matrix, the inverse transformation is given by $\mathbf{g} - \boldsymbol{\mu_g} = A^T\tilde{\mathbf{g}}$. We can write this expression explicitly:

$$
\begin{pmatrix}
g_{11} - \mu_g \\
g_{21} - \mu_g \\
\vdots \\
g_{N1} - \mu_g \\
g_{21} - \mu_g \\
\vdots \\
g_{2N} - \mu_g \\
\vdots \\
g_{N1} - \mu_g \\
\vdots \\
g_{NN} - \mu_g
\end{pmatrix}
=
\begin{pmatrix}
a_{11} & a_{21} & \cdots & a_{N1} \\
a_{12} & a_{22} & \cdots & a_{N2} \\
\vdots & \vdots & & \vdots \\
a_{1N} & a_{2N} & \cdots & a_{NN} \\
a_{1,N+1} & a_{2,N+1} & \cdots & a_{N,N+1} \\
\vdots & \vdots & & \vdots \\
a_{1,2N} & a_{2,2N} & \cdots & a_{N,2N} \\
\vdots & \vdots & & \vdots \\
a_{1,N^2-N+1} & a_{2,N^2-N+1} & \cdots & a_{N,N^2-N+1} \\
\vdots & \vdots & & \vdots \\
a_{1,N^2} & a_{2,N^2} & \cdots & a_{N,N^2}
\end{pmatrix}
\begin{pmatrix}
\tilde{g}_{11} \\
\tilde{g}_{21} \\
\vdots \\
\tilde{g}_{N1} \\
\tilde{g}_{21} \\
\vdots \\
\tilde{g}_{2N} \\
\vdots \\
\tilde{g}_{N1} \\
\vdots \\
\tilde{g}_{NN}
\end{pmatrix}
$$

$$
\Rightarrow
\begin{cases}
g_{11} - \mu_g &= a_{11}\tilde{g}_{11} + a_{21}\tilde{g}_{21} + \ldots + a_{N1}\tilde{g}_{NN} \\
g_{21} - \mu_g &= a_{12}\tilde{g}_{11} + a_{22}\tilde{g}_{21} + \ldots + a_{N2}\tilde{g}_{NN} \\
\cdots & \quad\cdots \\
g_{N1} - \mu_g &= a_{1N}\tilde{g}_{N1} + a_{2N}\tilde{g}_{21} + \ldots + a_{NN}\tilde{g}_{NN} \\
g_{21} - \mu_g &= a_{1,N+1}\tilde{g}_{N1} + a_{2,N+1}\tilde{g}_{21} + \ldots + a_{N,N+1}\tilde{g}_{NN} \\
\cdots & \quad\cdots \\
g_{2N} - \mu_g &= a_{1,2N}\tilde{g}_{N1} + a_{2,2N}\tilde{g}_{21} + \ldots + a_{N,2N}\tilde{g}_{NN} \\
g_{N1} - \mu_g &= a_{1,N^2-N+1}\tilde{g}_{N1} + a_{2,N^2-N+1}\tilde{g}_{21} + \ldots + a_{N,N^2-N+1}\tilde{g}_{NN} \\
\cdots & \quad\cdots \\
g_{NN} - \mu_g &= a_{1N^2}\tilde{g}_{11} + a_{2N^2}\tilde{g}_{21} + \ldots + a_{NN^2}\tilde{g}_{NN}
\end{cases}
$$

We can rearrange these equations into matrix form:

$$
\begin{pmatrix}
g_{11} - \mu_g & g_{21} - \mu_g & \cdots & g_{N1} - \mu_g \\
g_{21} - \mu_g & g_{22} - \mu_g & \cdots & g_{N2} - \mu_g \\
\vdots & \vdots & & \vdots \\
g_{N1} - \mu_g & g_{2N} - \mu_g & \cdots & g_{NN} - \mu_g
\end{pmatrix}
$$

$$
= \tilde{g}_{11}
\begin{pmatrix}
a_{11} & a_{1,N+1} & \cdots & a_{1,N^2-N+1} \\
a_{12} & a_{1,N+2} & \cdots & a_{1,N^2-N+2} \\
\vdots & \vdots & & \vdots \\
a_{1N} & a_{1,2N} & \cdots & a_{1,N^2}
\end{pmatrix}
+ \tilde{g}_{21}
\begin{pmatrix}
a_{21} & a_{2,N+1} & \cdots & a_{2,N^2-N+1} \\
a_{22} & a_{2,N+2} & \cdots & a_{2,N^2-N+2} \\
\vdots & \vdots & & \vdots \\
a_{2N} & a_{2,2N} & \cdots & a_{2,N^2}
\end{pmatrix}
+
$$

$$+\ldots+\tilde{g}_{NN}\begin{pmatrix} a_{N1} & a_{N,N+1} & \cdots & a_{N,N^2-N+1} \\ a_{N2} & a_{N,N+2} & \cdots & a_{N,N^2-N+2} \\ \vdots & \vdots & & \vdots \\ a_{NN} & a_{N,2N} & \cdots & a_{N,N^2} \end{pmatrix} \tag{3.47}$$

This expression makes it obvious that the eigenimages in terms of which the K-L transform expands an image are formed from the eigenvectors of its spatial autocorrelation matrix, by writing them in matrix form; i.e. by using the first N elements of an eigenvector to form the first column of the corresponding eigenimage, the next N elements to form the next column and so on. The coefficients of this expansion are the elements of the transformed image.

Example 3.11

Consider a 3×3 image with column representation g. Write down an expression for the K-L transform of the image in terms of the elements of g and the elements a_{ij} of the transformation matrix A. Calculate an approximation to the image g by setting the last six rows of A to zero. Show that the approximation will be a 9×1 vector with the first three elements equal to those of the full transformation of g and the remaining six elements zero.

Assume that μ_g is the average grey value of image g. Then the transformed image will have the form:

$$\begin{pmatrix} \tilde{g}_{11} \\ \tilde{g}_{21} \\ \tilde{g}_{31} \\ \tilde{g}_{12} \\ \tilde{g}_{22} \\ \tilde{g}_{32} \\ \tilde{g}_{13} \\ \tilde{g}_{23} \\ \tilde{g}_{33} \end{pmatrix} = \begin{pmatrix} a_{11} & a_{12} & \cdots & a_{19} \\ a_{21} & a_{22} & \cdots & a_{29} \\ a_{31} & a_{32} & \cdots & a_{39} \\ a_{41} & a_{42} & \cdots & a_{49} \\ \vdots & \vdots & & \vdots \\ \vdots & \vdots & & \vdots \\ \vdots & \vdots & & \vdots \\ \vdots & \vdots & & \vdots \\ a_{91} & a_{92} & \cdots & a_{99} \end{pmatrix} \begin{pmatrix} g_{11} - \mu_g \\ g_{21} - \mu_g \\ g_{31} - \mu_g \\ g_{12} - \mu_g \\ g_{22} - \mu_g \\ g_{32} - \mu_g \\ g_{13} - \mu_g \\ g_{23} - \mu_g \\ g_{33} - \mu_g \end{pmatrix}$$

$$= \begin{pmatrix} a_{11}(g_{11} - \mu_g) + a_{12}(g_{21} - \mu_g) + \ldots + a_{19}(g_{33} - \mu_g) \\ a_{21}(g_{11} - \mu_g) + a_{22}(g_{21} - \mu_g) + \ldots + a_{29}(g_{33} - \mu_g) \\ a_{31}(g_{11} - \mu_g) + a_{32}(g_{21} - \mu_g) + \ldots + a_{39}(g_{33} - \mu_g) \\ a_{41}(g_{11} - \mu_g) + a_{42}(g_{21} - \mu_g) + \ldots + a_{49}(g_{33} - \mu_g) \\ \vdots \\ a_{91}(g_{11} - \mu_g) + a_{92}(g_{21} - \mu_g) + \ldots + a_{99}(g_{33} - \mu_g) \end{pmatrix} \tag{3.48}$$

If we set $a_{41} = a_{42} = \ldots = a_{49} = a_{51} = \ldots = a_{59} = \ldots = a_{99} = 0$, clearly the last six rows of the above vector will be 0 and the truncated transformation of the image will be vector

$$\tilde{\mathbf{g}}' = (\tilde{g}_{11} \quad \tilde{g}_{21} \quad \tilde{g}_{31} \quad 0 \quad 0 \quad 0 \quad 0 \quad 0 \quad 0)^T \tag{3.49}$$

According to formula (3.47) the approximation of the image is then:

$$
\begin{pmatrix} g_{11} & g_{12} & g_{13} \\ g_{21} & g_{22} & g_{23} \\ g_{31} & g_{32} & g_{33} \end{pmatrix} = \begin{pmatrix} \mu_g & \mu_g & \mu_g \\ \mu_g & \mu_g & \mu_g \\ \mu_g & \mu_g & \mu_g \end{pmatrix} + \tilde{g}_{11} \begin{pmatrix} a_{11} & a_{14} & a_{17} \\ a_{12} & a_{15} & a_{18} \\ a_{13} & a_{16} & a_{19} \end{pmatrix}
$$

$$
+ \tilde{g}_{21} \begin{pmatrix} a_{21} & a_{24} & a_{27} \\ a_{22} & a_{25} & a_{28} \\ a_{23} & a_{26} & a_{29} \end{pmatrix} + \tilde{g}_{31} \begin{pmatrix} a_{31} & a_{34} & a_{37} \\ a_{32} & a_{35} & a_{38} \\ a_{33} & a_{36} & a_{39} \end{pmatrix} \tag{3.50}
$$

Example 3.12 (B)

Show that if A is an $N^2 \times N^2$ matrix the ith row of which is vector $\mathbf{u}_i^T$ and C_2 an $N^2 \times N^2$ matrix with all its elements zero except the element at position $(2,2)$ which is equal to c_2, then:

$$A^T C_2 A = c_2 \mathbf{u}_2 \mathbf{u}_2^T$$

Assume that u_{ij} indicates the jth component of vector $\mathbf{u}_i$. Then:

$$
A^T C_2 A = \begin{pmatrix} u_{11} & u_{21} & \cdots & u_{N^2 1} \\ u_{12} & u_{22} & \cdots & u_{N^2 2} \\ u_{13} & u_{23} & \cdots & u_{N^2 3} \\ \vdots & \vdots & & \vdots \\ u_{1N^2} & u_{2N^2} & \cdots & u_{N^2 N^2} \end{pmatrix} \begin{pmatrix} 0 & 0 & \cdots & 0 \\ 0 & c_2 & \cdots & 0 \\ 0 & 0 & \cdots & 0 \\ \vdots & \vdots & & \vdots \\ 0 & 0 & \cdots & 0 \end{pmatrix}
$$

$$
\begin{pmatrix} u_{11} & u_{12} & \cdots & u_{1N^2} \\ u_{21} & u_{22} & \cdots & u_{2N^2} \\ u_{31} & u_{32} & \cdots & u_{3N^2} \\ \vdots & \vdots & & \vdots \\ u_{N^2 1} & u_{N^2 2} & \cdots & u_{N^2 N^2} \end{pmatrix}
$$

$$
= \begin{pmatrix} u_{11} & u_{21} & \cdots & u_{N^21} \\ u_{12} & u_{22} & \cdots & u_{N^22} \\ u_{13} & u_{23} & \cdots & u_{N^23} \\ \vdots & \vdots & & \vdots \\ u_{1N^2} & u_{2N^2} & \cdots & u_{N^2N^2} \end{pmatrix} \begin{pmatrix} 0 & 0 & \cdots & 0 \\ c_2 u_{21} & c_2 u_{22} & \cdots & c_2 u_{2N^2} \\ 0 & 0 & \cdots & 0 \\ \vdots & \vdots & & \vdots \\ 0 & 0 & \cdots & 0 \end{pmatrix}
$$

$$
= \begin{pmatrix} c_2 u_{21}^2 & c_2 u_{21} u_{22} & \cdots & c_2 u_{21} u_{2N^2} \\ c_2 u_{22} u_{21} & c_2 u_{22}^2 & \cdots & c_2 u_{22} u_{2N^2} \\ \vdots & \vdots & & \vdots \\ c_2 u_{2N^2} u_{21} & c_2 u_{2N^2} & \cdots & c_2 u_{2N^2}^2 \end{pmatrix}
$$

$$
= c_2 \mathbf{u}_2 \mathbf{u}_2^T \tag{3.51}
$$

Example 3.13 (B)

Assuming a 3×3 image, and accepting that we approximate it retaining only the first three eigenvalues of its autocovariance matrix, show that:

$$
E\{\tilde{\mathbf{g}}\tilde{\mathbf{g}}'^{T}\} \;=\; C'_{\tilde{\mathbf{g}}\tilde{\mathbf{g}}} \tag{3.52}
$$

Using the result of Example 3.12 concerning the truncated transform of the image $\tilde{\mathbf{g}}'$, we have:

$$
E\{\tilde{\mathbf{g}}\tilde{\mathbf{g}}'^{T}\} \;=\; E\left\{ \begin{pmatrix} \tilde{g}_{11} \\ \tilde{g}_{21} \\ \tilde{g}_{31} \\ \tilde{g}_{12} \\ \tilde{g}_{22} \\ \tilde{g}_{32} \\ \tilde{g}_{13} \\ \tilde{g}_{23} \\ \tilde{g}_{33} \end{pmatrix} \begin{pmatrix} \tilde{g}_{11} & \tilde{g}_{21} & \tilde{g}_{31} & 0 & 0 & 0 & 0 & 0 & 0 \end{pmatrix} \right\}
$$

$$
= E\left\{ \begin{pmatrix} \tilde{g}_{11}^2 & \tilde{g}_{11}\tilde{g}_{21} & \tilde{g}_{11}\tilde{g}_{31} & 0 & 0 & 0 & 0 & 0 & 0 \\ \tilde{g}_{21}\tilde{g}_{11} & \tilde{g}_{21}^2 & \tilde{g}_{21}\tilde{g}_{31} & 0 & 0 & 0 & 0 & 0 & 0 \\ \tilde{g}_{31}\tilde{g}_{11} & \tilde{g}_{31}\tilde{g}_{21} & \tilde{g}_{31}^2 & 0 & 0 & 0 & 0 & 0 & 0 \\ & \vdots & & & & & & & \\ \tilde{g}_{33}\tilde{g}_{11} & \tilde{g}_{33}\tilde{g}_{21} & \tilde{g}_{33}\tilde{g}_{31} & 0 & 0 & 0 & 0 & 0 & 0 \end{pmatrix} \right\}
$$

$$
= \begin{pmatrix}
E\{\tilde{g}_{11}^2\} & E\{\tilde{g}_{11}\tilde{g}_{21}\} & E\{\tilde{g}_{11}\tilde{g}_{31}\} & 0 & 0 & 0 & 0 & 0 & 0 \\
E\{\tilde{g}_{21}\tilde{g}_{11}\} & E\{\tilde{g}_{21}^2\} & E\{\tilde{g}_{21}\tilde{g}_{31}\} & 0 & 0 & 0 & 0 & 0 & 0 \\
E\{\tilde{g}_{31}\tilde{g}_{11}\} & E\{\tilde{g}_{31}\tilde{g}_{21}\} & E\{\tilde{g}_{31}^2\} & 0 & 0 & 0 & 0 & 0 & 0 \\
\vdots & & & & & & & & \\
E\{\tilde{g}_{33}\tilde{g}_{11}\} & E\{\tilde{g}_{33}\tilde{g}_{21}\} & E\{\tilde{g}_{33}\tilde{g}_{31}\} & 0 & 0 & 0 & 0 & 0 & 0
\end{pmatrix}
$$

The transformed image $\tilde{\mathbf{g}}$ *is constructed in such a way that all the off-diagonal elements of its correlation matrix are equal to 0. Therefore, we have:*

$$
E\{\tilde{\mathbf{g}}\tilde{\mathbf{g}}'^T\} = \begin{pmatrix}
E\{\tilde{g}_{11}^2\} & 0 & 0 & 0 & 0 & 0 & 0 & 0 & 0 \\
0 & E\{\tilde{g}_{21}^2\} & 0 & 0 & 0 & 0 & 0 & 0 \\
0 & 0 & E\{\tilde{g}_{31}^2\} & 0 & 0 & 0 & 0 & 0 & 0 \\
0 & 0 & 0 & 0 & 0 & 0 & 0 & 0 \\
\vdots & & & & & & & & \\
0 & 0 & 0 & 0 & 0 & 0 & 0 & 0
\end{pmatrix} = C'_{\tilde{\mathbf{g}}\tilde{\mathbf{g}}}
$$

B3.2: What is the error of the approximation of an image using the Karhunen–Loeve transform?

We shall show now that the Karhunen–Loeve transform not only expresses an image in terms of uncorrelated data, but also, if truncated after a certain term, it can be used to approximate the image in the **least mean square error** sense. Suppose that the transform is:

$$
\tilde{\mathbf{g}} = A\mathbf{g} - A\boldsymbol{\mu}_{\mathbf{g}} \Rightarrow \mathbf{g} = A^T\tilde{\mathbf{g}} + \boldsymbol{\mu}_{\mathbf{g}} \tag{3.53}
$$

We assume that we have ordered the eigenvalues of $C_{\mathbf{gg}}$ in descending order. Suppose that we decide to neglect the last few eigenvalues and, say, we retain the first K most significant ones. $C_{\tilde{\mathbf{g}}\tilde{\mathbf{g}}}$ is an $N^2 \times N^2$ matrix and its truncated version, $C'_{\tilde{\mathbf{g}}\tilde{\mathbf{g}}}$, has the last $N^2 - K$ diagonal elements 0. The transformation matrix A^T was an $N^2 \times N^2$ matrix, the columns of which were the eigenvectors of $C_{\mathbf{gg}}$. Neglecting the $N^2 - K$ eigenvalues is like omitting $N^2 - K$ eigenvectors, so the new transformation matrix A'^T has the last $N^2 - K$ columns 0. The approximated image then is:

$$
\mathbf{g}' = A'^T\tilde{\mathbf{g}}' + \boldsymbol{\mu}_{\mathbf{g}} \tag{3.54}
$$

The error of the approximation is $\mathbf{g} - \mathbf{g}' = A^T\tilde{\mathbf{g}} - A'^T\tilde{\mathbf{g}}'$. The norm of this matrix is:

$$\|\mathbf{g} - \mathbf{g}'\| = \text{trace}[(\mathbf{g} - \mathbf{g}')(\mathbf{g} - \mathbf{g}')^T] \qquad (3.55)$$

Therefore, the mean square error is:

$$E\{\|\mathbf{g} - \mathbf{g}'\|\} = E\{\text{trace}[(\mathbf{g} - \mathbf{g}')(\mathbf{g} - \mathbf{g}')^T]\} \qquad (3.56)$$

We can exchange the order of taking the expectation value and taking the trace because trace means nothing more than the summing of the diagonal elements of the matrix:

$$
\begin{aligned}
E\{\|\mathbf{g}-\mathbf{g}'\|\} &= \text{trace}\left[E\left\{(\mathbf{g}-\mathbf{g}')(\mathbf{g}-\mathbf{g}')^T\right\}\right] \\
&= \text{trace}\left[E\left\{(A^T\tilde{\mathbf{g}}-A'^T\tilde{\mathbf{g}}')(A^T\tilde{\mathbf{g}}-A'^T\tilde{\mathbf{g}}')^T\right\}\right] \\
&= \text{trace}\left[E\left\{(A^T\tilde{\mathbf{g}}-A'^T\tilde{\mathbf{g}}')(\tilde{\mathbf{g}}^T A - \tilde{\mathbf{g}}'^T A')\right\}\right] \\
&= \text{trace}\left[E\left\{A^T\tilde{\mathbf{g}}\tilde{\mathbf{g}}^T A - A^T\tilde{\mathbf{g}}\tilde{\mathbf{g}}'^T A' - A'^T\tilde{\mathbf{g}}'\tilde{\mathbf{g}}^T A + A'^T\tilde{\mathbf{g}}'\tilde{\mathbf{g}}'^T A'\right\}\right]
\end{aligned}
$$
$$(3.57)$$

Matrices A and A' are fixed, so the expectation operator does not affect them. Therefore:

$$
\begin{aligned}
E\{\|\mathbf{g} - \mathbf{g}'\|\} = \ & \text{trace}[A^T E\{\tilde{\mathbf{g}}\tilde{\mathbf{g}}^T\}A - A^T E\{\tilde{\mathbf{g}}\tilde{\mathbf{g}}'^T\}A' \\
& -A'^T E\{\tilde{\mathbf{g}}'\tilde{\mathbf{g}}^T\}A + A'^T E\{\tilde{\mathbf{g}}'\tilde{\mathbf{g}}'^T\}A']
\end{aligned}
\qquad (3.58)
$$

In this expression we recognize $E\{\tilde{\mathbf{g}}\tilde{\mathbf{g}}^T\}$ and $E\{\tilde{\mathbf{g}}'\tilde{\mathbf{g}}'^T\}$ as the correlation matrices of the two transforms: $C_{\tilde{\mathbf{g}}\tilde{\mathbf{g}}}$ and $C'_{\tilde{\mathbf{g}}\tilde{\mathbf{g}}}$.

The $\tilde{\mathbf{g}}\tilde{\mathbf{g}}'^T$ matrix is the product of a vector and its transpose but with the last $N^2 - K$ components of the transpose replaced by 0. The expectation operator will make all the off-diagonal elements of $\tilde{\mathbf{g}}\tilde{\mathbf{g}}'$ zero anyway (since the transform is such that its autocorrelation matrix has 0 all the off-diagonal elements). The fact that the last $N^2 - K$ elements of $\tilde{\mathbf{g}}'^T$ are 0 too will also make the last $N^2 - K$ diagonal elements 0 (see Example 3.13). So, the result is:

$$E\{\tilde{\mathbf{g}}\tilde{\mathbf{g}}'^T\} = C'_{\tilde{\mathbf{g}}\tilde{\mathbf{g}}} \qquad (3.59)$$

Similar reasoning leads to:

$$E\{\tilde{\mathbf{g}}'\tilde{\mathbf{g}}^T\} = C'_{\tilde{\mathbf{g}}\tilde{\mathbf{g}}} \qquad (3.60)$$

So:

$$E\{\|\mathbf{g} - \mathbf{g}'\|\} = \text{trace}[A^T C_{\tilde{\mathbf{g}}\tilde{\mathbf{g}}} A - A^T C'_{\tilde{\mathbf{g}}\tilde{\mathbf{g}}} A' - A'^T C'_{\tilde{\mathbf{g}}\tilde{\mathbf{g}}} A + A'^T C'_{\tilde{\mathbf{g}}\tilde{\mathbf{g}}} A'] \quad (3.61)$$

Consider the sum: $-A^T C'_{\tilde{\mathbf{g}}\tilde{\mathbf{g}}} A' + A'^T C'_{\tilde{\mathbf{g}}\tilde{\mathbf{g}}} A' = -(A - A')^T C'_{\tilde{\mathbf{g}}\tilde{\mathbf{g}}} A'$. We can partition A in two sections, a $K \times N^2$ submatrix A_1 and an $(N^2 - K) \times N^2$ submatrix A_2. A' consists of A_1 and an $(N^2 - K) \times N^2$ submatrix with all its elements zero:

$$A = \begin{pmatrix} A_1 \\ --- \\ A_2 \end{pmatrix} \qquad A' = \begin{pmatrix} A_1 \\ --- \\ \mathbf{0} \end{pmatrix} \Rightarrow$$

$$A - A' = \begin{pmatrix} \mathbf{0} \\ --- \\ A_2 \end{pmatrix} \quad \text{and} \quad (A - A')^T = (\ \underbrace{\mathbf{0}}_{N^2 \times K} \ | \ \underbrace{A_2^T}_{N^2 \times (N^2 - K)} \ ) \quad (3.62)$$

Then $(A - A')^T C'_{\tilde{\mathbf{g}}\tilde{\mathbf{g}}} A' = (\ \mathbf{0} \ | \ A_2^T \ ) C'_{\tilde{\mathbf{g}}\tilde{\mathbf{g}}} A'$.
$C'_{\tilde{\mathbf{g}}\tilde{\mathbf{g}}}$ can be partitioned into four submatrices:

$$C'_{\tilde{\mathbf{g}}\tilde{\mathbf{g}}} = \begin{pmatrix} C_1 & | & \mathbf{0} \\ - & - & - \\ \mathbf{0} & | & \mathbf{0} \end{pmatrix} \quad (3.63)$$

where C_1 is $K \times K$ diagonal. Then the product is:

$$(\ \mathbf{0} \ | \ A_2^T \ ) \begin{pmatrix} C_1 & | & \mathbf{0} \\ - & - & - \\ \mathbf{0} & | & \mathbf{0} \end{pmatrix} = (\mathbf{0}) \quad (3.64)$$

Using this result in (3.61) we obtain:

$$E\{\|\mathbf{g} - \mathbf{g}'\|\} = \text{trace}[A^T C_{\tilde{\mathbf{g}}\tilde{\mathbf{g}}} A - A'^T C'_{\tilde{\mathbf{g}}\tilde{\mathbf{g}}} A] \quad (3.65)$$

Consider the term $A^T C_{\tilde{\mathbf{g}}\tilde{\mathbf{g}}} A$. We may assume that $C_{\tilde{\mathbf{g}}\tilde{\mathbf{g}}}$ is the sum of N^2 matrices, each one being $N^2 \times N^2$ and having only one non zero element:

$$C_{\tilde{\mathbf{g}}\tilde{\mathbf{g}}} = \begin{pmatrix} \lambda_1 & 0 & \cdots & 0 \\ 0 & 0 & \cdots & 0 \\ \vdots & \vdots & & \vdots \\ 0 & 0 & \cdots & 0 \end{pmatrix} + \begin{pmatrix} 0 & 0 & \cdots & 0 \\ 0 & \lambda_2 & \cdots & 0 \\ \vdots & \vdots & & \vdots \\ 0 & 0 & \cdots & 0 \end{pmatrix} + \ldots + \begin{pmatrix} 0 & 0 & \cdots & 0 \\ 0 & 0 & \cdots & 0 \\ \vdots & \vdots & & \vdots \\ 0 & 0 & \cdots & \lambda_{N^2} \end{pmatrix}$$

$$(3.66)$$

A is made up of rows of eigenvectors while A^T is made up of columns of eigenvectors. Then we can write:

$$A^T C_{\tilde{g}\tilde{g}} A = \sum_{i=1}^{N^2} (\, \mathbf{u_1} \quad \mathbf{u_2} \quad \cdots \quad \mathbf{u_{N^2}} \,) C_i \begin{pmatrix} \mathbf{u_1}^T \\ \mathbf{u_2}^T \\ \vdots \\ \mathbf{u_{N^2}}^T \end{pmatrix} \qquad (3.67)$$

where C_i is the matrix with its ith diagonal element non-zero and equal to λ_i. Generalizing then the result of Example 3.11, we have:

$$
\begin{aligned}
\text{trace}[A^T C_{\tilde{g}\tilde{g}} A] &= \text{trace}\left[\sum_{i=1}^{N^2} \lambda_i \mathbf{u_i} \mathbf{u_i}^T \right] \\
&= \text{trace}\left[\sum_{i=1}^{N^2} \lambda_i \begin{pmatrix} u_{i1}^2 & u_{i1}u_{i2} & \cdots & u_{i2}u_{iN^2} \\ u_{i2}u_{i1} & u_{i2}^2 & \cdots & u_{i2}u_{iN^2} \\ \vdots & \vdots & & \vdots \\ u_{iN^2}u_{i1} & u_{iN^2}u_{i2} & \cdots & u_{iN^2}^2 \end{pmatrix} \right] \\
&= \sum_{i=1}^{N^2} \lambda_i \text{trace}\begin{pmatrix} u_{i1}^2 & u_{i1}u_{i2} & \cdots & u_{i2}u_{iN^2} \\ u_{i2}u_{i1} & u_{i2}^2 & \cdots & u_{i2}u_{iN^2} \\ \vdots & \vdots & & \vdots \\ u_{iN^2}u_{i1} & u_{iN^2}u_{i2} & \cdots & u_{iN^2}^2 \end{pmatrix} \\
&= \sum_{i=1}^{N^2} \lambda_i (u_{i1}^2 + u_{i2}^2 + \ldots + u_{iN^2}^2) \\
&= \sum_{i=1}^{N^2} \lambda_i \qquad (3.68)
\end{aligned}
$$

To obtain this result we made use of the fact that $\mathbf{u_i}$ is an eigenvector, and therefore $u_{i1}^2 + u_{i2}^2 + \ldots + u_{iN^2}^2 = 1$.

Applying this to equation (3.65) we eventually get:

$$\text{Mean square error} = \sum_{i=1}^{N^2} \lambda_i - \sum_{i=1}^{K} \lambda_i = \sum_{i=K+1}^{N^2} \lambda_i \qquad (3.69)$$

Thus, when an image is approximated by its truncated Karhunen–Loeve expansion, the mean square error committed is equal to the sum of the omitted eigenvalues of the covariance matrix. Since λ_i are arranged in decreasing order, this shows that the mean square error is a minimum.

Example 3.14

The autocovariance matrix of a 2×2 image is given by:

$$C = \begin{pmatrix} 3 & 0 & -1 & 0 \\ 0 & 3 & 0 & -1 \\ -1 & 0 & 3 & 0 \\ 0 & -1 & 0 & 3 \end{pmatrix}$$

Calculate the transformation matrix A for the image, which when used for the inverse transform will approximate the image with mean square error equal to 2.

We must find the eigenvalues of the matrix by solving the equation:

$$\begin{vmatrix} 3-\lambda & 0 & -1 & 0 \\ 0 & 3-\lambda & 0 & -1 \\ -1 & 0 & 3-\lambda & 0 \\ 0 & -1 & 0 & 3-\lambda \end{vmatrix} = 0 \Rightarrow$$

$$(3-\lambda)\left[(3-\lambda)^3 - (3-\lambda)\right] - (-1)^2\left[(3-\lambda)^2 - (-1)^2\right] = 0 \Rightarrow$$

$$(3-\lambda)^2\left[(3-\lambda)^2 - 1\right] - \left[(3-\lambda)^2 - 1\right] = 0 \Rightarrow$$

$$\left[(3-\lambda)^2 - 1\right]^2 = 0 \Rightarrow$$

$$(3-\lambda-1)^2(3-\lambda+1)^2 = 0 \Rightarrow$$

$$(2-\lambda)^2(4-\lambda)^2 = 0 \Rightarrow$$

$$\lambda_1 = 4, \ \lambda_2 = 4, \ \lambda_3 = 2, \ \lambda_4 = 2 \qquad (3.70)$$

The corresponding eigenvectors for $\lambda = 4$ are:

$$\begin{pmatrix} 3 & 0 & -1 & 0 \\ 0 & 3 & 0 & -1 \\ -1 & 0 & 3 & 0 \\ 0 & -1 & 0 & 3 \end{pmatrix}\begin{pmatrix} x_1 \\ x_2 \\ x_3 \\ x_4 \end{pmatrix} = 4\begin{pmatrix} x_1 \\ x_2 \\ x_3 \\ x_4 \end{pmatrix} \Rightarrow \begin{vmatrix} 3x_1 - x_3 = 4x_1 \\ 3x_2 - x_4 = 4x_2 \\ -x_1 + 3x_3 = 4x_3 \\ -x_2 + 3x_4 = 4x_4 \end{vmatrix} \Rightarrow$$

$$x_3 = -x_1, \ x_4 = -x_2, \ x_1 = -x_3, \ x_2 = -x_4 \quad (3.71)$$

Choose: $x_1 = x_3 = 0$, $x_2 = \frac{1}{\sqrt{2}}$, $x_4 = -\frac{1}{\sqrt{2}}$

Or choose: $x_1 = \frac{1}{\sqrt{2}}$, $x_3 = -\frac{1}{\sqrt{2}}$, $x_2 = x_4 = 0$.

The first two eigenvectors, therefore, are: $\left(0 \ \ \frac{1}{\sqrt{2}} \ \ 0 \ \ -\frac{1}{\sqrt{2}}\right)$ and $\left(\frac{1}{\sqrt{2}} \ \ 0 \ \ -\frac{1}{\sqrt{2}} \ \ 0\right)$, which are orthogonal to each other. For $\lambda = 2$ we have:

$$
\begin{pmatrix} 3 & 0 & -1 & 0 \\ 0 & 3 & 0 & -1 \\ -1 & 0 & 3 & 0 \\ 0 & -1 & 0 & 3 \end{pmatrix} \begin{pmatrix} x_1 \\ x_2 \\ x_3 \\ x_4 \end{pmatrix} = 2 \begin{pmatrix} x_1 \\ x_2 \\ x_3 \\ x_4 \end{pmatrix} \Rightarrow \left. \begin{matrix} 3x_1 - x_3 = 2x_1 \\ 3x_2 - x_4 = 2x_2 \\ -x_1 + 3x_3 = 2x_3 \\ -x_2 + 3x_4 = 2x_4 \end{matrix} \right| \Rightarrow
$$

$$
x_1 = x_3, \; x_2 = x_4, \; x_1 = x_3, \; x_2 = x_4 \quad (3.72)
$$

Choose: $x_1 = x_3 = 0$, $x_2 = x_4 = \frac{1}{\sqrt{2}}$.

We do not need to calculate the fourth eigenvector because we are interested in an approximate transformation matrix. By setting some eigenvectors to $(0 \quad 0 \quad 0 \quad 0)$ *the mean square error we commit when reconstructing the image is equal to the sum of the corresponding eigenvalues. In this case, if we consider as transformation matrix the matrix* $\tilde{A}$:

$$
\tilde{A} = \begin{pmatrix} 0 & \frac{1}{\sqrt{2}} & 0 & -\frac{1}{\sqrt{2}} \\ \frac{1}{\sqrt{2}} & 0 & -\frac{1}{\sqrt{2}} & 0 \\ 0 & \frac{1}{\sqrt{2}} & 0 & \frac{1}{\sqrt{2}} \\ 0 & 0 & 0 & 0 \end{pmatrix} \quad (3.73)
$$

the error will be equal to $\lambda_4 = 2$.

Example 3.15

Show the different stages of the Karhunen–Loeve transform of the image in Example 2.14.

Figure 3.2 shows the 64 eigenimages of the original image of Example 2.14 according to the Karhunen–Loeve transform.

    The eight images shown in Figure 3.3 are the reconstructed images when 8, 16, 24, 32, 40, 48, 56 and 64 terms were used for the reconstruction.

    The sums of the mean squared errors for each reconstructed image are:

$$
\text{Error for image (a):} \quad 5360 \quad \left(\sum_{i=9}^{64} \lambda_i = 5365 \right)
$$

$$
\text{Error for image (b):} \quad 3846 \quad \left(\sum_{i=17}^{64} \lambda_i = 3850 \right)
$$

$$
\text{Error for image (c):} \quad 2715 \quad \left(\sum_{i=25}^{64} \lambda_i = 2718 \right)
$$

$$\text{Error for image (d):} \quad 1850 \quad \left(\sum_{i=33}^{64} \lambda_i = 1852\right)$$

$$\text{Error for image (e):} \quad 1194 \quad \left(\sum_{i=41}^{64} \lambda_i = 1195\right)$$

$$\text{Error for image (f):} \quad 715 \quad \left(\sum_{i=49}^{64} \lambda_i = 715\right)$$

$$\text{Error for image (g):} \quad 321 \quad \left(\sum_{i=57}^{64} \lambda_i = 321\right)$$

$$\text{Error for image (h):} \quad 0$$

Note that the mean squared errors of the reconstructions agree very well with the sum of the omitted eigenvalues in each case.

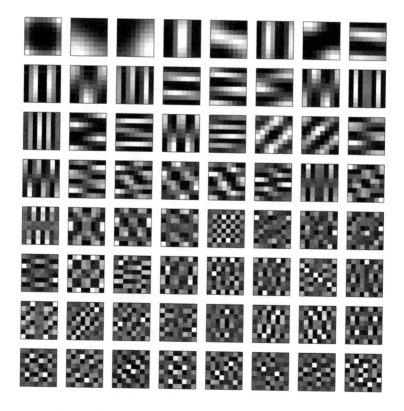

Figure 3.2: The 64 eigenimages, each scaled separately to have values from 0 to 255

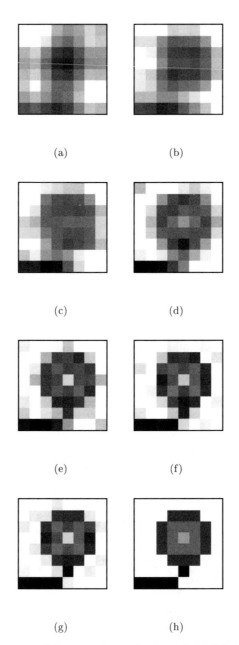

(a)   (b)

(c)   (d)

(e)   (f)

(g)   (h)

Figure 3.3: Reconstructed image when the first $8, 16, 24, 32, 40, 48, 56$ and 64 eigenimages shown in *Figure* 3.2 were used (from top left to bottom right respectively).

Example 3.16

The autocovariance matrix of a 2×2 image is given by:

$$C = \begin{pmatrix} 4 & 0 & -1 & 0 \\ 0 & 4 & 0 & -1 \\ -1 & 0 & 4 & 0 \\ 0 & -1 & 0 & 4 \end{pmatrix}$$

Calculate the transformation matrix A for the image, which, when used for the inverse transform, will approximate the image with mean-square-error equal to 6.

Find the eigenvalues of the matrix first:

$$\begin{vmatrix} 4 - \lambda & 0 & -1 & 0 \\ 0 & 4 - \lambda & 0 & -1 \\ -1 & 0 & 4 - \lambda & 0 \\ 0 & -1 & 0 & 4 - \lambda \end{vmatrix} = 0 \Rightarrow$$

$$(4 - \lambda) \begin{vmatrix} 4 - \lambda & 0 & -1 \\ 0 & 4 - \lambda & 0 \\ -1 & 0 & 4 - \lambda \end{vmatrix} - 1 \begin{vmatrix} 0 & 4 - \lambda & -1 \\ -1 & 0 & 0 \\ 0 & -1 & 4 - \lambda \end{vmatrix} = 0 \Rightarrow$$

$$(4 - \lambda)[(4 - \lambda)^2 - (4 - \lambda)] - [(4 - \lambda)^2 - 1] = 0 \Rightarrow$$

$$[(4 - \lambda)^2 - 1]^2 = 1 \Rightarrow (4 - \lambda - 1)^2 (4 - \lambda + 1)^2 = 0 \Rightarrow$$

$$\lambda_1 = 5$$
$$\lambda_2 = 5$$
$$\lambda_3 = 3$$
$$\lambda_4 = 3$$

Since we allow error of image reconstruction of 6, we do not need to calculate the eigenvectors that correspond to $\lambda = 3$.
   Eigenvectors for $\lambda = 5$:

$$\begin{pmatrix} 4 & 0 & -1 & 0 \\ 0 & 4 & 0 & -1 \\ -1 & 0 & 4 & 0 \\ 0 & -1 & 0 & 4 \end{pmatrix} \begin{pmatrix} x_1 \\ x_2 \\ x_3 \\ x_4 \end{pmatrix} = 5 \begin{pmatrix} x_1 \\ x_2 \\ x_3 \\ x_4 \end{pmatrix} \Rightarrow \begin{array}{l} 4x_1 - x_3 = 5x_1 \Rightarrow x_1 = x_3 \\ 4x_2 - x_4 = 5x_2 \Rightarrow x_2 = -x_4 \\ -x_1 - 4x_3 = 5x_3 \Rightarrow x_1 = -x_3 \\ -x_2 - 4x_4 = 5x_4 \Rightarrow x_2 = -x_4 \end{array}$$

Choose $x_1 = x_3 = 0$, $x_2 = \frac{1}{\sqrt{2}}$, $x_4 = -\frac{1}{\sqrt{2}}$. For λ_2 choose an orthogonal eigenvector, i.e. $x_2 = x_4 = 0$, $x_1 = \frac{1}{\sqrt{2}}$, $x_3 = -\frac{1}{\sqrt{2}}$. Then the transformation matrix which allows reconstruction with mean square error 6 (equal to the sum of the omitted eigenvalues) is:

$$
A = \begin{pmatrix} 0 & \frac{1}{\sqrt{2}} & 0 & -\frac{1}{\sqrt{2}} \\ \frac{1}{\sqrt{2}} & 0 & -\frac{1}{\sqrt{2}} & 0 \\ 0 & 0 & 0 & 0 \\ 0 & 0 & 0 & 0 \end{pmatrix}
$$

What is the "take home" message of this chapter?

If we view an image as an instantiation of a whole lot of images which are the result of a random process, then we can try to represent it as the linear superposition of some eigenimages which are appropriate for representing the whole ensemble of images. For an $N \times N$ image there may be as many as N^2 such eigenimages, while in the SVD approach there are only N. The difference is that with these N^2 eigenimages we are supposed to be able to represent the whole ensemble of images, while in the case of SVD the N eigenimages are appropriate only for representing the one image. If the set of eigenimages is arranged in decreasing order of the corresponding eigenvalues, truncating the expansion of the image in terms of them, approximates the image with the minimum **mean** square error, over the whole ensemble of images. In the SVD case similar truncation leads to the minimum square error approximation.

The crux of K-L expansion is the assumption of ergodicity. This assumption states that the spatial statistics of a single image are the same as the ensemble statistics over the whole set of images. If a restricted type of image is considered, this assumption is clearly unrealistic: Images are not simply the outcomes of a random process; there is always a deterministic underlying component which makes the assumption invalid. So, in such case, the K-L transform effectively puts more emphasis on the random component of the image; i.e. the noise, rather than the component of interest. However, if many different images are considered, the average grey value over the ensemble even of the deterministic component may be the same from pixel to pixel, and the assumption of ergodicity may be nearly valid. On the other hand, if one has available a collection of images representative of the type of image of interest, the assumption of ergodicity may not need to be made: The K-L transform can be calculated using ensemble statistics and used to define a basis of images taylor-made for the particular type of image.

Chapter 4

Image Enhancement

What is image enhancement?

Image enhancement is the process by which we try to improve an image so that it looks **subjectively** better. We do not really know how the image should look, but we can tell whether it has been improved or not, by considering, for example, whether more detail can be seen, or whether unwanted flickering has been removed, or the contrast is better etc.

How can we enhance an image?

The approach largely depends on what we wish to achieve. In general, there are two major approaches: those which reason about the statistics of the grey values of the image, and those which reason about the spatial frequency content of the image.

Which methods of the image enhancement reason about the grey level statistics of an image?

- Methods that manipulate the histogram of the image for the purpose of increasing its contrast.
- The method of principal component analysis of a multispectral image for obtaining a grey level version of it with the maximum possible contrast.
- Methods based on rank order filtering of the image for the purpose of removing noise.

What is the histogram of an image?

The histogram of an image is a discrete function that is formed by counting the number of pixels in the image that have a certain grey value. When this function is normalized to sum up to 1 for all the grey level values, it can be treated as a probability density function that expresses how probable is for a certain grey value to be found in the image. Seen this way, the grey value of a pixel becomes a random variable which takes values according to the outcome of an underlying random experiment.

When is it necessary to modify the histogram of an image?

Suppose that we cannot see much detail in the image. The reason is most likely that pixels which represent different objects or parts of objects tend to have grey level values which are very similar to each other. This is demonstrated with the example histograms shown in *Figure* 4.1. The histogram of the "bad" image is very narrow, while the histogram of the "good" image is more spread.

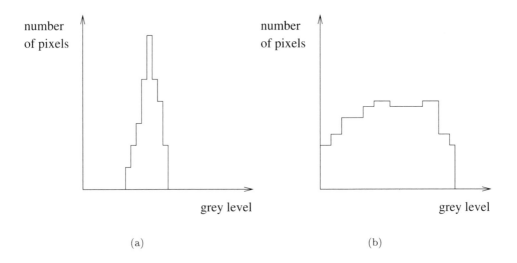

Figure 4.1: (a) is the histogram of a "bad" image while (b) is the histogram of a "good" image.

How can we modify the histogram of an image?

Suppose that the grey levels in the original image are given by the values a variable r obtains and in the new image by the values a variable s obtains. We would like to find a transformation $s = T(r)$ such that the probability density function $p_r(r)$ in *Figure* 4.1a is transformed into a probability density function $p_s(s)$ which looks like that in *Figure* 4.1b, say.

Since $p_r(r)$ is the probability density function of random variable r, the number of pixels with grey level values in the range r to $r + dr$ is $p_r(r)dr$. The transformation we seek will transform this range to $[s, s + ds]$. The total number of pixels in this range will remain the same but in the enhanced image this number will be $p_s(s)ds$:

$$p_s(s)ds = p_r(r)dr \qquad (4.1)$$

This equation can be used to define the transformation T that must be applied to variable r to obtain variable s, provided we define function $p_s(s)$.

What is histogram equalization?

Histogram equalization is the process by which we make all grey values in an image equally probable, i.e. we set $p_s(s) = c$, where c is a constant. Transformation $s = T(r)$ can be calculated from equation (4.1) by substitution of $p_s(s)$ and integration. We integrate from 0 up to an arbitrary value of the corresponding variable, making use of the fact that equation (4.1) is valid for any range of values. These limits are equivalent of saying that we equate the **distribution** functions of the two random variables s and r:

$$\int_0^s c\,ds = \int_0^r p_r(r)dr \Rightarrow s = \frac{1}{c}\int_0^r p_r(x)dx \qquad (4.2)$$

Here, in order to avoid confusion we replaced the dummy variable of integration by x. *Figures* 4.2a-4.2d show an example of applying this transformation to a low contrast image. Notice how narrow the histogram 4.2b of the original image 4.2a is. After histogram equalization, the histogram in 4.2d is much more spread, but contrary to our expectations, it is not flat, i.e. it does not look "equalized".

Why do histogram equalization programs usually not produce images with flat histograms?

In the above analysis, we tacitly assumed that variables r and s can take continuous values. In reality, of course, the grey level values are discrete. In the continuous domain there is an infinite number of numbers in any interval $[r, r + dr]$. In digital images we have only a finite number of pixels in each range. As the range is stretched, and the number of pixels in it is preserved, there is only this finite number of pixels with which the stretched range is populated. The histogram that results is spread over the whole range of grey values, but it is far from flat.

Is it possible to enhance an image to have an absolutely flat histogram?

Yes, if we randomly re-distribute the pixels across neighbouring grey values. This method is called histogram equalization with *random additions*. We can better follow it if we consider a very simple example. Let us assume that we have N_1 pixels with value g_1 and N_2 pixels with value g_2. Let us say that we wish to stretch this histogram so that we have $(N_1 + N_2)/3$ pixels with grey value $\tilde{g}_1$, $(N_1 + N_2)/3$ pixels with grey value $\tilde{g}_2$ and $(N_1 + N_2)/3$ pixels with grey value $\tilde{g}_3$. Let us also assume that we have worked out the transformation that leads from g_i to $\tilde{g}_i$. After we apply this transformation, we may find that we have $\tilde{N}_1$ pixels with grey value $\tilde{g}_1$, $\tilde{N}_2$ pixels with grey value $\tilde{g}_2$, $\tilde{N}_3$ pixels with grey value $\tilde{g}_3$, and that $\tilde{N}_1 > (N_1 + N_2)/3$, $\tilde{N}_2 < (N_1 + N_2)/3$ and $\tilde{N}_3 < (N_1 + N_2)/3$. We may pick at random $(N_1 + N_2)/3 - \tilde{N}_3$ pixels with value $\tilde{g}_2$ and give them value $\tilde{g}_3$. Then we may pick at random $\tilde{N}_1 - (N_1 + N_2)/3$

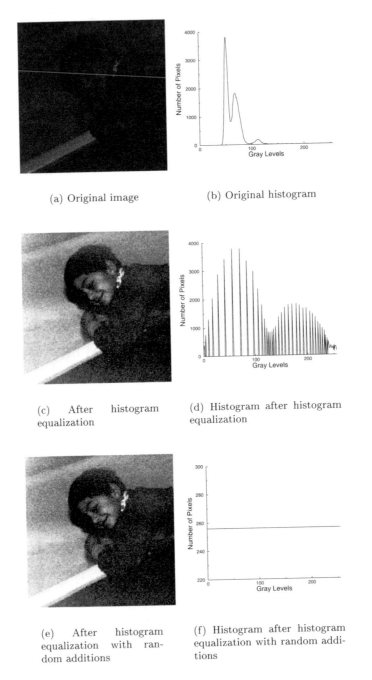

(a) Original image

(b) Original histogram

(c) After histogram equalization

(d) Histogram after histogram equalization

(e) After histogram equalization with random additions

(f) Histogram after histogram equalization with random additions

Figure 4.2: Enhancing the image of a bathtub cleaner by histogram equalization

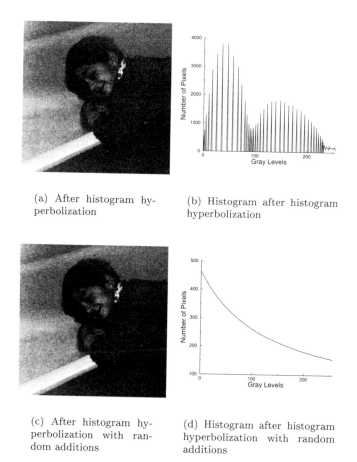

(a) After histogram hyperbolization

(b) Histogram after histogram hyperbolization

(c) After histogram hyperbolization with random additions

(d) Histogram after histogram hyperbolization with random additions

Figure 4.3: Histogram hyperbolization with $\alpha = 0.5$ applied to the image of *Figure* 4.2a

with value $\tilde{g}_1$ and give them value $\tilde{g}_2$. The result will be a perfectly flat histogram. An example of applying this method can be seen in *Figures* 4.2e and 4.3c.

What if we do not wish to have an image with a flat histogram?

We may define $p_s(s)$ to be any function we wish. Then:

$$\int_0^s p_s(y)dy = \int_0^r p_r(x)dx$$

where in order to avoid confusion we have used x and y as the dummy variables of integration. This equation defines s directly in terms of r. Since $p_s(y)$ is known (the desired histogram), one can solve the integral on the left hand side to find a function

f_1 of s. Similarly, the integral on the right hand side can be performed to yield a function f_2 of r; i.e.

$$f_1(s) = f_2(r) \Rightarrow s = f_1^{-1} f_2(r) \tag{4.3}$$

In practice, one may define an intermediate transformation by:

$$w \equiv \int_0^r p_r(x) dx \tag{4.4}$$

This transformation gives the values w of the equalized histogram of the given image. Clearly:

$$\int_0^s p_s(y) dy = w \tag{4.5}$$

This defines another transformation such that $w = T_2(s)$ while we actually need s in terms of w. So this is a three-step process:

1. Equalize the histogram of the given image.
2. Specify the desired histogram and obtain the transformation $w = T_2(s)$.
3. Apply the inverse of the above transformation to the equalized histogram.

An example of applying this method to image 4.2a can be seen in *Figures* 4.3a and 4.3b. This histogram has been produced by setting $p_s(s) = \alpha e^{-\alpha s}$ where α is some positive constant. The effect is to give more emphasis to low grey level values and less to the high ones. This effect is barely visible in *Figure* 4.3b because it is masked by the discretization effect. However, in *Figure* 4.3d it can be seen clearly because the method of random additions was used.

Why should one wish to perform something other than histogram equalization?

One may wish to emphasize certain grey values more than others, in order to compensate for a certain effect; for example, to compensate for the way the human eye responds to the different degrees of brightness. This is a reason for doing histogram hyperbolization: it produces a more pleasing picture.

Example 4.1

The histogram of an image can be approximated by the probability density function

$$p_r(r) = Ae^{-r}$$

where r is the grey-level variable taking values between 0 and b, and A is a normalizing factor. Calculate the transformation $s = T(r)$, where s is the grey level value in the transformed image, such that the transformed image has probability density function

$$p_s(s) = Bse^{-s^2}$$

where s takes values between 0 and b, and B is some normalizing factor.

The transformation $s = T(r)$ can be calculated using equation (4.1).

$$Bse^{-s^2}ds = Ae^{-r}dr$$

We integrate both sides of this equation to obtain the relationship between the distribution functions of variables s and r. To avoid confusion we use as dummy variables of integration y on the left hand side and x on the right hand side:

$$B\int_0^s ye^{-y^2}dy = A\int_0^r e^{-x}dx \tag{4.6}$$

The left hand side of (4.6) is:

$$\int_0^s ye^{-y^2}dy = \frac{1}{2}\int_0^s e^{-y^2}dy^2 = -\frac{1}{2}e^{-y^2}\Big|_0^s = \frac{1-e^{-s^2}}{2} \tag{4.7}$$

The right hand side of (4.6) is:

$$\int_0^s e^{-x}dx = -e^{-x}\Big|_0^r = 1 - e^{-r} \tag{4.8}$$

We substitute from (4.7) and (4.8) into (4.6) to obtain:

$$\frac{1-e^{-s^2}}{2} = \frac{A}{B}(1-e^{-r}) \Rightarrow$$

$$e^{-s^2} = 1 - \frac{2A}{B}(1-e^{-r}) \Rightarrow$$

$$-s^2 = ln\left[1 - \frac{2A}{B}(1-e^{-r})\right] \Rightarrow$$

$$s = \sqrt{-ln\left[1 - \frac{2A}{B}(1-e^{-r})\right]}$$

What if the image has inhomogeneous contrast?

The approach described above is global, i.e. we modify the histogram which refers to the whole image. However, the image may have variable quality at various parts. For example, it may have a wide shadow band right in the middle, with its top and bottom parts being adequately visible. In that case we can apply the above techniques locally: We scan the image with a window inside which we modify the histogram but we alter only the value of the grey level of the central pixel. Clearly such a method is costly and various algorithms have been devised to make it more efficient.

Figure 4.4a shows a classical example of an image that requires local enhancement. The picture was taken indoors looking towards windows with plenty of ambient light coming through. All outdoor sections are fine, but in the indoor part the film was under-exposed. The result of global histogram equalization shown in *Figure* 4.4b is not bad, but it makes the outdoor parts over-exposed in order to allow us to see the details of the interior. The result of the local histogram equalization on the other hand, shown in *Figure* 4.4c, is overall a much more balanced picture. The window size used for this was 40×40, with the original image being of size 400×400. Notice that no part of the picture gives the impression of being over-exposed or under-exposed. There are parts of the image, however, that look damaged: at the bottom of the picture and a little at the top. They correspond to parts of the original film which received too little light to record anything. They correspond to flat black patches, and by trying to enhance them we simply enhance the film grain or the instrument noise. This effect is more prominent in the picture of the hanging train of Wupertal shown in *Figure* 4.5. Local histogram equalization (the result of which is shown in *Figure* 4.5c) attempts to improve parts of the picture that are totally black, in effect trying to amplify non-existing information. However, those parts of the image with some information content are enhanced in a pleasing way.

A totally different effect becomes evident in *Figure* 4.6c which shows the local histogram enhancement of a picture taken at Karlstejn castle in the Czech Republic, shown in *Figure* 4.6a. The castle at the back consists of flat grey walls. The process of local histogram equalization amplifies every small variation of the wall to such a degree that the wall looks like the rough surface of a rock. Further, on the left of the picture we observe again the effect of trying to enhance a totally black area. In this case, the result of global histogram equalization looks much more acceptable, in spite of the fact that if we were to judge from the original image, we would have thought that local histogram equalization would produce a better result.

Is there an alternative to histogram manipulation?

Yes, one may use the mean and standard deviation of the distribution of pixels inside a window. Let us say that the mean grey value inside a window centred at (x, y) is $m(x, y)$, the variance of the pixels inside the window is $\sigma(x, y)$, and the value of pixel (x, y) is $f(x, y)$. We can enhance the variance inside each such window by using a transformation of the form:

$$g(x, y) = A[f(x, y) - m(x, y)] + m(x, y) \tag{4.9}$$

where A is some scalar.

We would like areas which have low variance to have their variance amplified most. So we choose the amplification factor A inversely proportional to $\sigma(x, y)$:

$$A = \frac{kM}{\sigma(x, y)}$$

where k is a constant, and M is the average grey value of the image.

Figure 4.4d shows the results of applying this process to image 4.4a with $k = 3$ and window size 5×5. Note that although details in the image have become explicit, the picture overall is too dark and not particularly pleasing. *Figures* 4.5d and 4.6d show the results of applying the same process to the images 4.5a and 4.6a respectively, with the additional post-processing of histogram equalization.

(a) Original image

(b) After global histogram equalization

(c) After local histogram equalization

(d) After local enhancement

Figure 4.4: Enhancing the image of a young train driver.

(a) Original image   (b) After global histogram equalization

(c) After local histogram equalization   (d) After local enhancement

Figure 4.5: Enhancing the image of the hanging train of Wupertal.

(a) Original image   (b) After global histogram equalization

(c) After local histogram equalization   (d) After local enhancement

Figure 4.6: Enhancing the image at the Karlstejn castle.

How can we improve the contrast of a multispectral image?

A *multispectral* or *multiband* or colour image consists of several arrays of the same scene, one for each spectral component. Each of these bands is a grey level image giving the intensity of light at the particular spectral component at the position of each pixel. Suppose for simplicity that we have three spectral bands, Red, Green and Blue. Then each picture consists of three bands, three grey level images. Alternatively, we may say that each pixel carries three values, one for each spectral band. We can plot these triplets in a 3D coordinate space, called RGB because we measure the grey value of a pixel in each of the three bands along the three axes. The pixels of the colour image plotted in this space form a cluster.

If we were to use only one of these bands, we would like to choose the one that shows the most detail; i.e. the one with the maximum contrast, the one in which the values of the pixels are most spread.

It is possible that the maximum spread of the values of the pixels is not along any of the axes, but along another line (see *Figure* 4.7a). To identify this line we must perform *principal component analysis* or take the *Karhunen–Loeve transformation* of the image.

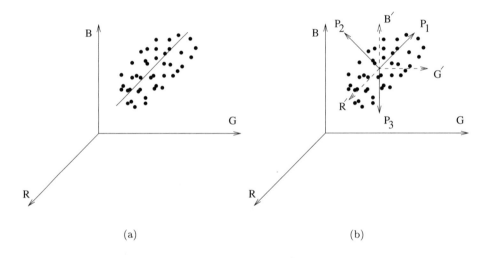

(a)   (b)

Figure 4.7: The pixels of a colour image form a cluster in the colour space. The maximum spread of this cluster may be along a line not parallel with any of the colour axes.

What is principal component analysis?

Principal component analysis (or Karhunen–Loeve transformation) identifies a linear transformation of the coordinate system such that the three axes of the new coordinate system coincide with the directions of the three largest spreads of the point distribution. In this new set of axes the data are uncorrelated. This means that if we form a grey image by using the values of the first co-ordinate of each pixel, it will contain totally uncorrelated information from the information that will be contained in the grey image formed by the second coordinate of each pixel and the information contained in the image formed by the third coordinate of each pixel.

What is the relationship of the Karhunen–Loeve transformation discussed here and the one discussed in Chapter 3?

They both analyse an ensemble of random outcomes into their uncorrelated components. However, in Chapter 3 the whole image was considered as the outcome of a random experiment, with the other random outcomes in the ensemble not available. Their lack of availability was compensated by the assumed ergodicity. So, although the ensemble statistics were computed over the single available image using spatial statistics, they were assumed to be averages computed over all random outcomes, i.e. all versions of the image. Here the values of a **single** pixel are considered to be the outcomes of a random experiment and we have at our disposal the whole ensemble of random outcomes made up from all the image pixels.

How can we perform principal component analysis?

To perform principal component analysis we must diagonalize the covariance matrix of our data. The autocovariance function of the outputs of the assumed random experiment is:

$$C(i,j) \equiv E\{(x_i(k,l) - x_{i0})(x_j(k,l) - x_{j0})\}$$

where $x_i(k,l)$ is the value of pixel (k,l) at band i, x_{i0} is the mean of band i, $x_j(k,l)$ is the value of the same pixel in band j, x_{j0} is the mean of band j, and the expectation value is over all outcomes of the random experiment, i.e. over all pixels of the image:

$$C(i,j) = \frac{1}{N^2} \sum_{k=1}^{N} \sum_{l=1}^{N} (x_i(k,l) - x_{i0})(x_j(k,l) - x_{j0}) \qquad (4.10)$$

Since we have three bands, variables i and j take only three values to indicate R,G and B and the covariance matrix is a 3×3 matrix. For data that are uncorrelated, C is diagonal; i.e. $C(i,j) = 0$ for $i \neq j$. To achieve this we must transform our data using the transformation matrix A made up from the eigenvectors of the covariance matrix of the untransformed data. The process is as follows:

1. Find the mean of the distribution of points in the colour space, say point (R_0, G_0, B_0).

2. Subtract the mean grey level value from each corresponding band. This is equivalent to translating the RGB coordinate system to be centred at the centre of the pixel distribution (see axes $R'G'B'$ in *Figure* 4.7b).

3. Find the autocorrelation matrix $C(i,j)$ of the initial distribution (where i and j take the values R, G and B).

4. Find the eigenvalues of $C(i,j)$ and arrange them in **decreasing** order. Form the eigenvector matrix A, having the eigenvectors as rows.

5. Transform the distribution using matrix A. Each triplet $\mathbf{x} = \begin{pmatrix} R \\ G \\ B \end{pmatrix}$ is trans-

formed into $\mathbf{y} = \begin{pmatrix} P_1 \\ P_2 \\ P_3 \end{pmatrix}$ by: $\mathbf{y} = A\mathbf{x}$; i.e. $y_k = \sum_i a_{ki} x_i$.

This is a linear transformation. The new "colours" are linear combinations of the intensity values of the initial colours, arranged so that the first principal component contains most of the information for the image (see *Figure* 4.7b).

What are the advantages of using principal components to express an image?

The advantages of using principal components are:

1. The information conveyed by each band is maximal for the number of bits used because the bands are uncorrelated and no information contained in one band can be predicted by the knowledge of the other bands.

2. If we want to use a monochrome version of the image, we can restrict ourselves to the first principal component only and be sure that it has the maximum contrast and contains the maximum possible information conveyed by a single band of the image.

An example of principal component analysis is shown in *Figure* 4.8. Although at first glance not much difference is observed between *Figures* 4.8a, 4.8b, 4.8c and 4.8d, at a more careful examination, we can see that the first principal component combines the best parts of all three bands: For example, the face of the boy has more contrast in 4.8b and 4.8c than in 4.8a, while his right leg has more contrast with his trousers in 4.8a and 4.8b than in 4.8c. In 4.8d we have good contrast in both these places. Similarly, the contrast between the trousers and the ground is non-existent in 4.8a and 4.8b but it is obvious in 4.8c. Image 4.8d shows it as well.

What are the disadvantages of principal component analysis?

The grey values in the bands created from principal component analysis have no physical meaning, as they do not correspond to any physical colours. As a result, the grey value of a pixel cannot be used for the classification of a pixel. This is particularly relevant to remote sensing applications, where often pixels are classified according to their grey values. In a **principal component band**, pixels that represent water, for example, may appear darker or brighter than other pixels in the image depending on the image content, while the degree of greyness of water pixels in the various **spectral bands** is always consistent, well understood by remote sensing scientists, and often used to identify them.

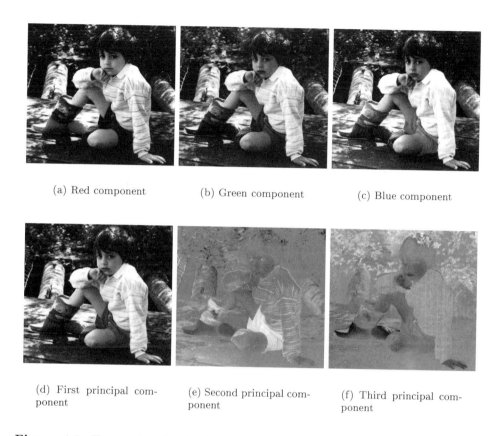

(a) Red component (b) Green component (c) Blue component

(d) First principal component
(e) Second principal component
(f) Third principal component

Figure 4.8: Example of principal component analysis of a colour image.

Example 4.2

Is it possible for matrix C below to represent the autocovariance matrix of a three-band image?

$$C = \begin{pmatrix} -1 & 0 & 1 \\ 0 & 1 & -2 \\ -2 & 2 & 0 \end{pmatrix}$$

This matrix cannot represent the autocovariance matrix of an image because from equation (4.10) it is obvious that C must be symmetric with positive elements along its diagonal.

Example 4.3

A three-band image consists of a green, blue and red band with mean
$3, 2$ an 3 respectively. The autocovariance matrix of this image is given
by:

$$B = \begin{pmatrix} 2 & 0 & 1 \\ 0 & 2 & 0 \\ 1 & 0 & 2 \end{pmatrix}$$

A pixel has intensity values $5, 3$ and 4 in the three bands respectively.
What will be the transformed values of the same pixel in the three
principal component bands?

First we must find the eigenvalues of matrix B:

$$\begin{vmatrix} 2 - \lambda & 0 & 1 \\ 0 & 2 - \lambda & 0 \\ 1 & 0 & 2 - \lambda \end{vmatrix} = 0$$

$$\Rightarrow (2 - \lambda)^3 - (2 - \lambda) = 0 \Rightarrow (2 - \lambda) \left[(2 - \lambda)^2 - 1 \right] = 0$$

$$\Rightarrow (2 - \lambda)(1 - \lambda)(3 - \lambda) = 0$$

Therefore, $\lambda_1 = 3$, $\lambda_2 = 2$, $\lambda_3 = 1$. The corresponding eigenvectors are:

$$\begin{pmatrix} 2 & 0 & 1 \\ 0 & 2 & 0 \\ 1 & 0 & 2 \end{pmatrix} \begin{pmatrix} x_1 \\ x_2 \\ x_3 \end{pmatrix} = 3 \begin{pmatrix} x_1 \\ x_2 \\ x_3 \end{pmatrix} \Rightarrow \begin{array}{l} 2x_1 + x_3 = 3x_1 \\ 2x_2 = 3x_2 \\ x_1 + 2x_3 = 3x_3 \end{array} \Bigg| \Rightarrow \begin{array}{l} x_1 = x_3 \\ x_2 = 0 \end{array} \Bigg| \Rightarrow$$

$$\Rightarrow \mathbf{u_1} = \begin{pmatrix} \frac{1}{\sqrt{2}} \\ 0 \\ \frac{1}{\sqrt{2}} \end{pmatrix}$$

$$\begin{pmatrix} 2 & 0 & 1 \\ 0 & 2 & 0 \\ 1 & 0 & 2 \end{pmatrix} \begin{pmatrix} x_1 \\ x_2 \\ x_3 \end{pmatrix} = 2 \begin{pmatrix} x_1 \\ x_2 \\ x_3 \end{pmatrix} \Rightarrow \begin{array}{l} 2x_1 + x_3 = 2x_1 \\ 2x_2 = 2x_2 \\ x_1 + 2x_3 = 2x_3 \end{array} \Bigg| \Rightarrow \begin{array}{l} x_2 \text{ anything} \\ x_1 = x_3 = 0 \end{array} \Bigg| \Rightarrow$$

$$\Rightarrow \mathbf{u_2} = \begin{pmatrix} 0 \\ 1 \\ 0 \end{pmatrix}$$

$$\begin{pmatrix} 2 & 0 & 1 \\ 0 & 2 & 0 \\ 1 & 0 & 2 \end{pmatrix} \begin{pmatrix} x_1 \\ x_2 \\ x_3 \end{pmatrix} = \begin{pmatrix} x_1 \\ x_2 \\ x_3 \end{pmatrix} \Rightarrow \begin{array}{l} 2x_1 + x_3 = x_1 \\ 2x_2 = x_2 \\ x_1 + 2x_3 = x_3 \end{array} \Bigg| \Rightarrow \begin{array}{l} x_1 = x_3 \\ x_2 = 0 \end{array} \Bigg| \Rightarrow$$

$$\Rightarrow \mathbf{u_3} = \begin{pmatrix} -\frac{1}{\sqrt{2}} \\ 0 \\ \frac{1}{\sqrt{2}} \end{pmatrix}$$

The transformation matrix A is:

$$\begin{pmatrix} \frac{1}{\sqrt{2}} & 0 & \frac{1}{\sqrt{2}} \\ 0 & 1 & 0 \\ -\frac{1}{\sqrt{2}} & 0 & \frac{1}{\sqrt{2}} \end{pmatrix}$$

Pixel $\begin{pmatrix} 5 \\ 3 \\ 4 \end{pmatrix}$ will transform to:

$$\begin{pmatrix} p_1 \\ p_2 \\ p_3 \end{pmatrix} = \begin{pmatrix} \frac{1}{\sqrt{2}} & 0 & \frac{1}{\sqrt{2}} \\ 0 & 1 & 0 \\ -\frac{1}{\sqrt{2}} & 0 & \frac{1}{\sqrt{2}} \end{pmatrix} \begin{pmatrix} 5 \\ 3 \\ 4 \end{pmatrix} = \begin{pmatrix} \frac{9}{\sqrt{2}} \\ 3 \\ -\frac{1}{\sqrt{2}} \end{pmatrix}$$

Example 4.4

A 4×4 three-band image is given:

$$R = \begin{pmatrix} 3 & 3 & 5 & 6 \\ 3 & 4 & 4 & 5 \\ 4 & 5 & 5 & 6 \\ 4 & 5 & 5 & 6 \end{pmatrix} \quad G = \begin{pmatrix} 3 & 2 & 3 & 4 \\ 1 & 5 & 3 & 6 \\ 4 & 5 & 3 & 6 \\ 2 & 4 & 4 & 5 \end{pmatrix} \quad B = \begin{pmatrix} 4 & 2 & 3 & 4 \\ 1 & 4 & 2 & 4 \\ 4 & 3 & 3 & 5 \\ 2 & 3 & 5 & 5 \end{pmatrix}$$

Calculate its three principal components and verify that they are uncorrelated.

First we calculate the mean of each band:

$$\begin{aligned} R_0 &= \frac{1}{16}(3+3+5+6+3+4+4+5+4+5+5+6+4+5+5+6) \\ &= \frac{73}{16} = 4.5625 \\ G_0 &= \frac{1}{16}(3+2+3+4+1+5+3+6+4+5+3+6+2+4+4+5) \\ &= \frac{60}{16} = 3.75 \\ B_0 &= \frac{1}{16}(4+2+3+4+1+4+2+4+4+3+3+5+2+3+5+5) \\ &= \frac{50}{16} = 3.375 \end{aligned}$$

Next we calculate the elements of the covariance matrix as:

$$C_{RR} = \frac{1}{16}\sum_{k=1}^{4}\sum_{l=1}^{4}(R(k,l) - R_0)^2 = 0.996094$$

$$C_{RG} = \frac{1}{16}\sum_{k=1}^{4}\sum_{l=1}^{4}(R(k,l) - R_0)(G(k,l) - G_0) = 0.953125$$

$$C_{RB} = \frac{1}{16}\sum_{k=1}^{4}\sum_{l=1}^{4}(R(k,l) - R_0)(B(k,l) - B_0) = 0.726563$$

$$C_{GG} = \frac{1}{16}\sum_{k=1}^{4}\sum_{l=1}^{4}(G(k,l) - G_0)^2 = 1.9375$$

$$C_{GB} = \frac{1}{16}\sum_{k=1}^{4}\sum_{l=1}^{4}(G(k,l) - G_0)(B(k,l) - B_0) = 1.28125$$

$$C_{BB} = \frac{1}{16}\sum_{k=1}^{4}\sum_{l=1}^{4}(B(k,l) - B_0)^2 = 1.359375$$

Therefore, the covariance matrix is:

$$C = \begin{pmatrix} 0.996094 & 0.953125 & 0.726563 \\ 0.953125 & 1.937500 & 1.281250 \\ 0.726563 & 1.28125 & 1.359375 \end{pmatrix}$$

The eigenvalues of this matrix are:

$$\lambda_1 = 3.528765 \quad \lambda_2 = 0.435504 \quad \lambda_3 = 0.328700$$

The corresponding eigenvectors are:

$$\mathbf{u_1} = \begin{pmatrix} 0.427670 \\ 0.708330 \\ 0.561576 \end{pmatrix} \quad \mathbf{u_2} = \begin{pmatrix} 0.876742 \\ -0.173808 \\ -0.448457 \end{pmatrix} \quad \mathbf{u_3} = \begin{pmatrix} 0.220050 \\ -0.684149 \\ 0.695355 \end{pmatrix}$$

The transformation matrix therefore is:

$$A = \begin{pmatrix} 0.427670 & 0.708330 & 0.561576 \\ 0.876742 & -0.173808 & -0.448457 \\ 0.220050 & -0.684149 & 0.695355 \end{pmatrix}$$

We can find the principal components by using this matrix to transform the values of every pixel. For example, for the first few pixels we find:

$$\begin{pmatrix} 5.654302 \\ 0.314974 \\ 1.389125 \end{pmatrix} = \begin{pmatrix} 0.427670 & 0.708330 & 0.561576 \\ 0.876742 & -0.173808 & -0.448457 \\ 0.220050 & -0.684149 & 0.695355 \end{pmatrix}\begin{pmatrix} 3 \\ 3 \\ 4 \end{pmatrix}$$

$$\begin{pmatrix} 3.822820 \\ 1.385694 \\ 0.682562 \end{pmatrix} = \begin{pmatrix} 0.427670 & 0.708330 & 0.561576 \\ 0.876742 & -0.173808 & -0.448457 \\ 0.220050 & -0.684149 & 0.695355 \end{pmatrix} \begin{pmatrix} 3 \\ 2 \\ 2 \end{pmatrix}$$

$$\begin{pmatrix} 5.948065 \\ 2.516912 \\ 1.133874 \end{pmatrix} = \begin{pmatrix} 0.427670 & 0.708330 & 0.561576 \\ 0.876742 & -0.173808 & -0.448457 \\ 0.220050 & -0.684149 & 0.695355 \end{pmatrix} \begin{pmatrix} 5 \\ 3 \\ 3 \end{pmatrix}$$

We use the first element of each transformed triplet to form the first principal component of the image, the second element for the second principal component, and the third for the third one. In this way we derive:

$$P_1 = \begin{pmatrix} 5.654302 & 3.822820 & 5.948065 & 7.645640 \\ 2.552915 & 7.498631 & 4.958820 & 8.634631 \\ 6.790301 & 7.364725 & 5.948065 & 9.623876 \\ 4.250490 & 6.656395 & 7.779546 & 8.915546 \end{pmatrix}$$

$$P_2 = \begin{pmatrix} 0.314974 & 1.385694 & 2.516912 & 2.771389 \\ 2.007960 & 0.844099 & 2.088630 & 1.547035 \\ 1.017905 & 2.169300 & 2.516912 & 1.975317 \\ 2.262436 & 2.343106 & 1.446188 & 2.149123 \end{pmatrix}$$

$$P_3 = \begin{pmatrix} 1.389125 & 0.682562 & 1.133874 & 1.365131 \\ 0.671360 & 0.240878 & 0.218468 & -0.223219 \\ 0.925027 & -0.234424 & 1.133874 & 0.692187 \\ 0.902617 & 0.449725 & 1.840433 & 1.376336 \end{pmatrix}$$

To confirm that these new bands contain uncorrelated data we shall calculate their autocovariance matrix. First we find the mean of each band: P_{10}, P_{20}, P_{30}. Then we compute:

$$C_{P_1 P_1} = \frac{1}{16} \sum_{i=1}^{4} \sum_{j=1}^{4} (P_1(i,j) - P_{10})^2 = 3.528765$$

$$C_{P_1 P_2} = \frac{1}{16} \sum_{i=1}^{4} \sum_{j=1}^{4} (P_1(i,j) - P_{10})(P_2(i,j) - P_{20}) = 0.0$$

$$C_{P_1 P_3} = \frac{1}{16} \sum_{i=1}^{4} \sum_{j=1}^{4} (P_1(i,j) - P_{10})(P_3(i,j) - P_{30}) = 0.0$$

$$C_{P_2 P_2} = \frac{1}{16} \sum_{i=1}^{4} \sum_{j=1}^{4} (P_2(i,j) - P_{20})^2 = 0.435504$$

$$C_{P_2 P_3} = \frac{1}{16} \sum_{i=1}^{4} \sum_{j=1}^{4} (P_2(i,j) - P_{20})(P_3(i,j) - P_{30}) = 0.0$$

$$C_{P_3 P_3} = \frac{1}{16} \sum_{i=1}^{4} \sum_{j=1}^{4} (P_3(i,j) - P_{30})^2 = 0.328700$$

We see that this covariance matrix is diagonal, so it refers to uncorrelated data.

Example 4.5

For the image in Example 4.4 show that the first principal component has more contrast than any of the original bands.

The contrast of an image can be characterized by the range of grey values it has. We can see that the contrast of the original image was 3 in the red band, 5 in the green band and 4 in the blue band. The range of values in the first principal component is 9.623876 − 2.552915 = 7.070961. This is larger than any of the previous ranges.

Some of the images with enhanced contrast appear very noisy. Can we do anything about that?

Indeed, this is the case for images 4.4c, 4.5c and 4.6c, where there are large uniformly coloured regions, which happen to cover entirely the window inside which local histogram equalization takes place. Then the grey values of these pixels are stretched to the full range of $0 - 255$, and the noise is significantly enhanced. We have to use then some noise reduction techniques to post-process the image. The technique we use depends on the type of noise that is present in the image.

What are the types of noise present in an image?

There are various types of noise. However, they fall into two major classes: *additive* and *multiplicative* noise. An example of multiplicative noise is variable illumination. This is perhaps the most common type of noise in images. Additive noise is often assumed to be *impulse noise* or *Gaussian noise*. *Figure* 4.9a shows an image corrupted with impulse noise and *Figure* 4.9b shows an image corrupted with additive zero-mean Gaussian noise.

Impulse noise alters at random the value of some pixels. In a binary image this means that some black pixels become white and some white pixels become black. This is why this noise is also called *salt and pepper noise*. Additive zero-mean Gaussian noise means that a value drawn from a zero-mean Gaussian probability density function is added to the true value of every pixel.

(a) Image with impulse noise   (b) Image with additive Gaussian noise

(c) Median filtering of (a)   (d) Median filtering of (b)

(e) Smoothing of (a) by averaging   (f) Smoothing of (b) by averaging

Figure 4.9: Examples of filtering to remove noise.

We use *rank order filtering* to remove impulse noise and *smoothing* to reduce
Gaussian noise.

What is a rank order filter?

A rank order filter is a filter the output value of which depends on the ranking of the
pixels according to their grey values inside the filter window. The most common rank
order filter is the *median filter*.

Figure 4.10 shows the result of trying to remove the noise from output images 4.4c
and 4.5c by median filtering.

(a)   (b)

**Figure 4.10: Improving images 4.4c and 4.5c by using median filtering
(median filter size $= 3 \times 3$).**

What is median filtering?

The median is the grey level value which divides a distribution in two equally num-
bered populations. For example, if we use a 5×5 window, we have 25 grey level
values which we order in an increasing sequence. Then the median is the thirteenth
value. This has the effect of forcing points with distinct intensities to be more like
their neighbours, thus eliminating intensity spikes which appear isolated.

Figure 4.9c shows image 4.9a processed with a median filter and with a window of
size 5×5, while *Figure* 4.9d shows image 4.9b (which contains Gaussian noise) having
been processed in the same way. It is clear that the median filter removes the impulse
noise almost completely.

What if the noise in an image is not impulse?

The most common type of noise in images is Gaussian. We can remove Gaussian noise by *smoothing* the image. For example, we may replace the value of each pixel by the average value inside a small window around the pixel. *Figures* 4.9e and 4.9f show the result of applying this process to images 4.9a and 4.9b respectively. The size of the window used is the same as for the median filtering of the same images, i.e. 5×5. We note that this type of filtering is much more effective for the Gaussian noise, but produces bad results in the case of impulse noise. This is a simple form of *lowpass filtering* of the image.

Why does lowpass filtering reduce noise?

Usually, the noise which is superimposed on the image is uncorrelated. This means that it has a flat spectrum. On the other hand, most images have spectra which have higher values in the low frequencies and gradually reducing values for high frequencies. After a certain frequency, the spectrum of a noisy signal is dominated by the noise component (see *Figure* 4.11). So, if we use a lowpass filter, we kill off all the noise-dominated high-frequency components. At the same time, of course, we kill also the useful information of the image buried in these high frequencies. The result is a clean, but blurred image. The process is as follows:

1. Find the Fourier transform of the image.

2. Multiply it with a function which does not alter frequencies below a certain cutoff frequency but which kills off all higher frequencies. In the 2-dimensional frequency space, this ideal lowpass filter is schematically depicted in *Figure* 4.12.

3. Take the inverse Fourier transform of this product.

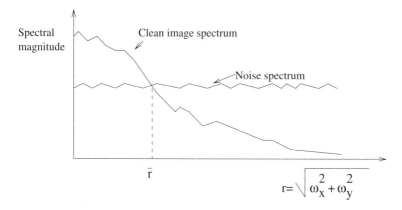

Figure 4.11: After a certain frequency, the spectrum of the noisy image is dominated by the noise. Ideally the cutoff frequency of the lowpass filter should be at $r_0 = \bar{r}$.

Multiplication of the two frequency spectra is equivalent to convolution of the actual functions. So, what we can do instead of the above procedure, is to find the 2-dimensional function in real space which has as its Fourier transform the ideal lowpass filter, and convolve our image with that function. This would be ideal, but it does not work in practice because the function the Fourier transform of which is the ideal lowpass filter is infinite in extent.

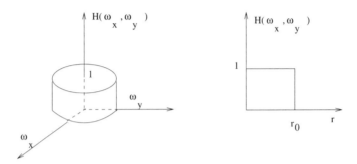

Figure 4.12: The ideal lowpass filter in 2D in the frequency domain. On the right a cross-section of this filter with cutoff frequency r_0 $(r \equiv \sqrt{\omega_x^2 + \omega_y^2})$.

What if we are interested in the high frequencies of an image?

It is possible that we may want to enhance the small details of a picture instead of ironing them out. Such a process is called *sharpening* and it enhances small fluctuations in the intensity of the image, noise included.

One way to achieve this is to calculate at each pixel the local gradient of the intensity using numerical difference formulae. If one wants to be more sophisticated, one can use the filtering approach we discussed in the context of smoothing. Only now, of course, the filter should be highpass and allow the high-frequency components to survive while killing the low-frequency components.

What is the ideal highpass filter?

The ideal highpass filter in the frequency domain is schematically depicted in *Figure 4.13*.

Filtering with such a filter in the frequency domain is equivalent to convolving in real space with the function that has this filter as its Fourier transform. There is no finite function which corresponds to the highpass filter and one has to resort to various approximations.

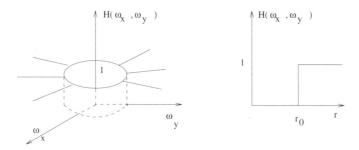

Figure 4.13: The spectrum of the ideal highpass filter is 1 everywhere, except inside a circle of radius r_0 in the frequency domain, where it is 0. On the right, a cross-section of such a filter. Here $r \equiv \sqrt{\omega_x^2 + \omega_y^2}$.

How can we improve an image which suffers from variable illumination?

This is a problem which can be dealt with if we realize that every image function $f(x, y)$ is the product of two factors: an illumination function $i(x, y)$ and a reflectance function $r(x, y)$ that is intrinsic to the imaged surface:

$$f(x, y) = i(x, y)r(x, y) \tag{4.11}$$

Illumination is generally of uniform nature and yields low-frequency components in the Fourier transform of the image. Different materials (objects) on the other hand, imaged next to each other, cause sharp changes of the reflectance function, which causes sharp transitions in the intensity of an image. These sharp changes are associated with high-frequency components. We can try to separate these two factors by first taking the logarithm of equation (4.11) so that the two effects are additive rather than multiplicative: $\ln f(x, y) = \ln i(x, y) + \ln r(x, y)$

Then we filter this logarithmic image by what is called a *homomorphic filter*. Such a filter will enhance the high frequencies and suppress the low frequencies so that the variation in the illumination will be reduced while edges (and details) will be sharpened. The cross-section of a homomorphic filter looks like the one shown in *Figure* 4.14.

Figures 4.15a and 4.16a show two images with smoothly varying illumination from left to right. The results after homomorphic filtering shown in *Figures* 4.15b and 4.16b constitute a clear improvement with the effect of variable illumination greatly reduced and several details particularly in the darker parts of the images made visible. These results were obtained by applying to the logarithm of the original image, a filter with the following transfer function:

$$H(\omega_x, \omega_y) = \frac{1}{1 + e^{-s(\sqrt{\omega_x^2 + \omega_y^2} - \omega_0)}} + A$$

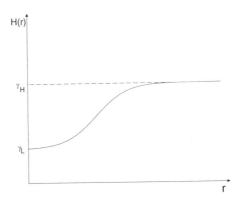

Figure 4.14: A cross-section of a homomorphic filter as a function of polar frequency, $r \equiv \sqrt{\omega_x^2 + \omega_y^2}$.

(a) Original image   (b) After homomorphic filtering

Figure 4.15: Example of homomorphic filtering Haroula at the keyboard.

with $s = 1$, $\omega_0 = 128$ and $A = 10$. The parameters of this filter are related as follows to the parameters γ_H and γ_L of *Figure* 4.14:

$$\gamma_L = \frac{1}{1 + e^{s\omega_0}} + A, \qquad \gamma_H = 1 + A$$

(a) Original image   (b) After homomorphic filtering

Figure 4.16: Homomorphic filtering Mrs Petrou at home.

Can any of the objectives of image enhancement be achieved by the linear methods we learned in Chapter 2?

Yes, often image smoothing and image sharpening is done by convolving the image with a suitable filter. This is the reason we prefer the filters to be finite in extent: Finite convolution filters can be implemented as linear operators applied on the image.

Example 4.6

You have a 3×3 image which can be represented by a 9×1 vector. Derive a matrix which, when it operates on this image, smoothes its columns by averaging every three successive pixels, giving them weights $\frac{1}{4}, \frac{1}{2}, \frac{1}{4}$. To deal with the border pixels assume that the image is repeated periodically in all directions.

Let us say that the original image is

$$\begin{pmatrix} g_{11} & g_{12} & g_{13} \\ g_{21} & g_{22} & g_{23} \\ g_{31} & g_{32} & g_{33} \end{pmatrix}$$

and its smoothed version is

$$\begin{pmatrix} \tilde{g}_{11} & \tilde{g}_{12} & \tilde{g}_{13} \\ \tilde{g}_{21} & \tilde{g}_{22} & \tilde{g}_{23} \\ \tilde{g}_{31} & \tilde{g}_{32} & \tilde{g}_{33} \end{pmatrix}$$

Let as also say that the smoothing matrix we wish to identify is A, with elements a_{ij}:

$$\begin{pmatrix} \tilde{g}_{11} \\ \tilde{g}_{12} \\ \tilde{g}_{13} \\ \tilde{g}_{21} \\ \tilde{g}_{22} \\ \tilde{g}_{23} \\ \tilde{g}_{31} \\ \tilde{g}_{32} \\ \tilde{g}_{33} \end{pmatrix} = \begin{pmatrix} a_{11} & a_{12} & \dots & a_{19} \\ a_{21} & a_{22} & \dots & a_{29} \\ a_{31} & a_{32} & \dots & a_{39} \\ a_{41} & a_{42} & \dots & a_{49} \\ a_{51} & a_{52} & \dots & a_{59} \\ a_{61} & a_{62} & \dots & a_{69} \\ a_{71} & a_{72} & \dots & a_{79} \\ a_{81} & a_{82} & \dots & a_{89} \\ a_{91} & a_{92} & \dots & a_{99} \end{pmatrix} \begin{pmatrix} g_{11} \\ g_{12} \\ g_{13} \\ g_{21} \\ g_{22} \\ g_{23} \\ g_{31} \\ g_{32} \\ g_{33} \end{pmatrix}$$

From the above equation we have:

$$\begin{aligned} \tilde{g}_{11} = \quad & a_{11}g_{11} + a_{12}g_{21} + a_{13}g_{31} + a_{14}g_{12} + a_{15}g_{22} \\ & + a_{16}g_{32} + a_{17}g_{13} + a_{18}g_{23} + a_{19}g_{33} \end{aligned} \tag{4.12}$$

From the definition of the smoothing mask, we have:

$$\tilde{g}_{11} = \frac{1}{4}g_{31} + \frac{1}{2}g_{11} + \frac{1}{4}g_{21} \tag{4.13}$$

Comparison of equations (4.12) and (4.13) shows that we must set:

$$a_{11} = \frac{1}{2}, a_{12} = \frac{1}{4}, a_{13} = \frac{1}{4}, a_{14} = a_{15} = \dots = a_{19} = 0$$

Working in a similar way for a few more elements, we can see that the matrix we wish to identify has the form:

$$A = \begin{pmatrix} \frac{1}{2} & \frac{1}{4} & \frac{1}{4} & 0 & 0 & 0 & 0 & 0 & 0 \\ \frac{1}{4} & \frac{1}{2} & \frac{1}{4} & 0 & 0 & 0 & 0 & 0 & 0 \\ \frac{1}{4} & \frac{1}{4} & \frac{1}{2} & 0 & 0 & 0 & 0 & 0 & 0 \\ 0 & 0 & 0 & \frac{1}{2} & \frac{1}{4} & \frac{1}{4} & 0 & 0 & 0 \\ 0 & 0 & 0 & \frac{1}{4} & \frac{1}{2} & \frac{1}{4} & 0 & 0 & 0 \\ 0 & 0 & 0 & \frac{1}{4} & \frac{1}{4} & \frac{1}{2} & 0 & 0 & 0 \\ 0 & 0 & 0 & 0 & 0 & 0 & \frac{1}{2} & \frac{1}{4} & \frac{1}{4} \\ 0 & 0 & 0 & 0 & 0 & 0 & \frac{1}{4} & \frac{1}{2} & \frac{1}{4} \\ 0 & 0 & 0 & 0 & 0 & 0 & \frac{1}{4} & \frac{1}{4} & \frac{1}{2} \end{pmatrix}$$

What is the "take home" message of this chapter?

With image enhancement we try to make images look better according to subjective criteria. Most of the methods used are non-linear, and therefore they cannot be described within the framework developed in the previous chapters. Some of the methods used are not only non-linear, but also inhomogeneous.

Contrast enhancement of a grey image can be achieved by manipulating the grey values of the pixels so that they become more diverse. This can be done by defining a transformation that converts the distribution of the grey values to a pre-specified shape. The choice of this shape may be totally arbitrary. If, however, we are dealing with a multiband image, we can define a line in the multidimensional space (where we measure the grey values of a pixel in the different bands along the corresponding axes) such that when we project the data on this line, their projected values are maximally spread (see *Figure* 4.7a). This line defines the first principal axis of the data ($\equiv$ pixel values) in the multidimensional space. We can create a grey image by assigning to each pixel its projected value along the first principal axis. This is called the first principal component of the multiband image. From all grey bands we can construct from the multiband image, the first principal component has the maximum contrast and contains the most information. However, the grey values of the pixels do not have physical meaning as they are linear combinations of the grey values of the original bands. The basic difference between histogram manipulation and principal component analysis is that in the former the values the pixels carry are changed arbitrarily, while in the latter the joint probability density function of the pixel values remains the same; we simply "read" the data in a more convenient way, without introducing information that is not already there (compare, for example, the enhanced images 4.6c and 4.8d).

We can also enhance an image in a desirable way by manipulating its Fourier spectrum: We can preferentially kill frequency bands we do not want. This can be achieved with the help of convolution filters that have pre-specified spectra. It is necessary, therefore, to understand how we can develop and use filters appropriate for each task. This is the topic of the next chapter.

Chapter 5

Two-Dimensional Filters

What is this chapter about?

Manipulation of images often entails omitting or enhancing details of certain spatial frequencies. This is equivalent to multiplying the Fourier transform of the image with a certain function that "kills" or modifies certain frequency components. When we do that, we say that we *filter the image*, and the function we use is called a *filter*.

This chapter explores some of the basic properties of 2D filters and presents some methods by which the operation we wish to apply to the Fourier transform of the image can be converted into a simple convolution operation applied to the image directly, allowing us to avoid using the Fourier transform itself.

How do we define a 2D filter?

A 2D filter is defined in terms of its Fourier transform $H(\mu, \nu)$, called the *system function*. By taking the inverse Fourier transform of $H(\mu, \nu)$ we can calculate the filter in the real domain. This is called the *unit sample response* of the filter and is denoted by $h(k, l)$.

How are the system function and the unit sample response of the filter related?

$H(\mu, \nu)$ is defined as a continuous function of (μ, ν). The unit sample response $h(k, l)$ is defined as the inverse Fourier transform of $H(\mu, \nu)$, but since it has to be used for the convolution of a digital image, it is defined at discrete points only. Then the equations relating these two functions are:

$$h(k, l) = \frac{1}{2\pi} \int_{-\pi}^{\pi} \int_{-\pi}^{\pi} H(\mu, \nu) e^{j(\mu k + \nu l)} \, d\mu d\nu \qquad (5.1)$$

$$H(\mu, \nu) = \frac{1}{2\pi} \sum_{n=-\infty}^{+\infty} \sum_{m=-\infty}^{+\infty} h(n, m) e^{-j(\mu n + \nu m)} \qquad (5.2)$$

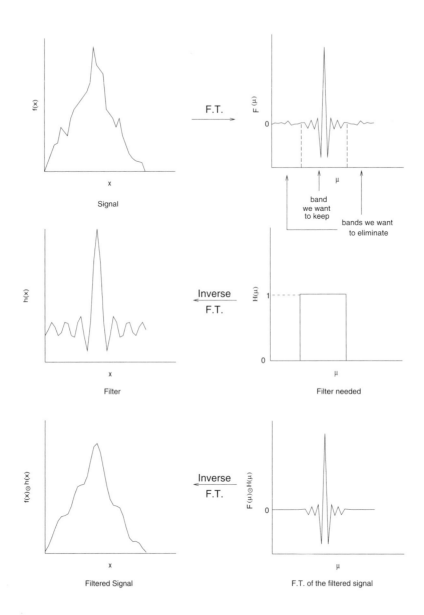

Figure 5.1: Top row: a signal and its Fourier transform. Middle row: the unit sample response function of a filter on the left, and the filter's system function on the right. Bottom row: On the left the filtered signal that can be obtained by convolving the signal at the top with the filter in the middle. On the right the Fourier transform of the filtered signal obtained by multiplying the Fourier transform of the signal at the top, with the Fourier transform (system function) of the filter in the middle.

If we are interested in real filters only, these equations can be modified as follows:

$$h(k,l) \quad = \quad \frac{1}{2\pi} \int_{-\pi}^{\pi} \int_{-\pi}^{\pi} H(\mu,\nu) \cos(\mu k + \nu l) d\mu d\nu \tag{5.3}$$

$$H(\mu,\nu) \quad = \quad \frac{1}{2\pi} \sum_{n=-\infty}^{+\infty} \sum_{m=-\infty}^{+\infty} h(n,m) \cos(\mu n + \nu m) \tag{5.4}$$

Why are we interested in the filter function in the real domain?

We can achieve the enhancement of the image as desired, by simply convolving it with $h(k,l)$ instead of multiplying its Fourier transform with $H(\mu,\nu)$. *Figure* 5.1 shows this schematically in the 1D case. The 2D case is totally analogous.

Are there any conditions which $h(k,l)$ must fulfil so that it can be used as a convolution filter?

Yes, $h(k,l)$ must be zero for $k > K$ and $l > L$, for some finite values K and L; i.e. the filter with which we want to convolve the image must be a finite array of numbers. The ideal lowpass, bandpass and highpass filters do not fulfil this condition.

B5.1: What is the unit sample response of the ideal lowpass filter?

The ideal lowpass filter which cuts to zero all frequencies above a certain frequency R, say, is defined as:

$$H(\mu,\nu) \quad = \quad \begin{cases} 1 & \text{for } \sqrt{\mu^2 + \nu^2} \leq R \\ 0 & \text{otherwise} \end{cases} \tag{5.5}$$

We can use this definition of $H(\mu,\nu)$ to calculate the corresponding unit sample response from equation (5.3):

$$h(k,l) \quad = \quad \frac{1}{2\pi} \int_{-\pi}^{\pi} \int_{-\pi}^{\pi} \cos(\mu k + \nu l) H(\mu,\nu) d\mu d\nu$$

We shall introduce polar coordinates (r,θ) in the (μ,ν) frequency space:

$$\left. \begin{array}{l} \mu \equiv r\cos\theta \\ \nu \equiv r\sin\theta \end{array} \right\} \Rightarrow \mu^2 + \nu^2 = r^2, \ d\mu d\nu = r dr d\theta$$

Then

$$h(k,l) = \frac{1}{2\pi} \int_{0}^{2\pi} \int_{0}^{R} \cos(rk\cos\theta + rl\sin\theta) r dr d\theta \tag{5.6}$$

We can write:

$$k\cos\theta + l\sin\theta = \sqrt{k^2+l^2}\left[\frac{k}{\sqrt{k^2+l^2}}\cos\theta + \frac{l}{\sqrt{k^2+l^2}}\sin\theta\right]$$

$$= \sqrt{k^2+l^2}[\sin\phi\cos\theta + \cos\phi\sin\theta] = \sqrt{k^2+l^2}\sin(\theta+\phi)$$

where angle ϕ has been defined so that:

$$\sin\phi \equiv \frac{k}{\sqrt{k^2+l^2}} \quad\text{and}\quad \cos\phi \equiv \frac{l}{\sqrt{k^2+l^2}}$$

We define a new variable $t \equiv \theta + \phi$. Then equation (5.6) can be written as:

$$h(k,l) = \frac{1}{2\pi}\int_\phi^{2\pi+\phi}\int_0^R \cos(r\sqrt{k^2+l^2}\sin t)rdrdt$$

$$= \frac{1}{2\pi}\int_\phi^{2\pi}\int_0^R \cos(r\sqrt{k^2+l^2}\sin t)rdrdt$$

$$+\frac{1}{2\pi}\int_{2\pi}^{2\pi+\phi}\int_0^R \cos(r\sqrt{k^2+l^2}\sin t)rdrdt$$

In the second term we change the variable to $\tilde{t} = t-2\pi \Rightarrow t = \tilde{t}+2\pi \Rightarrow \sin t = \sin\tilde{t}$. Therefore, we can write:

$$h(k,l) = \frac{1}{2\pi}\int_\phi^{2\pi}\int_0^R \cos(r\sqrt{k^2+l^2}\sin t)rdrdt$$

$$+\frac{1}{2\pi}\int_0^\phi\int_0^R \cos(r\sqrt{k^2+l^2}\sin\tilde{t})rdrd\tilde{t}$$

$$= \frac{1}{2\pi}\int_0^{2\pi}\int_0^R \cos(r\sqrt{k^2+l^2}\sin t)rdrdt$$

This can be written as:

$$h(k,l) = \frac{1}{2\pi}\int_0^\pi\int_0^R \cos(r\sqrt{k^2+l^2}\sin t)rdrdt$$

$$+\frac{1}{2\pi}\int_\pi^{2\pi}\int_0^R \cos(r\sqrt{k^2+l^2}\sin t)rdrdt$$

In the second term, we define a new variable of integration: $\tilde{t} \equiv t - \pi \Rightarrow t = \tilde{t} + \pi \Rightarrow \sin t = -\sin\tilde{t} \Rightarrow \cos(r\sqrt{k^2+l^2}\sin t) = \cos(r\sqrt{k^2+l^2}\sin\tilde{t})$ and $dt = d\tilde{t}$. Then:

$$h(k,l) = \frac{1}{2\pi} \int_0^\pi \int_0^R \cos(r\sqrt{k^2+l^2}\sin t)r\,dr\,dt$$

$$+ \frac{1}{2\pi} \int_0^\pi \int_0^R \cos(r\sqrt{k^2+l^2}\sin\tilde{t})r\,dr\,d\tilde{t}$$

$$= \frac{1}{\pi} \int_0^R \left\{ \int_0^\pi \cos(r\sqrt{k^2+l^2}\sin t)dt \right\} r\,dr \tag{5.7}$$

We know that the Bessel function of the first kind of zero order is defined as:

$$J_0(z) \equiv \frac{1}{\pi} \int_0^\pi \cos(z\sin\theta)d\theta \tag{5.8}$$

If we use definition (5.8) in equation (5.7) we obtain:

$$h(k,l) = \int_0^R rJ_0(r\sqrt{k^2+l^2})dr$$

We define a new variable of integration $x \equiv r\sqrt{k^2+l^2} \Rightarrow dr = \frac{1}{\sqrt{k^2+l^2}}dx$. Then:

$$h(k,l) = \frac{1}{k^2+l^2} \int_0^{R\sqrt{k^2+l^2}} xJ_0(x)dx \tag{5.9}$$

From the theory of Bessel functions, it is known that:

$$\int x^{p+1}J_p(x)dx = x^{p+1}J_{p+1}(x) \tag{5.10}$$

We apply formula (5.10) with $p = 0$ in equation (5.9):

$$h(k,l) = \frac{1}{k^2+l^2} xJ_1(x)\Big|_0^{R\sqrt{k^2+l^2}} \Rightarrow$$

$$h(k,l) = \frac{R}{\sqrt{k^2+l^2}} J_1(R\sqrt{k^2+l^2}) \tag{5.11}$$

This function is a function of infinite extent, defined at each point (k,l) of integer coordinates. It corresponds, therefore, to an array of infinite dimensions. The implication is that this filter cannot be implemented as a linear convolution filter of the image.

Example 5.1 (B)

What is the impulse response of the ideal bandpass filter?

The ideal bandpass filter is defined as:

$$H(\mu, \nu) = \begin{cases} 1 & \text{for } R_1 \leq \sqrt{\mu^2 + \nu^2} \leq R_2 \\ 0 & \text{otherwise} \end{cases}$$

The only difference, therefore, with the ideal lowpass filter derived in box **B5.1** *is in the limits of equation (5.9):*

$$
\begin{aligned}
h(k, l) &= \frac{1}{k^2 + l^2} \int_{R_1\sqrt{k^2+l^2}}^{R_2\sqrt{k^2+l^2}} x J_0(x) dx \\
&= \frac{1}{k^2 + l^2} x J_1(x) \Big|_{R_1\sqrt{k^2+l^2}}^{R_2\sqrt{k^2+l^2}} \\
&= \frac{1}{\sqrt{k^2 + l^2}} \left(R_2 J_1(R_2\sqrt{k^2 + l^2}) - R_1 J_1(R_1\sqrt{k^2 + l^2}) \right)
\end{aligned}
$$

This is a function defined for all values (k, l). Therefore the ideal bandpass filter is an infinite impulse response filter.

Example 5.2 (B)

What is the impulse response of the ideal highpass filter?

The ideal highpass filter is defined as:

$$H(\mu, \nu) = \begin{cases} 0 & \text{for } \sqrt{\mu^2 + \nu^2} \leq R \\ 1 & \text{otherwise} \end{cases}$$

The only difference, therefore, with the ideal lowpass filter derived in box **B5.1** *is in the limits of equation (5.9):*

$$
\begin{aligned}
h(k, l) &= \frac{1}{k^2 + l^2} \int_{R\sqrt{k^2+l^2}}^{\infty} x J_0(x) dx \\
&= \frac{1}{k^2 + l^2} x J_1(x) \Big|_{R\sqrt{k^2+l^2}}^{\infty}
\end{aligned}
\tag{5.12}
$$

Bessel function $J_1(x)$ tends to 0 for $x \to \infty$. However, its asymptotic behaviour is $\lim_{x \to \infty} J_1(x) = \frac{1}{\sqrt{x}}$. This means that $J_1(x)$ does not tend to 0 fast enough to compensate for the factor x which multiplies it, i.e. $\lim_{x \to \infty} x J_1(x) \to \infty$. Therefore, there is no real domain function that has as Fourier transform the ideal highpass filter. In practice, of course, the highest frequency we may possibly be interested in is $\frac{1}{N}$ where N is the number of pixels in the image, so the issue of ∞ upper limit in equation (5.12) does not arise and the ideal highpass filter becomes the same as the ideal bandpass filter.

What is the relationship between the 1D and the 2D ideal lowpass filters?

The 1D ideal lowpass filter is given by:

$$h(k) = \frac{\sin k}{k} \tag{5.13}$$

The 2D ideal lowpass filter is given by:

$$h(k,l) = \frac{J_1(\sqrt{k^2 + l^2})}{\sqrt{k^2 + l^2}}$$

where $J_1(x)$ is the first-order Bessel function of the first kind. *Figure 5.2* shows the plot of $h(k)$ versus k and the plot of $h(k,l)$ versus k for $l = 0$. It can be seen that although the two filters look similar, they differ in significant details: their zero crossings are at different places, and the amplitudes of their side-lobes are different.

This implies that we cannot take an ideal or optimal (according to some criteria) 1D filter replace its variable by the polar radius (i.e. replace k by $\sqrt{k^2 + l^2}$ in equation (5.13)) and create the corresponding "ideal or optimal" filter in 2D. However, although the 2D filter we shall create this way will not be the ideal or optimal one according to the corresponding criteria in 2D, it will be a good suboptimal filter with qualitatively the same behaviour as the optimal one.

How can we implement a filter of infinite extent?

A filter which is of infinite extent in real space can be implemented in a recursive way, and that is why it is called a *recursive filter*. Filters which are of finite extent in real space are called *non-recursive* filters. Filters are usually represented and manipulated with the help of their z-transforms.

How is the z-transform of a digital 1D filter defined?

A filter of finite extent is essentially a finite string of numbers $\{x_1, x_2, x_3, \ldots, x_n\}$. Sometimes an arrow is used to denote the element of the string that corresponds to

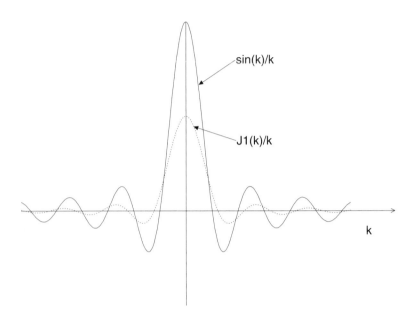

Figure 5.2: The cross-section of the 2D ideal lowpass filter is similar to but different from the cross-section of the 1D ideal lowpass filter.

the zeroth position. The z-transform of such a string is defined as:

$$X(z) = \sum_{k=l}^{m} x_k z^{-k} \tag{5.14}$$

where l and m are defined according to which term of the string of numbers $\{x_k\}$ is assumed to be the $k = 0$ term. (For example, if x_3 is the $k = 0$ term then $l = -2$ and $m = n - 3$ in the above sequence).

If the filter is of infinite extent, the sequence of numbers which represents it is of infinite extent too and its z-transform is given by an infinite sum of the form:

$$X(z) = \sum_{k=-\infty}^{\infty} x_k z^{-k} \quad \text{or} \quad \sum_{k=0}^{\infty} x_k z^{-k} \quad \text{or} \quad \sum_{k=-\infty}^{0} x_k z^{-k}$$

In such a case we can usually write this sum in closed form as the **ratio** of two polynomials in z, as opposed to writing it as a single polynomial in z which is the case for the z-transform of the finite filter.

Why do we use z-transforms?

The reason we use z-transforms is because digital filters can easily be realized in hardware in terms of their z-transforms. The z-transform of a sequence together with its region of convergence uniquely defines the sequence. Further, it obeys the convolution theorem: The z-transform of the convolution of two sequences is the product of the z-transforms of the two sequences.

How is the z-transform defined in 2D?

For a finite 2D array of dimensions $M \times N$, the z-transform is a finite polynomial in two complex variables z_1 and z_2:

$$H(z_1, z_2) = \sum_{i=0}^{M} \sum_{j=0}^{N} c_{ij} z_1^i z_2^j \tag{5.15}$$

where c_{ij} are the elements of the array. For an infinite array, the double summation is of infinite extent and it can be written as a **ratio** of two finite summations:

$$H(z_1, z_2) = \frac{\sum_{i=0}^{M_a} \sum_{j=0}^{N_a} a_{ij} z_1^i z_2^j}{\sum_{i=0}^{M_b} \sum_{j=0}^{N_b} b_{ij} z_1^i z_2^j} \tag{5.16}$$

where M_a, N_a, M_b and N_b are some integers. Conventionally we choose $b_{00} = 1$.

B5.2: Why does the extent of a filter determine whether the filter is recursive or not?

When we convolve an image with a digital filter, we essentially multiply the z-transform of the image with the z-transform of the filter:

$$\underbrace{R(z_1, z_2)}_{\substack{\text{z-transform of} \\ \text{output image}}} = \underbrace{H(z_1, z_2)}_{\substack{\text{z-transform} \\ \text{of filter}}} \underbrace{D(z_1, z_2)}_{\substack{\text{z-transform of} \\ \text{input image}}} \tag{5.17}$$

If we substitute (5.16) in (5.17) and bring the denominator to the left hand side of the equation, we have:

$$R(z_1, z_2) \sum_{i=0}^{M_b} \sum_{j=0}^{N_b} b_{ij} z_1^i z_2^j = \left(\sum_{i=0}^{M_a} \sum_{j=0}^{N_a} a_{ij} z_1^i z_2^j \right) D(z_1, z_2)$$

In the sum on the left hand side we separate the $k = 0$, $l = 0$ term, which by convention has $b_{00} = 1$:

$$R(z_1, z_2) + \left(\underbrace{\sum_{i=0}^{M_b} \sum_{j=0}^{N_b}}_{\text{not both 0}} b_{ij} z_1^i z_2^j\right) R(z_1, z_2) = \left(\sum_{i=0}^{M_a} \sum_{j=0}^{N_a} a_{ij} z_1^i z_2^j\right) D(z_1, z_2)$$

Therefore:

$$R(z_1, z_2) = \left(\sum_{i=0}^{M_a} \sum_{j=0}^{N_a} a_{ij} z_1^i z_2^j\right) D(z_1, z_2) - \left(\underbrace{\sum_{i=0}^{M_b} \sum_{j=0}^{N_b}}_{\text{not both 0}} b_{ij} z_1^i z_2^j\right) R(z_1, z_2) \tag{5.18}$$

Remember that $R(z_1, z_2)$ is a sum in $z_1^m z_2^n$ with coefficients say r_{mn}. It is clear from the above equation that the value of r_{mn} can be calculated in terms of the previously calculated values of r_{mn} since the series $R(z_1, z_2)$ appears on the right hand side of the equation too. That is why such a filter is called recursive. In the case of a finite filter all b_{kl}'s are zero (except b_{00} which is 1) and so the coefficients r_{mn} of $R(z_1, z_2)$ are expressed in terms of a_{ij} and the coefficients which appear in $D(z_1, z_2)$ only (i.e. we have no recursion).

Example 5.3 (B)

A 256×256 image is to be processed by an infinite impulse response filter. The z-transform of the filter can be written as the ratio of a third-degree polynomial in each of the variables z_1 and z_2 over another third-degree polynomial in the same variables. Calculate the values of the output image in terms of the values of the input image and the filter coefficients.

In equation (5.18) we have $M_a = N_a = M_b = N_b = 3$. Let us say that the z-transform of the input image is:

$$D(z_1, z_2) = \sum_{k=0}^{255} \sum_{l=0}^{255} d_{kl} z_1^k z_2^l \tag{5.19}$$

and of the output image is:

$$R(z_1, z_2) = \sum_{k=0}^{255} \sum_{l=0}^{255} r_{kl} z_1^k z_2^l \tag{5.20}$$

We wish to derive expressions for r_{ij} in terms of a_{ij}, b_{ij} and d_{ij}. If we substitute (5.19) and (5.20) into equation (5.18) we obtain:

$$\sum_{k=0}^{255} \sum_{l=0}^{255} r_{kl} z_1^k z_2^l = \sum_{i=0}^{3} \sum_{j=0}^{3} \sum_{k=0}^{255} \sum_{l=0}^{255} (a_{ij} d_{kl} - b_{ij} r_{kl}) z_1^{k+i} z_2^{l+j} \tag{5.21}$$

To avoid having to treat separately the terms with 0 indices, we allowed in the above expression the term $i = j = 0$, but we must remember that now $b_{00} \equiv 0$. To be able to relate terms of the same order on the two sides of this equation, we must define new variables of summation on the right hand side. We define:

$$\left. \begin{array}{l} k + i \equiv u \\ k - i \equiv v \end{array} \right| \Rightarrow \begin{array}{l} k = \frac{u+v}{2} \\ i = \frac{u-v}{2} \end{array} \qquad \left. \begin{array}{l} l + j \equiv t \\ l - j \equiv s \end{array} \right| \Rightarrow \begin{array}{l} l = \frac{t+s}{2} \\ j = \frac{t-s}{2} \end{array} \right|$$

We must find the limits of summation for the new variables. The figure below shows how the area over which we sum using variables k and i is transformed when the new variables are used. The transformed area is defined by the transformed equations of the lines that define the original area, i.e. lines $i = 0$, $i = 3$, $k = 0$ and $k = 255$, which become respectively: $v = u$, $v = u - 6$, $v = -u$ and $v = 510 - u$.

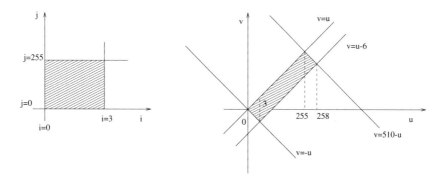

Figure 5.3: The transformation of the area of summation.

Equation (5.21) then becomes:

$$\sum_{k=0}^{255} \sum_{l=0}^{255} r_{kl} z_1^k z_2^l = \sum_{j=0}^{3} \sum_{l=0}^{255} \left\{ \sum_{u=0}^{2} \sum_{v=-u}^{u} + \sum_{u=3}^{255} \sum_{v=u-6}^{u} + \sum_{u=256}^{258} \sum_{v=u-6}^{510-u} \right\}$$

$$(a_{\frac{u-v}{2} j} d_{\frac{u+v}{2} l} - b_{\frac{u-v}{2} j} r_{\frac{u+v}{2} l}) z_1^u z_2^{l+j} \tag{5.22}$$

In a similar way we can find the limits of summation of variables s and t to obtain:

$$\sum_{k=0}^{255}\sum_{l=0}^{255} r_{kl} z_1^k z_2^l =$$

$$\left\{\sum_{t=0}^{2}\sum_{s=-t}^{t}\sum_{u=0}^{2}\sum_{v=-u}^{u} + \sum_{t=0}^{2}\sum_{s=-t}^{t}\sum_{u=3}^{255}\sum_{v=u-6}^{u} + \sum_{t=0}^{2}\sum_{s=-t}^{t}\sum_{u=256}^{258}\sum_{v=u-6}^{510-u}\right.$$

$$+\sum_{t=3}^{255}\sum_{s=t-6}^{t}\sum_{u=0}^{2}\sum_{v=-u}^{u} + \sum_{t=3}^{255}\sum_{s=t-6}^{t}\sum_{u=3}^{255}\sum_{v=u-6}^{u} + \sum_{t=3}^{255}\sum_{s=t-6}^{t}\sum_{u=256}^{258}\sum_{v=u-6}^{510-u}$$

$$\left.+\sum_{t=256}^{258}\sum_{s=t-6}^{510-t}\sum_{u=0}^{2}\sum_{v=-u}^{u} + \sum_{t=256}^{258}\sum_{s=t-6}^{510-t}\sum_{u=3}^{255}\sum_{v=u-6}^{u} + \sum_{t=256}^{258}\sum_{s=t-6}^{510-t}\sum_{u=256}^{258}\sum_{v=u-6}^{510-u}\right\}$$

$$\left(\left(a_{\frac{u-v}{2}\frac{t-s}{2}}d_{\frac{u+v}{2}\frac{t+s}{2}} - b_{\frac{u-v}{2}\frac{t-s}{2}}r_{\frac{u+v}{2}\frac{t-s}{2}}\right)z_1^u z_2^t\right) \tag{5.23}$$

We further define the following quantities:

$$A_{ut} \equiv \sum_{s=-t}^{t}\sum_{v=-u}^{u}\left(a_{\frac{u-v}{2}\frac{t-s}{2}}d_{\frac{u+v}{2}\frac{t+s}{2}} - b_{\frac{u-v}{2}\frac{t-s}{2}}r_{\frac{u+v}{2}\frac{t-s}{2}}\right)$$

$$B_{ut} \equiv \sum_{s=-t}^{t}\sum_{v=u-6}^{u}\left(a_{\frac{u-v}{2}\frac{t-s}{2}}d_{\frac{u+v}{2}\frac{t+s}{2}} - b_{\frac{u-v}{2}\frac{t-s}{2}}r_{\frac{u+v}{2}\frac{t-s}{2}}\right)$$

$$C_{ut} \equiv \sum_{s=-t}^{t}\sum_{v=u-6}^{510-u}\left(a_{\frac{u-v}{2}\frac{t-s}{2}}d_{\frac{u+v}{2}\frac{t+s}{2}} - b_{\frac{u-v}{2}\frac{t-s}{2}}r_{\frac{u+v}{2}\frac{t-s}{2}}\right)$$

$$D_{ut} \equiv \sum_{s=t-6}^{t}\sum_{v=-u}^{u}\left(a_{\frac{u-v}{2}\frac{t-s}{2}}d_{\frac{u+v}{2}\frac{t+s}{2}} - b_{\frac{u-v}{2}\frac{t-s}{2}}r_{\frac{u+v}{2}\frac{t-s}{2}}\right)$$

$$E_{ut} \equiv \sum_{s=t-6}^{t}\sum_{v=u-6}^{u}\left(a_{\frac{u-v}{2}\frac{t-s}{2}}d_{\frac{u+v}{2}\frac{t+s}{2}} - b_{\frac{u-v}{2}\frac{t-s}{2}}r_{\frac{u+v}{2}\frac{t-s}{2}}\right)$$

$$F_{ut} \equiv \sum_{s=t-6}^{t}\sum_{v=u-6}^{510-u}\left(a_{\frac{u-v}{2}\frac{t-s}{2}}d_{\frac{u+v}{2}\frac{t+s}{2}} - b_{\frac{u-v}{2}\frac{t-s}{2}}r_{\frac{u+v}{2}\frac{t-s}{2}}\right)$$

$$G_{ut} \equiv \sum_{s=t-6}^{510-t}\sum_{v=-u}^{u}\left(a_{\frac{u-v}{2}\frac{t-s}{2}}d_{\frac{u+v}{2}\frac{t+s}{2}} - b_{\frac{u-v}{2}\frac{t-s}{2}}r_{\frac{u+v}{2}\frac{t-s}{2}}\right)$$

$$H_{ut} \equiv \sum_{s=t-6}^{510-t}\sum_{v=u-6}^{u}\left(a_{\frac{u-v}{2}\frac{t-s}{2}}d_{\frac{u+v}{2}\frac{t+s}{2}} - b_{\frac{u-v}{2}\frac{t-s}{2}}r_{\frac{u+v}{2}\frac{t-s}{2}}\right)$$

$$I_{ut} \equiv \sum_{s=t-6}^{510-t}\sum_{v=u-6}^{510-u}\left(a_{\frac{u-v}{2}\frac{t-s}{2}}d_{\frac{u+v}{2}\frac{t+s}{2}} - b_{\frac{u-v}{2}\frac{t-s}{2}}r_{\frac{u+v}{2}\frac{t-s}{2}}\right) \tag{5.24}$$

Then equation (5.23) can be written as:

$$
\sum_{u=0}^{255}\sum_{t=0}^{255} r_{ut}\,z_1^u z_2^t = \sum_{t=0}^{2}\sum_{u=0}^{2} A_{ut}\,z_1^u z_2^t + \sum_{t=0}^{2}\sum_{u=3}^{255} B_{ut}\,z_1^u z_2^t + \sum_{t=0}^{2}\sum_{u=256}^{258} C_{ut}\,z_1^u z_2^t
$$

$$
+ \sum_{t=3}^{255}\sum_{u=0}^{2} D_{ut}\,z_1^u z_2^t + \sum_{t=3}^{255}\sum_{u=3}^{255} E_{ut}\,z_1^u z_2^t + \sum_{t=3}^{255}\sum_{u=256}^{258} F_{ut}\,z_1^u z_2^t
$$

$$
+ \sum_{t=256}^{258}\sum_{u=0}^{2} G_{ut}\,z_1^u z_2^t + \sum_{t=256}^{258}\sum_{u=3}^{255} H_{ut}\,z_1^u z_2^t + \sum_{t=256}^{258}\sum_{u=256}^{258} I_{ut}\,z_1^u z_2^t
$$

$$
(5.25)
$$

Here we changed the dummy summation variables on the left hand side to u and t as well. To derive expressions for r_{ut} in terms of a_{ut}, b_{ut} and d_{ut}, we must equate the coefficients of the terms of the same degree in z_1 and z_2 that appear on the two sides of equation (5.25). We must remember that b_{00} was redefined in (5.21) to be 0. To make sure that we deal with all the cases, we draw the diagram in Figure 5.4, where the two axes correspond to the summing variables on the right hand side of equation (5.25). Each rectangle in this diagram corresponds to a combination of ranges of values of the two variables that have to be considered separately.

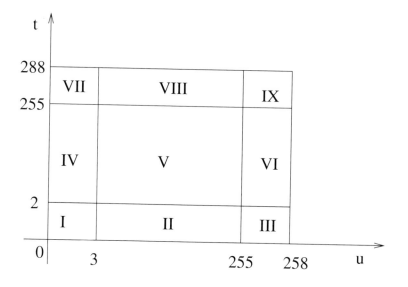

Figure 5.4: Different recursive formulae apply to different ranges of index values.

Case I: $0 \leq u \leq 2$ *and* $0 \leq t \leq 2$

$$r_{ut} = A_{ut} = \sum_{s=-t}^{t} \sum_{v=-u}^{u} \left(a_{\frac{u-v}{2} \frac{t-s}{2}} d_{\frac{u+v}{2} \frac{t+s}{2}} - b_{\frac{u-v}{2} \frac{t-s}{2}} r_{\frac{u+v}{2} \frac{t+s}{2}} \right) \tag{5.26}$$

We define new summation variables:
$i \equiv \frac{u-v}{2}$ *with limits* $i = u$ *(corresponding to* $v = -u$*) and* $i = 0$ *(corresponding to*
$v = u$*).*
Then $v = u - 2i \Rightarrow \frac{u+v}{2} = \frac{2u-2i}{2} = u - i$
$j \equiv \frac{t-s}{2}$ *with limits* $j = t$ *(corresponding to* $s = -t$*) and* $j = 0$ *(corresponding to*
$s = t$*).*
Then $s = t - 2j \Rightarrow \frac{t+s}{2} = \frac{2t-2j}{2} = t - j$
Expression (5.26) can then be written as:

$$r_{ut} = \sum_{i=0}^{u} \sum_{j=0}^{t} \left(a_{ij} d_{u-i,t-j} - b_{ij} r_{u-i,t-j} \right) \tag{5.27}$$

We consider first the case $u = t = 0$ *and remember that* $b_{00} = 0$:

$$r_{00} = a_{00} d_{00}$$

Then consider the case $u = 0$, $t = 1$ *for which* j *takes the values 0 and 1:*

$$r_{01} = a_{00} d_{01} + a_{01} d_{00} - b_{01} r_{00}$$

For the case $u = 0$, $t = 2$, j *takes values 0, 1 and 2:*

$$r_{02} = a_{00} d_{02} + a_{01} d_{01} - b_{01} r_{01} + a_{02} d_{00} - b_{02} r_{00}$$

Notice that although some r *values appear on the right hand side of these equations, they are always with smaller indices than those of* r *on the left hand side, and therefore their values have already been computed.*

Case II: $2 < u \leq 255$ *and* $0 \leq t \leq 2$

$$r_{ut} = B_{ut} = \sum_{s=-t}^{t} \sum_{v=u-6}^{u} \left(a_{\frac{u-v}{2} \frac{t-s}{2}} d_{\frac{u+v}{2} \frac{t+s}{2}} - b_{\frac{u-v}{2} \frac{t-s}{2}} r_{\frac{u+v}{2} \frac{t-s}{2}} \right) \tag{5.28}$$

The only difference with the previous case is that the limits of the new summation variable i *will be from* $i = 3$ *(corresponding to* $v = u - 6$*) to* $i = 0$ *(corresponding to* $v = u$*):*

$$r_{ut} = \sum_{i=0}^{3} \sum_{j=0}^{t} \left(a_{ij} d_{u-i,t-j} - b_{ij} r_{u-i,t-j} \right) \tag{5.29}$$

Case III: $255 < u \le 258$ and $0 \le t \le 2$

In this case there is no corresponding term on the left of the equation since the maximum value u takes on the left is 255. We have therefore:

$$0 = C_{ut} \equiv \sum_{s=-t}^{t} \sum_{v=u-6}^{510-u} \left(a_{\frac{u-v}{2}\,\frac{t-s}{2}} d_{\frac{u+v}{2}\,\frac{t+s}{2}} - b_{\frac{u-v}{2}\,\frac{t-s}{2}} r_{\frac{u+v}{2}\,\frac{t-s}{2}} \right) \quad (5.30)$$

Or

$$\sum_{i=u-255}^{3} \sum_{j=0}^{t} \left(a_{ij} d_{u-i,t-j} - b_{ij} r_{u-i,t-j} \right) = 0 \quad (5.31)$$

Consider $u = 256$, $t = 0$:

$$\sum_{i=1}^{3} a_{i0} d_{256-i,0} - b_{i0} r_{256-i,0} = 0 \Rightarrow$$

$$a_{10} d_{255,0} - b_{10} r_{255,0} + a_{20} d_{254,0} - b_{10} r_{254,0} + a_{30} d_{253,0} - b_{30} r_{253,0} = 0 \quad (5.32)$$

The values of $r_{255,0}$, $r_{254,0}$ *and* $r_{253,0}$ *have already been computed by equation (5.29). Equation (5.32), therefore, serves only as a check that these values have been computed correctly. In general, this is the role of equation (5.31) and the equations that will be derived for cases VI, VII, VIII and IX.*

Case IV: $0 \le u \le 2$ and $2 \le t \le 255$

$$r_{ut} = D_{ut} = \sum_{s=t-6}^{t} \sum_{v=-u}^{u} \left(a_{\frac{u-v}{2}\,\frac{t-s}{2}} d_{\frac{u+v}{2}\,\frac{t+s}{2}} - b_{\frac{u-v}{2}\,\frac{t-s}{2}} r_{\frac{u+v}{2}\,\frac{t-s}{2}} \right) \quad (5.33)$$

Or

$$r_{ut} = \sum_{i=0}^{u} \sum_{j=0}^{3} \left(a_{ij} d_{u-i,t-j} - b_{ij} r_{u-i,t-j} \right) \quad (5.34)$$

Case V: $2 < u \le 255$ and $2 < t \le 255$

$$r_{ut} = E_{ut} = \sum_{s=t-6}^{t} \sum_{v=u-6}^{u} \left(a_{\frac{u-v}{2}\,\frac{t-s}{2}} d_{\frac{u+v}{2}\,\frac{t+s}{2}} - b_{\frac{u-v}{2}\,\frac{t-s}{2}} r_{\frac{u+v}{2}\,\frac{t-s}{2}} \right) \quad (5.35)$$

Or

$$r_{ut} = \sum_{i=0}^{3} \sum_{j=0}^{3} \left(a_{ij} d_{u-i,t-j} - b_{ij} r_{u-i,t-j} \right) \quad (5.36)$$

Case VI: $255 < u \le 258$ and $2 < t \le 255$

$$0 = F_{ut} = \sum_{s=t-6}^{t} \sum_{v=u-6}^{510-u} \left(a_{\frac{u-v}{2}\,\frac{t-s}{2}} d_{\frac{u+v}{2}\,\frac{t+s}{2}} - b_{\frac{u-v}{2}\,\frac{t-s}{2}} r_{\frac{u+v}{2}\,\frac{t-s}{2}} \right) \quad (5.37)$$

Or

$$\sum_{i=3}^{u-255} \sum_{j=0}^{3} (a_{ij}d_{u-i,t-j} - b_{ij}r_{u-i,t-j}) = 0 \tag{5.38}$$

Case VII: $0 < u \le 2$ and $255 < t \le 258$

$$0 = G_{ut} = \sum_{s=t-6}^{510-t} \sum_{v=-u}^{u} \left(a_{\frac{u-v}{2}\frac{t-s}{2}} d_{\frac{u+v}{2}\frac{t+s}{2}} - b_{\frac{u-v}{2}\frac{t-s}{2}} r_{\frac{u+v}{2}\frac{t-s}{2}} \right) \tag{5.39}$$

Or

$$\sum_{i=0}^{u} \sum_{j=3}^{t-255} (a_{ij}d_{u-i,t-j} - b_{ij}r_{u-i,t-j}) = 0 \tag{5.40}$$

Case VIII: $2 < u \le 255$ and $255 < t \le 258$

$$0 = H_{ut} = \sum_{s=t-6}^{510-t} \sum_{v=u-6}^{u} \left(a_{\frac{u-v}{2}\frac{t-s}{2}} d_{\frac{u+v}{2}\frac{t+s}{2}} - b_{\frac{u-v}{2}\frac{t-s}{2}} r_{\frac{u+v}{2}\frac{t-s}{2}} \right) \tag{5.41}$$

Or

$$\sum_{i=0}^{3} \sum_{j=3}^{t-255} (a_{ij}d_{u-i,t-j} - b_{ij}r_{u-i,t-j}) = 0 \tag{5.42}$$

Case VIII: $2 < u \le 255$ and $255 < t \le 258$

$$0 = I_{ut} = \sum_{s=t-6}^{510-t} \sum_{v=u-6}^{510-u} \left(a_{\frac{u-v}{2}\frac{t-s}{2}} d_{\frac{u+v}{2}\frac{t+s}{2}} - b_{\frac{u-v}{2}\frac{t-s}{2}} r_{\frac{u+v}{2}\frac{t-s}{2}} \right) \tag{5.43}$$

Or

$$\sum_{i=0}^{3} \sum_{j=3}^{t-255} (a_{ij}d_{u-i,t-j} - b_{ij}r_{u-i,t-j}) = 0 \tag{5.44}$$

Is there any fundamental difference between 1D and 2D recursive filters?

In one dimension, if we express the z-transform of a function as a ratio of two polynomials, we can factorize these polynomials to obtain the poles and zeroes of the transform. Once the poles are known, the inverse z-transform can be computed for each possible region of convergence of the z-transform.

In two dimensions, factorization is much more difficult and even if it is achieved, the zeroes of the polynomial of the denominator (poles) are not usually isolated discrete points but rather ranges of continuous values of the variables. Further, the (z_1, z_2) space is 4-dimensional, and this makes the visualization of regions of convergence of the polynomials impossible.

How do we know that a filter does not amplify noise?

A filter must be *stable* in order not to amplify noise. A digital filter is said to be stable if the output remains bounded for all bounded inputs. A digital filter $h(n, m)$ is stable if and only if:

$$\sum_{n=-\infty}^{\infty} \sum_{m=-\infty}^{\infty} |h(n, m)| < \infty$$

The above criterion is obviously fulfilled if the filter is of finite extent or non-recursive. For infinite impulse response (recursive) filters, one has to check the behaviour of the filter carefully, with the help of special theorems on the subject.

Is there an alternative to using infinite impulse response filters?

Yes! We can compromise in our requirements on the frequency response of the filter. In particular, we may adopt the following approach:

1. Decide upon the desired system function (frequency response) of the filter.
2. Choose a **finite** filter (non-recursive) which approximates as well as possible the desired system function.

This approach leads to the *approximation theory.*

Why do we need approximation theory?

Because the required frequency response of a filter cannot be realized exactly by a finite impulse response filter.

How do we know how good an approximate filter is?

We use as a measure of the quality of the approximation the *Chebyshev norm* of the difference between the desired and the approximated response, defined as the maximum deviation between the two functions.

Suppose $F(\mu, \nu)$ is the ideal frequency response we require and $H(\mu, \nu)$ is the frequency response of the approximated finite filter. Then the closeness of $H(\mu, \nu)$ to $F(\mu, \nu)$ is measured by:

$$\text{Error} = ||F(\mu, \nu) - H(\mu, \nu)|| \equiv max_{(\mu,\nu)}|F(\mu, \nu) - H(\mu, \nu)| \qquad (5.45)$$

What is the best approximation to an ideal given system function?

The best approximation to an ideal system function $F(\mu, \nu)$ is that function $H(\mu, \nu)$ which minimizes the error; i.e.

$$\text{Best approximation error} = min_{H(...)}\{max_{(\mu,\nu)}|F(\mu, \nu) - H(\mu, \nu)|\} \qquad (5.46)$$

Why do we judge an approximation according to the Chebyshev norm instead of the square error?

The total square error $(\int\int [F(\mu,\nu) - H(\mu,\nu)]^2 d\mu d\nu)$ is an **average** measure of how close $F(\mu,\nu)$ and $H(\mu,\nu)$ are. Two different approximations may have identical square errors, but one may introduce very undesirable features to the filter at specific frequencies (see *Figure* 5.5). The Chebyshev norm on the other hand guarantees that at no frequency will the deviation of the two functions exceed a certain limit.

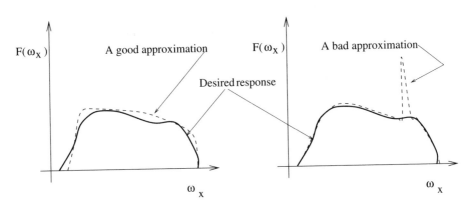

Figure 5.5: In the frequency domain, a good approximation is one that never deviates too much away from the desired response.

How can we obtain an approximation to a system function?

There are three techniques of designing 2-dimensional filters. All three are direct adaptations of corresponding techniques in one dimension. They are:

1. Windowing
2. Frequency sampling
3. Linear programming

Of the above techniques only the last one gives the best approximation in the sense that it minimizes the maximum deviation between the exact and the approximate functions.

What is windowing?

The most obvious way to make an infinite impulse response filter finite, is to truncate it at some desired size. This method is called *windowing* because it is as if we put a window on the top of the filter and look only at what is shown inside the window. However, this method creates filters with sharp edges.

What is wrong with windowing?

A function which contains discontinuities (like a sharply truncated filter) has a Fourier transform with lots of ripples at high frequencies as lots of higher harmonics are required to contribute in order to produce those sharp edges. So, the process of truncation distorts badly the system function of the filter at high frequencies.

How can we improve the result of the windowing process?

Truncating the filter function at certain frequencies is equivalent to multiplying it with a window function of the form:

$$w(x,y) = \begin{cases} 1 & \text{for } -a \le x \le a \text{ and } -b \le y \le b \\ 0 & \text{elsewhere} \end{cases}$$

where a and b are some parameters. To avoid sharp edges of the truncated filter this window has to be replaced by a smooth function. Several such smooth windows have been developed for the case of 1D signals.

Can we make use of the windowing functions that have been developed for 1D signals, to define a windowing function for images?

Once we have decided upon a good window in the 1-dimensional case, the extension to two dimensions is not hazard-free. One must have in mind that if $w(x)$ is a good 1-dimensional window, its circularly symmetric version $w(r) \equiv w(\sqrt{x^2 + y^2})$ does not have the same frequency response as $w(x)$. In other words, the Fourier transform of the circularly symmetric window is not the circular version of the Fourier transform of the 1-dimensional window. This was discussed extensively earlier in this chapter (see also *Figure* 5.2).

In general, however, if $w(x)$ is a good 1-dimensional window, $w_2(\sqrt{x^2 + y^2})$ is a good circularly symmetric 2-dimensional window function. In both cases of course, one should be very careful when one discretizes the window to sample it densely enough so that aliasing problems are avoided.

The windowing method might yield some reasonably good 2-dimensional filters, but these filters are not optimal approximations to the desired filter as no effort has been put into optimizing the approximation.

What is the formal definition of the approximation problem we are trying to solve?

Suppose that $F(\mu, \nu)$ is an ideal, unrealizable filter response. The goal is to find a set of n coefficients $c_1, c_2, \ldots, c_n$ such that the generalized polynomial $P(\mu, \nu) \equiv \sum_{i=1}^{n} c_i g_i(\mu, \nu)$ minimizes the Chebyshev norm of the error function Δ defined by:

$$\Delta \equiv F(\mu, \nu) - \sum_{i=1}^{n} c_i g_i(\mu, \nu) \tag{5.47}$$

Here $g_i(\mu, \nu)$ are known functions of the frequencies μ and ν, like for example $\cos(\mu + \nu)$, $\sin(2\mu + \nu)$ etc.

In (5.47) μ and ν take continuous values. We shall consider, however, a finite set of samples of Δ by choosing M points in the (μ, ν) space:

$$\Delta_m \equiv F(\mu_m, \nu_m) - \sum_{i=1}^{n} c_i g_i(\mu_m, \nu_m) \qquad m = 1, 2, \ldots, M \qquad (5.48)$$

Our purpose is to choose the coefficients $c_1, c_2, \ldots, c_n$ so that the maximum value of the Δ_m's is as small as possible. This problem can be solved with the help of *linear programming*.

What is linear programming?

Linear programming is an optimization method designed to solve the following problem: Choose n_1 positive variables x_{1i} and n_2 variables x_{2j} in order to

minimize: $z = \mathbf{c}_1^T \mathbf{x_1} + \mathbf{c}_2^T \mathbf{x_2} + d$

subject to: $A_{11}\mathbf{x_1} + A_{12}\mathbf{x_2} \geq \mathbf{B_1}$ p_1 inequality constraints

              $A_{21}\mathbf{x_1} + A_{22}\mathbf{x_2} = \mathbf{B_2}$ p_2 equality constraints

              $x_{1i} \geq 0$ n_1 variables x_1

              x_{2j} free n_2 variables x_2

where $\mathbf{c_1}$, $\mathbf{c_2}$, $\mathbf{B_1}$, $\mathbf{B_2}$ are vectors and A_{11}, A_{11}, A_{12}, A_{21}, A_{22} are matrices. Vectors $\mathbf{x_1}$ and $\mathbf{x_2}$ are made up from variables x_{1i} and x_{2j} respectively.

To solve the above problem one can use a commercial package algorithm. It is, therefore, enough simply to formulate our problem in the above way. Then we can use one of the commercially available packages to solve it.

How can we formulate the filter design problem as a linear programming problem?

Suppose that $F(\mu, \nu)$ is the desired frequency response of the filter and suppose that $h(n, m)$ is the digitized finite filter which we want to use. The frequency response of this digitized filter is:

$$H(\mu, \nu) = \sum_{n=-N}^{N} \sum_{m=-N}^{N} h(n, m) e^{-j(\mu n + \nu m)} \qquad (5.49)$$

where $h(n, m)$ are real numbers. It can be shown (see Box **B5.3**) that if $h(\mu, \nu) = h(-\nu, \mu)$ the system function $H(\mu, \nu)$ is real and it can be written as:

$$H(\mu, \nu) = 2 \sum_{n=-N}^{N} \sum_{m=1}^{N} h(n, m) \cos(\mu n + \nu m) + 2 \sum_{n=1}^{N} h(n, 0) \cos(\mu n) + h(0, 0) \quad (5.50)$$

This equation expresses the Fourier transform of the filter in terms of the discrete filter values in real space. We want to specify these values; namely, we want to specify $h(0,0), h(1,0), \ldots, h(N,0), \ldots, h(-N,m), h(-N+1,m), \ldots, h(N,m)$ for $m = 1, 2, \ldots, N$ so that $H(\mu, \nu)$ approximates optimally the desired filter response $F(\mu, \nu)$. This means that we want to choose the above values of $h(i,j)$ so that the error:

$$\delta \equiv max_{(\mu,\nu)} |H(\mu,\nu) - F(\mu,\nu)| \tag{5.51}$$

is minimum.

Variables (μ, ν) take continuous values. We choose a set of p_1 specific points in the (μ, ν) space which we shall use as constraint points. For every point (μ_i, ν_i) then the value of $|H(\mu_i, \nu_i) - F(\mu_i, \nu_i)|$ will be less or equal to the value of δ owing to the definition of δ. So, we can write for every point:

$$\delta \geq |H(\mu_i, \nu_i) - F(\mu_i, \nu_i)| \Rightarrow \begin{cases} -\delta \leq H(\mu_i, \nu_i) - F(\mu_i, \nu_i) \\ \delta \geq H(\mu_i, \nu_i) - F(\mu_i, \nu_i) \end{cases}$$

or

$$\left. \begin{array}{l} \delta + H(\mu_i, \nu_i) \geq F(\mu_i, \nu_i) \\ \delta - H(\mu_i, \nu_i) \geq -F(\mu_i, \nu_i) \end{array} \right\} \quad \text{for } i = 1, 2, \ldots, p_1 \tag{5.52}$$

The problem is one of trying to minimizing δ under the $2p_1$ inequality constraints (5.52) and the assumption that $\delta > 0$, while the variables we have to choose $h(i,j)$ are free to be positive or negative.

From equation (5.50) it is clear that the number of variables we have to choose is:

$$(2N+1)N + N + 1 = 2N^2 + N + N + 1 = 2N^2 + 2N + 1$$

So, we have $2N^2 + 2N + 1$ free variables plus one constrained variable (δ itself). We have no equality constraints ($p_2 = 0$) and we have $2p_1$ inequality constraints. In the language of linear programming:

Minimize: $z = x_{11}$
under the constraints: $A_{11}x_{11} + A_{12}\mathbf{x_2} \geq \mathbf{B_1}$ $2p_1$ inequality constraints
                  $x_{11} \geq 0$ 1 non-negative variable   (5.53)
                  x_{2j} free $2N^2 + 2N + 1$ free variables

where

$$x_{11} = \delta$$

$$\mathbf{x_2} = \begin{pmatrix} h(0,0) \\ h(1,0) \\ h(2,0) \\ \vdots \\ h(N,0) \\ h(-N,1) \\ \vdots \\ h(N,1) \\ h(-N,2) \\ \vdots \\ h(N,2) \\ \vdots \\ h(-N,N) \\ \vdots \\ h(N,N) \end{pmatrix} \leftarrow \begin{array}{l} (2N^2 + 2N + 1) \times 1 \\ \text{matrix} \end{array} \qquad \mathbf{B_1} = \begin{pmatrix} F(\mu_1,\nu_1) \\ -F(\mu_1,\nu_1) \\ F(\mu_2,\nu_2) \\ -F(\mu_2,\nu_2) \\ \vdots \\ F(\mu_{p_1},\nu_p) \\ -F(\mu_{p_1},\nu_p) \end{pmatrix} \leftarrow \begin{array}{l} 2p_1 \times 1 \\ \text{matrix} \end{array}$$

$$A_{11} = \begin{pmatrix} 1 \\ 1 \\ 1 \\ \vdots \\ 1 \end{pmatrix} \leftarrow \begin{array}{l} 2p_1 \times 1 \\ \text{matrix} \end{array}$$

A_{12} is a $2p_1 \times (2N^2 + 2N + 1)$ matrix the elements of which can easily be found if we substitute equation (5.50) into (5.52) or (5.53). It turns out to be:

$$A_{12} = \begin{pmatrix} 1 & 2\cos(\mu_1) & 2\cos(2\mu_1) & \cdots & 2\cos(N\mu_1) & 2\cos(-N\mu_1 + \nu_1) & \cdots \\ -1 & -2\cos(\mu_1) & -2\cos(2\mu_1) & \cdots & -2\cos(N\mu_1) & -2\cos(-N\mu_1 + \nu_1) & \cdots \\ \vdots & \vdots & \vdots & \vdots & \vdots & \vdots & \vdots \\ 1 & 2\cos(\mu_{p_1}) & 2\cos(2\mu_{p_1}) & \cdots & 2\cos(N\mu_{p_1}) & 2\cos(-N\mu_{p_1} + \nu_{p_1}) & \cdots \\ -1 & -2\cos(\mu_{p_1}) & -2\cos(2\mu_{p_1}) & \cdots & -2\cos(N\mu_{p_1}) & -2\cos(-N\mu_{p_1} + \nu_{p_1}) & \cdots \end{pmatrix}$$

$$\begin{pmatrix} 2\cos(N\mu_1 + \nu_1) & \cdots & 2\cos(-N\mu_1 + N\nu_1) & \cdots & 2\cos(N\mu_1 + N\nu_1) \\ -2\cos(N\mu_1 + \nu_1) & \cdots & -2\cos(-N\mu_1 + N\nu_1) & \cdots & -2\cos(N\mu_1 + N\nu_1) \\ \vdots & \vdots & & \vdots & \vdots & \vdots \\ 2\cos(N\mu_{p_1} + \nu_{p_1}) & \cdots & 2\cos(-N\mu_{p_1} + N\nu_{p_1}) & \cdots & 2\cos(N\mu_{p_1} + N\nu_{p_1}) \\ -2\cos(N\mu_{p_1} + \nu_{p_1}) & \cdots & -2\cos(-N\mu_{p_1} + N\nu_{p_1}) & \cdots & -2\cos(N\mu_{p_1} + N\nu_{p_1}) \end{pmatrix}$$

Once the problem has been formulated in this way any of the commercially available linear programming packages can be used to solve it. The drawback of this straightforward application of linear programming is that lots of constraint points are required for a given number of required coefficients (i.e. for given N). The resultant filter, however, is optimal in the Chebyshev sense.

B5.3: Show that if the impulse response of a filter is symmetric in both its arguments ($h(n,m) = h(-n,-m)$), the system function is a real function that can be written in the form (5.50).

From (5.49) we have:

$$H(\mu,\nu) = \sum_{n=-N}^{N} \sum_{m=-N}^{N} h(n,m)[\cos(\mu n + \nu m) + j\sin(\mu n + \nu m)] \quad (5.54)$$

Let us consider the imaginary terms only and split the sum over m into the positive, the negative and the 0 terms:

$$\sum_{n=-N}^{N} \sum_{m=-N}^{N} h(n,m)\sin(\mu n + \nu m) =$$

$$\sum_{n=-N}^{N} \left[\sum_{m=-N}^{-1} h(n,m)\sin(\mu n + \nu m) + \sum_{m=1}^{N} h(n,m)\sin(\mu n + \nu m) + h(n,0)\sin(\mu n) \right]$$

In the negative sum inside the square bracket, change the variable of summation to $\tilde{m} \equiv -m$. Then:

$$\sum_{n=-N}^{N} \sum_{m=-N}^{N} h(m,n)\sin(\mu n + \nu m) = \sum_{n=-N}^{N} \left[\sum_{\tilde{m}=N}^{1} h(n,-\tilde{m})\sin(\mu n - \nu\tilde{m}) \right.$$

$$\left. + \sum_{m=1}^{N} h(n,m)\sin(\mu n + \nu m) + h(n,0)\sin(\mu n) \right]$$

We can also change the variable for the sum over n for negative values of n: $\tilde{n} \equiv -n$. Then:

$$\sum_{n=-N}^{N} \sum_{m=-N}^{N} h(m,n)\sin(\mu n + \nu m) = \underbrace{\sum_{\tilde{n}=N}^{1} \sum_{\tilde{m}=1}^{N} h(-\tilde{n},-\tilde{m})\sin(-\mu\tilde{n} - \nu\tilde{m})}$$

$$+ \underbrace{\sum_{\tilde{n}=1}^{N} \sum_{m=1}^{N} h(-\tilde{n},m)\sin(-\mu\tilde{n} + \nu m)} + \underbrace{\sum_{n=1}^{N} \sum_{\tilde{m}=1}^{N} h(n,-\tilde{m})\sin(\mu n - \nu\tilde{m})}$$

$$+ \sum_{n=1}^{N} \sum_{m=1}^{N} h(n,m) \sin(\mu n + \nu m) + \sum_{n=1}^{N} h(n,0) \sin(\mu n)$$

$$+ \sum_{\tilde{n}=1}^{N} h(-\tilde{n},0) \sin(-\mu \tilde{n}) + \sum_{\tilde{m}=1}^{N} h(0,-\tilde{m}) \sin(-\nu \tilde{m})$$

$$+ \sum_{m=1}^{N} h(0,m) \sin(\nu m) + h(0,0) \sin(\mu 0) = 0 \qquad (5.55)$$

In the above expression we have also separated the terms with $n = 0$ or $m = 0$. Since:

$$h(n,m) = h(-n,-m) \Rightarrow \begin{cases} h(0,m) = h(0,-m) \\ h(n,0) = h(-n,0) \\ h(n,-m) = h(-n,m) \end{cases} \qquad (5.56)$$

and because $\sin(-x) = -\sin x$, the terms in (5.55) that are underlined with the same type of line cancel each other. Therefore, the imaginary part in the equation (5.54) is zero and the Fourier transform of the filter is real, given by:

$$H(\mu,\nu) = \sum_{n=-N}^{N} \sum_{m=-N}^{N} h(n,m) \cos(\mu n + \nu m) \qquad (5.57)$$

We can further simplify this expression by changing variables again for the sums over negative indices (i.e. set $\tilde{n} = -n$ for negative n and $\tilde{m} = -m$ for negative m):

$$H(\mu,\nu) =$$

$$= \sum_{n=-N}^{N} \left[\sum_{\tilde{m}=N}^{1} h(n,-\tilde{m}) \cos(\mu n - \nu \tilde{m}) \right.$$

$$\left. + \sum_{m=1}^{N} h(n,m) \cos(\mu n + \nu m) + h(n,0) \cos(\mu n) \right]$$

$$= \sum_{\tilde{n}=N}^{1} \sum_{\tilde{m}=N}^{1} h(-\tilde{n},-\tilde{m}) \cos(-\mu \tilde{n} - \nu \tilde{m}) + \sum_{\tilde{n}=N}^{1} \sum_{m=1}^{N} h(-\tilde{n},m) \cos(-\mu \tilde{n} + \nu m)$$

$$+ \sum_{n=1}^{N} \sum_{\tilde{m}=N}^{1} h(n,-\tilde{m}) \cos(\mu n - \nu \tilde{m}) + \sum_{n=1}^{N} \sum_{m=1}^{N} h(n,m) \cos(\mu n + \nu m)$$

$$+ \sum_{\tilde{n}=N}^{1} h(-\tilde{n},0)\cos(-\mu\tilde{n}) + \sum_{n=1}^{N} h(n,0)\cos(\mu n)$$

$$+ \sum_{\tilde{m}=N}^{1} h(0,-\tilde{m})\cos(-\tilde{m}\nu) + \sum_{m=1}^{N} h(0,m)\cos(m\nu) + h(0,0)$$

Using (5.56) again and the fact that $\cos(-x) = \cos x$, we collect together the terms that are underlined with the same type of line to obtain:

$$H(\mu,\nu) = 2\sum_{n=1}^{N}\sum_{m=1}^{N} h(n,m)\cos(\mu n + \nu m) + 2\sum_{n=1}^{N}\sum_{m=1}^{N} h(-n,m)\cos(-\mu n + \nu m)$$

$$+2\sum_{m=1}^{N} h(0,m)\cos(m\nu) + h(0,0) + 2\sum_{n=1}^{N} h(n,0)\cos(\mu n)$$

Finally, by writing together the first three sums, we get:

$$H(\mu,\nu) = 2\sum_{n=-N}^{N}\sum_{m=1}^{N} h(n,m)\cos(\mu n + \nu m) + 2\sum_{n=1}^{N} h(n,0)\cos(\mu n) + h(0,0)$$

Is there any way by which we can reduce the computational intensity of the linear programming solution?

Yes, by breaking the problem into a series of smaller ones within the framework of an iterative algorithm. At every iteration step of this algorithm the optimal solution is found within a subset of the set of all the constraint points. The algorithm uses linear programming at each iteration step and at every step narrows the range over which the Chebyshev error is allowed to lie. So, it breaks a big linear programming problem into a series of small, more manageable ones.

What is the philosophy of the iterative approach?

Imagine that we have a metal sheet with which we want to clad a curved surface. Every time we try to bend the sheet to pass through some points of the surface, it bulges badly out at some other points owing to its rigidity. We find it difficult to hold it down everywhere simultaneously. So, we bend and distort it first so that it fits very well some points (see *Figure* 5.6). Then we find the point where it deviates most from the desired shape and deform it again so that it now fits another subset of points that includes the one with the previously maximum deviation. It is clear that now it will not fit all that well the original set of points, and since we included the most deviating one from the remaining points, our new fitting will be a little bit worse than before. In other words, as our fitting progresses, we gradually increase the fitting error. This is the penalty we pay in order to make our metal sheet fit better and better overall the shape we want. Such an algorithm is called *maximizing* because from one iteration step to the next it increases the lower limit of the fitting error.

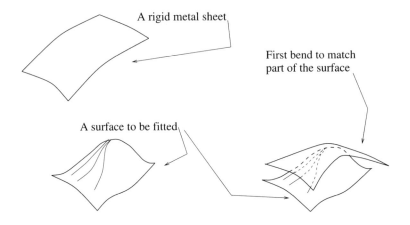

Figure 5.6: How to fit a surface gradually.

Are there any algorithms that work by decreasing the upper limit of the fitting error?

Yes, they are called *minimizing*. There are also algorithms called *mini-max* that work by simultaneously increasing the lower limit of the error and decreasing its upper limit. However, we are not going to discuss them here.

How does the maximizing algorithm work?

It works by making use of the concept of the *limiting set of equations* and the *La Vallee Poussin theorem*.

What is a limiting set of equations?

The set of equations (5.48) is called *limiting* if all Δ_m's are $\neq 0$ and their absolute values $|\Delta_m|$ cannot simultaneously be reduced for any choice of c's we make.

What does the La Vallee Poussin theorem say?

Suppose that equations (5.48) form a limiting set. We choose some c_i's which give the best approximation to F according to the Chebyshev sense. Call this approximation

$P^* \equiv \sum_{i=1}^{n} c_i^* g(\mu, \nu)$. Best approximation in the Chebyshev sense over the set X of M points $(\mu_1, \nu_1), (\mu_2, \nu_2), \ldots, (\mu_m, \nu_m)$ means that:

$$\underbrace{max|P^* - F|}_{\text{over set of points X}} \leq \underbrace{max|P - F|}_{\text{over same set of points X}}$$

where P is any other approximation; i.e. any other set of c_i's. The generalized Vallee Poussin theorem states that the above best Chebyshev error is greater than the minimum value of the error of the random approximation P:

$$min_X|P - F| \leq \underbrace{max_X|P^* - F|}_{\substack{\text{error of the best Chebyshev} \\ \text{approximation}}}$$

So, the theorem states that:

$$min_X|P - F| \leq max_X|P^* - F| \leq max_X|P - F| \tag{5.58}$$

i.e. the error of the best Chebyshev approximation is bounded from above and below, by the minimum and the maximum error of any other approximation.

What is the proof of the La Vallee Poussin theorem?

The right hand side of (5.58) is obvious by the definition of the best Chebyshev approximation. To prove the left inequality, assume first that it does not hold. Assume:

$$max_X|P^* - F| < min_X|P - F| \Rightarrow max_X|\Delta_m^*| < min_X|\Delta_m|$$

This implies that **all** Δ_m^*'s are less than **all** Δ_m's (since the maximum of the Δ_m^*'s is less than the minimum of the Δ_m's). This means that when we change the values of c_i's from c_i's to c_i^*'s, **all** the absolute values of the Δ_m's reduce simultaneously. This is in contradiction to the assumption that the set of equations over the points in X is limiting. So (5.58) holds.

What are the steps of the iterative algorithm?

Suppose that we choose a subset X_k of the set of points X (these points are the chosen discrete (μ, ν) points over which we want to determine the best approximation). In this subset of points we choose the best approximation

$$P_k(\mu, \nu) = \sum_{i=1}^{n} c_i^k g_i(\mu, \nu) \qquad \text{best in } X_k$$

and we define its error in X_k by:

$$\delta_k \equiv max_{X_k}|P_k - F|$$

and its error in the whole set X by:

$$E_k \equiv max_X |P_k - F|$$

Now determine a new set of points which is a subset of X, such that this P_k approximation in the X_{k+1} set has maximum error

$$max_{X_{k+1}} |P_k - F| = E_k$$

and minimum error

$$min_{X_{k+1}} |P_k - F| \geq \delta_k$$

and the set of equations

$$\Delta \equiv F(\mu, \nu) - \sum_{i=1}^{n} c_i g_i(\mu, \nu) \qquad \forall (\mu, \nu) \in X_{k+1}$$

is limiting. Then according to the La Vallee Poussin theorem:

$$min_{X_{k+1}} |P_k - F| \leq max_{X_{k+1}} |P_{k+1} - F| \leq max_{X_{k+1}} |P_k - F|$$

or

$$\delta_k \leq \delta_{k+1} \leq E_k \tag{5.59}$$

where P_{k+1} is the best approximation in the new set of points X_{k+1} and:

$$\delta_{k+1} = max_{X_{k+1}} |P_{k+1} - F|$$

From the way we choose every time the extra point to include in the subset X_k, it is obvious that from one step to the next we narrow the double inequality (5.59) by increasing its lower limit.

Can we approximate a filter by working fully in the frequency domain?

Yes, by calculating the values of the system function in some frequencies as functions of the values of the same function at other frequencies. In other words, express $H(\mu, \nu)$ in terms of $H(k, l)$ where k, l take up specific integer values and calculate the values of $H(k, l)$ for which $max|H(\mu, \nu) - F(\mu, \nu)|$ is minimum. We shall consider certain discrete values of μ, ν. These points will be the constraint points.

How can we express the system function of a filter at some frequencies as a function of its values at other frequencies?

Consider the equations that express the relationship between the impulse response of the filter and its system function:

$$H(\mu, \nu) = \frac{1}{2N + 1} \sum_{n=-N}^{N} \sum_{m=-N}^{N} h(n, m) e^{-j(\frac{\mu n + \nu m}{2N + 1})2\pi} \tag{5.60}$$

Note that here (μ, ν) take integer values and they no longer signify the angular frequencies like they did in all other expressions in this chapter so far. The inverse discrete Fourier transform of this system function is:

$$h(m,n) = \frac{1}{2N+1} \sum_{k=-N}^{N} \sum_{l=-N}^{N} H(k,l) e^{j\frac{2\pi}{2N+1}(nk+ml)} \tag{5.61}$$

Call $2N + 1 \equiv p$. Notice that if we assume that $H(k,l) = H(-k,-l)$ the filter will be real (as we want it) because the imaginary terms will cancel out; i.e.

$$h(m,n) = \frac{1}{p} \sum_{k=-N}^{N} \sum_{l=-N}^{N} H(k,l) \cos\left[\frac{2\pi}{p}(nk+ml)\right] \tag{5.62}$$

From (5.62) it is obvious that $h(m,n) = h(-m,-n)$ and using that in (5.60) we have:

$$H(\mu,\nu) = \frac{1}{p} \sum_{n=-N}^{N} \sum_{m=-N}^{N} h(n,m) \cos\left[\frac{2\pi}{p}(\mu n+\nu m)\right] \tag{5.63}$$

Since we want to express $H(\mu,\nu)$ in terms of other values of H, we must substitute from (5.62) into (5.63):

$$H(\mu,\nu) = \frac{1}{p^2} \sum_{n=-N}^{N} \sum_{m=-N}^{N} \sum_{k=-N}^{N} \sum_{l=-N}^{N} H(k,l) \cos\left[\frac{2\pi}{p}(nk+ml)\right] \cos\left[\frac{2\pi}{p}(\mu n+\nu m)\right] \tag{5.64}$$

It can be shown (see Box **B5.4**) that this expression can be written as:

$$H(\mu,\nu) = \frac{1}{(2N+1)^2} \left[\frac{1}{2} H(0,0)\Phi(0,0) + \sum_{l=1}^{N} H(0,l)\Phi(0,l) + \sum_{k=1}^{N} \sum_{l=-N}^{N} H(k,l)\Phi(k,l)\right] \tag{5.65}$$

where $\Phi(k,l)$ is some known function of the trigonometric functions.

Example 5.4

We want to construct a 3×3 real and symmetric (when both its arguments change sign) filter, with the following frequency response:

$$\begin{array}{c} \xrightarrow{\quad} \mu \\ \begin{array}{ccc} 0 & 1 & 0 \\ 1 & 0 & 1 \\ \nu \downarrow \quad 0 & 1 & 0 \end{array} \end{array}$$

i.e.

$$F(\mu,\nu) = 1 \quad \textbf{for} \quad (\mu,\nu) = (0,-1),(-1,0),(1,0),(0,1)$$
$$F(\mu,\nu) = 0 \quad \textbf{for} \quad (\mu,\nu) = (-1,-1),(1,-1),(0,0),(-1,1),(1,1)$$

Formulate the problem so that it can be solved by linear programming. State clearly the dimensions and elements of each array involved. (Suggestion: Use at most nine constraint points in Fourier space.)

Because of symmetry

$$
\begin{aligned}
h(-1,-1) &= h(1,1) \\
h(-1,0) &= h(1,0) \\
h(0,-1) &= h(0,1) \\
h(-1,1) &= h(1,-1)
\end{aligned}
$$

For a 3×3 filter, clearly $N = 1$ in equation (5.65). Then:

$$
\begin{aligned}
H(\mu,\nu) = {}& 2h(-1,1)\cos(\nu-\mu) + 2h(0,1)\cos\nu + 2h(1,1)\cos(\mu+\nu) \\
& + 2h(1,0)\cos\mu + h(0,0)
\end{aligned}
$$

$h(1,1), h(1,0), h(0,1), h(1,-1)$ and $h(0,0)$ are the unknowns we must specify. Since they can be positive or negative, these must be our free unknowns. Define the error as $\delta \equiv max_{\mu,\nu}|H(\mu,\nu) - F(\mu,\nu)|$. Then:

$$
\delta \geq |H - F| \Rightarrow
\begin{cases}
\delta \geq H - F & \Rightarrow \quad \delta - H \geq -F \\
\delta \geq -(H - F) & \Rightarrow \quad \delta + H \geq F
\end{cases}
$$

We use two inequalities like the above for every constraint point in frequency space, i.e. a total of 18 inequalities.

We want to minimize δ which must always be non-negative, so it must be our non-negative unknown. So: Minimize $Z = X_1$ under the constraints

$$
\begin{aligned}
\mathbf{A_{11}}X_1 + A_{12}\mathbf{X_2} &\geq \mathbf{B_1} \quad &&\textit{18 inequalities} \\
X_1 &\geq 0 \quad &&\textit{1 unknown} \\
\mathbf{X_2} &\textit{ free} \quad &&\textit{5 unknowns}
\end{aligned}
$$

where $X_1 = \delta$

$$
\mathbf{X_2} =
\begin{pmatrix}
h(0,0) \\
h(0,1) \\
h(-1,1) \\
h(1,0) \\
h(1,1)
\end{pmatrix}
\quad
\mathbf{A_{11}} =
\begin{pmatrix}
1 \\
1 \\
\vdots \\
1
\end{pmatrix}
$$

We consider the points in the Fourier space in the following order:

$$
\begin{matrix}
1 & 2 & 3 \\
4 & 5 & 6 \\
7 & 8 & 9 \\
(-1,1) & (0,-1) & (1,-1) \\
(-1,0) & (0,0) & (1,0) \\
(-1,1) & (0,1) & (1,1)
\end{matrix}
$$

Then

$$
\mathbf{B_1} =
\begin{pmatrix}
0 \\ 0 \\ -1 \\ 1 \\ 0 \\ 0 \\ -1 \\ 1 \\ 0 \\ 0 \\ -1 \\ 1 \\ 0 \\ 0 \\ -1 \\ 1 \\ 0 \\ 0
\end{pmatrix}
\qquad (\textit{Each element is the value of} - F(\mu,\nu) \ \textit{or} \ F(\mu,\nu))
$$

$$
A_{12} =
\begin{pmatrix}
-1 & -2\cos 1 & -2 & -2\cos 1 & -2\cos 2 \\
1 & 2\cos 1 & 2 & 2\cos 1 & 2\cos 2 \\
-1 & -2\cos 1 & -2\cos 1 & -2 & -2\cos 1 \\
1 & 2\cos 1 & 2\cos 1 & 2 & 2\cos 1 \\
-1 & -2\cos 1 & -2\cos 2 & -2\cos 1 & -2 \\
1 & 2\cos 1 & 2\cos 2 & 2\cos 1 & 2 \\
-1 & -2 & -2\cos 1 & -2\cos 1 & -2\cos 1 \\
1 & 2 & 2\cos 1 & 2\cos 1 & 2\cos 1 \\
-1 & -2 & -2 & -2 & -2 \\
1 & 2 & 2 & 2 & 2 \\
-1 & -2 & -2\cos 1 & -2\cos 1 & -2\cos 1 \\
1 & 2 & 2\cos 1 & 2\cos 1 & 2\cos 1 \\
-1 & -2\cos 1 & -2\cos 2 & -2\cos 1 & -2 \\
1 & 2\cos 1 & 2\cos 2 & 2\cos 1 & 2 \\
-1 & -2\cos 1 & -2\cos 1 & -2 & -2\cos 1 \\
1 & 2\cos 1 & 2\cos 1 & 2 & 2\cos 1 \\
-1 & -2\cos 1 & -2 & -2\cos 1 & -2\cos 2 \\
1 & 2\cos 1 & 2 & 2\cos 1 & 2\cos 2
\end{pmatrix}
$$

B5.4: Prove expression (5.65).

If we use the identity $\cos\alpha\cos\beta = \frac{1}{2}[\cos(\alpha+\beta)+\cos(\alpha-\beta)]$ and define $\frac{2\pi}{p} \equiv A$ equation (5.64) can be written as:

$$H(\mu,\nu) = \frac{1}{2p^2} \sum_{k=-N}^{N} \sum_{l=-N}^{N} \sum_{n=-N}^{N} \sum_{m=-N}^{N} H(k,l)\{\cos[An(\mu+k)+mA(l+\nu)]$$
$$+ \cos[nA(k-\mu)+mA(l-\nu)]\}$$

We define:

$$\begin{aligned} A(\mu+k) &\equiv x \\ A(l+\nu) &\equiv y \\ A(k-\mu) &\equiv u \\ A(l-\nu) &\equiv v \end{aligned}$$

Then:

$$H(\mu,\nu) = \frac{1}{2p^2} \sum_{k=-N}^{N} \sum_{l=-N}^{N} H(k,l) \sum_{n=-N}^{N} \sum_{m=-N}^{N} \{\cos(nx+my)+\cos(nu+mv)\} \quad (5.66)$$

Consider the first of the terms inside the curly brackets:

$$\sum_{n=-N}^{N} \sum_{m=-N}^{N} \cos(nx+my) = \sum_{n=-N}^{N} \sum_{m=-N}^{N} (\cos(nx)\cos(my)+\sin(my)\sin(nx))$$

$$= \sum_{n=-N}^{N} \cos(nx) \sum_{m=-N}^{N} \cos(my)$$

$$+ \sum_{m=-N}^{N} \sin(my) \sum_{n=-N}^{N} \sin(nx)$$

As sine is an odd function, $\sin(my)$ summed over a symmetric interval of values of m gives zero. So the second double sum in the above expression vanishes. On the other hand, $\cos(nx)$ being an even function allows us to write:

$$\sum_{n=-N}^{N} \sum_{m=-N}^{N} \cos(nx+my) = \left(1+2\sum_{n=1}^{N}\cos(nx)\right)\left(1+2\sum_{m=1}^{N}\cos(my)\right)$$

We use the identity:

$$\sum_{k=1}^{n} \cos(kx) = \cos\left(\frac{n+1}{2}x\right)\sin\frac{nx}{2}\frac{1}{\sin\frac{x}{2}}.$$

We apply it here to obtain:

$$\sum_{n=-N}^{N}\sum_{m=-N}^{N}\cos(nx+my) = \left(1+\frac{2\cos\left(\frac{N+1}{2}x\right)\sin\frac{Nx}{2}}{\sin\frac{x}{2}}\right)\left(1+\frac{2\cos\left(\frac{N+1}{2}y\right)\sin\frac{Ny}{2}}{\sin\frac{y}{2}}\right)$$

We also use the identity: $2\cos\alpha\sin\beta = \sin(\alpha+\beta) - \sin(\alpha-\beta)$. So:

$$\sum_{n=-N}^{N}\sum_{m=-N}^{N}\cos(nx+my) = \left(1+\frac{\sin\left(\frac{2N+1}{2}x\right)-\sin\frac{x}{2}}{\sin\frac{x}{2}}\right)\left(1+\frac{\sin\left(\frac{2N+1}{2}y\right)-\sin\frac{y}{2}}{\sin\frac{y}{2}}\right)$$

$$= \frac{\sin\left(\frac{2N+1}{2}x\right)}{\sin\frac{x}{2}}\frac{\sin\left(\frac{2N+1}{2}y\right)}{\sin\frac{y}{2}} \tag{5.67}$$

We define:

$$\phi(\alpha,\beta) \equiv \frac{\sin[\pi(\alpha+\beta)]}{\sin[\frac{2\pi}{p}(\alpha+\beta)]} \tag{5.68}$$

From the definition of p, x and A we have: $\frac{2N+1}{2}x = \frac{2N+1}{2}A(\mu+k) = \frac{p}{2}\frac{2\pi}{p}(\mu+\nu) = \pi(\mu+\nu)$. Similarly for $\frac{2N+1}{2}y$. Then equation (5.67) can be written as:

$$\sum_{n=-N}^{N}\sum_{m=-N}^{N}\cos(nx+my) = \phi(\mu,k)\phi(l,\nu) \tag{5.69}$$

Working in a similar way we obtain:

$$\sum_{n=-N}^{N}\sum_{m=-N}^{N}\cos(nu+mv) = \phi(k,-\mu)\phi(l,-\nu)$$

Then (5.66) can be written as:

$$H(\mu,\nu) = \frac{1}{2p^2}\sum_{k=-N}^{N}\sum_{l=-N}^{N}H(k,l)\left\{\phi(\mu,k)\phi(l,\nu) + \phi(k,-\mu)\phi(l,-\nu)\right\} \tag{5.70}$$

We define:

$$\Phi(k,l) \equiv \phi(\mu,k)\phi(l,\nu) + \phi(k,-\mu)\phi(l,-\nu) \tag{5.71}$$

From the definition of ϕ (equation (5.68)) it is clear that:

$$\Phi(k,l) = \Phi(-k,-l)$$

We also have $H(k,l) = H(-k,-l)$. Then equation (5.70) can be further simplified:

$$H(\mu,\nu) = \frac{1}{2p^2} \sum_{k=-N}^{N} \left[\sum_{l=-N}^{-1} H(k,l)\Phi(k,l) + \sum_{l=1}^{N} H(k,l)\Phi(k,l) + H(k,0)\Phi(k,0) \right]$$

$$= \frac{1}{2p^2} \sum_{k=-N}^{N} \left[\sum_{\tilde{l}=1}^{N} H(k,-\tilde{l})\Phi(k,-\tilde{l}) + \sum_{l=1}^{N} H(k,l)\Phi(k,l) + H(k,0)\Phi(k,0) \right]$$

In a similar manner we can split the summation over k into the 0 term, the positive and the negative terms and define a new variable $\tilde{k} = -k$ to sum over negative values of k:

$$
\begin{aligned}
H(\mu,\nu) \;=\; \frac{1}{2p^2} \bigg[& \underbrace{\sum_{\tilde{k}=1}^{N}\sum_{l=1}^{N} H(-\tilde{k},-l)\Phi(-\tilde{k},-l)} + \sum_{\tilde{k}=1}^{N}\sum_{l=1}^{N} H(-\tilde{k},l)\Phi(-\tilde{k},l) \\[2mm]
& + \underline{\sum_{\tilde{k}=1}^{N} H(-\tilde{k},0)\Phi(-\tilde{k},0)} + \sum_{k=1}^{N}\sum_{l=1}^{N} H(k,-l)\Phi(k,-l) \\[2mm]
& + \underbrace{\sum_{k=1}^{N}\sum_{l=1}^{N} H(k,l)\Phi(k,l)} \\[2mm]
& + \underline{\sum_{k=1}^{N} H(k,0)\Phi(k,0)} + \underline{\underline{\sum_{\tilde{l}=1}^{N} H(0,-l)\Phi(0,-l)}} \\[2mm]
& + \underline{\underline{\sum_{\tilde{l}=1}^{N} H(0,l)\Phi(0,l)}} + H(0,0)\Phi(0,0) \bigg]
\end{aligned}
$$

Owing to the symmetry properties of functions $\Phi(k,l)$ and $H(k,l)$, the terms that are underlined by the same type of line are equal, and therefore they can be combined to yield:

$$
\begin{aligned}
H(\mu,\nu) = \frac{1}{2p^2} \bigg[& 2\sum_{k=1}^{N}\sum_{l=1}^{N} H(k,l)\Phi(k,l) + 2\sum_{k=1}^{N}\sum_{l=1}^{N} H(k,-l)\Phi(k,-l) \\[2mm]
& + 2\sum_{k=1}^{N} H(k,0)\Phi(k,0) + 2\sum_{l=1}^{N} H(0,l)\Phi(0,l) + H(0,0)\Phi(0,0) \bigg] \quad (5.72)
\end{aligned}
$$

The first three sums on the right hand side of the above expression can be combined into one as they include summation over all positive values of l, all negative values of l and the term for $l = 0$:

$$H(\mu,\nu) = \frac{1}{2p^2} \left[2\sum_{k=1}^{N}\sum_{l=-N}^{N} H(k,l)\Phi(k,l) + 2\sum_{l=1}^{N} H(0,l)\Phi(0,l) + H(0,0)\Phi(0,0) \right]$$

Finally, upon substituting the value of p we obtain:

$$H(\mu,\nu) = \frac{1}{(2N+1)^2}\left[\frac{1}{2}H(0,0)\Phi(0,0) + \sum_{l=1}^{N} H(0,l)\Phi(0,l)\right.$$

$$\left. + \sum_{k=1}^{N}\sum_{l=-N}^{N} H(k,l)\Phi(k,l)\right] \qquad (5.73)$$

What exactly are we trying to do when we design the filter in the frequency domain only?

We try to find optimal values for the coefficients $H(k,l)$ for $k = 0,1,\ldots,N$, $l = 0,1,\ldots,N$ so that the error in approximating $F(\mu,\nu)$ by $H(\mu,\nu)$ at a fixed number of constraint points (μ_i,ν_i) is minimum.

In practice, what people usually do is to fix some values of the $H(k,l)$ and try to optimize the choice of the remaining values. For example, in the case of a lowpass filter, we can put $H(k,l) = 1$ for the frequencies we want to pass and $H(k,l) = 0$ for the undesired frequencies and then leave the values of $H(k,l)$ which refer to transitional frequencies to be determined (see *Figure* 5.7).

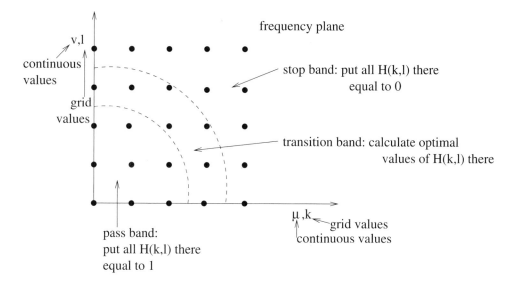

Figure 5.7: **Fixing some of the filter values in the frequency domain and letting others be free.**

How can we solve for the unknown values $H(k,l)$?

Equation (5.65) is similar to equation (5.50) and can be solved by using linear programming.

Does the frequency sampling method yield optimal solutions according to the Chebyshev criterion?

No, because the optimization is not done for all $H(k,l)$ since some of them are fixed.

Example 5.5

Formulate the problem of designing a filter in the frequency domain so that it can be solved by a linear programming package.

If we compare equations (5.65) and (5.50), we see that the problem is very similar to the problem of specifying the values of the filter in the real domain using linear programming: In that case we were seeking to specify $h(0,0), h(1,0), \ldots, h(N,0), \ldots, h(-N,m), h(-N + 1,m), \ldots, h(N,m)$ for $m = 1,2,\ldots,N$. Now we wish to specify $H(0,0), H(0,1), \ldots, H(0,N), H(k,-N), H(k,-N + 1), \ldots, H(k,N)$ for $k = 1,2,\ldots,N$.

In that case, we had function $\cos(\mu n + \nu m)$ appearing in equation (5.50). Now we have function $\Phi(\mu,\nu)$ instead. $\Phi(\mu,\nu)$, however, is also an even function with respect to both its arguments, and therefore it behaves in a similar way to the cosine function as far as the solution of the problem is concerned. The problem therefore can be formulated as follows:

Minimize:

$$z = x_{11}$$

under the constraints:

$$A_{11}x_{11} + A_{12}x_2 \geq \mathbf{B_1}$$

where:

$$x_{11} = \delta$$

and

$$\mathbf{x_2} = \begin{pmatrix} H(0,0) \\ H(1,0) \\ H(2,0) \\ \vdots \\ H(N,0) \\ H(-N,1) \\ \vdots \\ H(N,1) \\ H(-N,2) \\ \vdots \\ H(N,2) \\ \vdots \\ H(-N,N) \\ \vdots \\ H(N,N) \end{pmatrix} \leftarrow \begin{matrix} (2N^2 + 2N + 1) \times 1 \\ matrix \end{matrix} \qquad \mathbf{B_1} = \begin{pmatrix} F(\mu_1, \nu_1) \\ -F(\mu_1, \nu_1) \\ F(\mu_2, \nu_2) \\ -F(\mu_2, \nu_2) \\ \vdots \\ F(\mu_{p_1}, \nu_p) \\ -F(\mu_{p_1}, \nu_p) \end{pmatrix} \leftarrow \begin{matrix} 2p_1 \times 1 \\ matrix \end{matrix}$$

$$A_{11} = \begin{pmatrix} 1 \\ 1 \\ 1 \\ \vdots \\ 1 \end{pmatrix} \leftarrow \begin{matrix} 2p_1 \times 1 \\ matrix \end{matrix}$$

A_{12} is a $2p_1 \times (2N^2 + 2N + 1)$ matrix the elements of which can easily be found if we substitute equation (5.65) into the inequality constraint. It turns out to be:

$$A_{12} = \begin{pmatrix} 1 & 2\Phi(\mu_1,0) & 2\Phi(2\mu_1,0) & \dots & 2\Phi(N\mu_1,0) & 2\Phi(-N\mu_1,\nu_1) \\ -1 & -2\Phi(\mu_1,0) & -2\Phi(2\mu_1,0) & \dots & -2\Phi(N\mu_1,0) & -2\Phi(-N\mu_1,\nu_1) \\ \vdots & \vdots & & \vdots\ \vdots & \vdots & \vdots \\ \vdots & \vdots & & \vdots\ \vdots & \vdots & \vdots \\ 1 & 2\Phi(\mu_{p_1},0) & 2\Phi(2\mu_{p_1},0) & \dots & 2\Phi(N\mu_{p_1},0) & 2\Phi(-N\mu_{p_1},\nu_{p_1}) \\ -1 & -2\Phi(\mu_{p_1},0) & -2\Phi(2\mu_{p_1},0) & \dots & -2\Phi(N\mu_{p_1},0) & -2\Phi(-N\mu_{p_1},\nu_{p_1}) \end{pmatrix}$$

$$\begin{pmatrix} \dots & 2\Phi(N\mu_1,\nu_1) & \dots & 2\Phi(-N\mu_1,N\nu_1) & \dots & 2\Phi(N\mu_1,N\nu_1) \\ \dots & -2\Phi(N\mu_1,\nu_1) & \dots & -2\Phi(-N\mu_1,N\nu_1) & \dots & -2\Phi(N\mu_1,N\nu_1) \\ \vdots & & \vdots\ \vdots & & \vdots\ \vdots & \vdots \\ \vdots & & \vdots\ \vdots & & \vdots\ \vdots & \vdots \\ \dots & 2\Phi(N\mu_{p_1},\nu_{p_1}) & \dots & 2\Phi(-N\mu_{p_1},N\nu_{p_1}) & \dots & 2\Phi(N\mu_{p_1},N\nu_{p_1}) \\ \dots & -2\Phi(N\mu_{p_1},\nu_{p_1}) & \dots & -2\Phi(-N\mu_{p_1},N\nu_{p_1}) & \dots & -2\Phi(N\mu_{p_1},N\nu_{p_1}) \end{pmatrix}$$

What is the "take home" message of this chapter?

If we specify exactly the desired frequency response of a filter to be zero outside a finite range, we obtain a filter of infinite extent in the real domain. A digital filter of infinite extent can be realized as a recursive filter with the help of its z-transform. z-transforms in two dimensions are much more difficult to be inverted than 1-dimensional z-transforms. Also, recursive filters have to fulfil certain criteria to be stable. The theory of 2-dimensional digital filters, their stability criteria and their design problems are beyond the scope of this book.

To avoid using infinite impulse response filters we resort to approximations. We specify exactly the frequency response of the filter we require and then try to find the finite impulse response filter which approximates the desired frequency behaviour as well as possible.

Sometimes it is important to have exactly the desired frequency response to some frequencies, while we do not care much about some other frequencies. Then we fix the frequency response of the filter at the specific frequencies and try to find the value of the filter at the remaining frequencies in an approximate way.

In filter design the criterion of approximation we use is the Chebyshev norm instead of the least square error. This is because we are interested in the worse artifact we create at a single frequency, rather than in an overall good approximation which may have a very unevenly distributed error among the various frequencies.

Chapter 6

Image Restoration

What is image restoration?

Image restoration is the improvement of an image using **objective** criteria and prior knowledge as to what the image should look like.

What is the difference between image enhancement and image restoration?

In image enhancement we try to improve the image using **subjective** criteria, while in image restoration we are trying to reverse a specific damage suffered by the image, using **objective** criteria.

Why may an image require restoration?

An image may be degraded because the grey values of individual pixels may be altered, or it may be distorted because the position of individual pixels may be shifted away from their correct position. The second case is the subject of *geometric restoration*.

Geometric restoration is also called *image registration* because it helps in finding corresponding points between two images of the same region taken from different viewing angles. Image registration is very important in remote sensing when aerial photographs have to be registered against the map, or two aerial photographs of the same region have to be registered with each other.

How may geometric distortion arise?

Geometric distortion may arise because of the lens or because of the irregular movement of the sensor during image capture. In the former case, the distortion looks regular like those shown in *Figure* 6.1. The latter case arises, for example, when an aeroplane photographs the surface of the Earth with a line scan camera. As the aeroplane wobbles, the captured image may be **inhomogeneously** distorted, with pixels displaced by as much as $4-5$ interpixel distances away from their true positions.

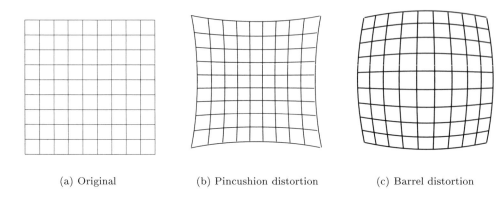

(a) Original   (b) Pincushion distortion   (c) Barrel distortion

Figure 6.1: Examples of geometric distortions caused by the lens.

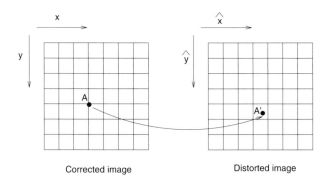

Corrected image   Distorted image

Figure 6.2: In this figure the pixels correspond to the nodes of the grids. Pixel A of the corrected grid corresponds to inter-pixel position A' of the original image.

How can a geometrically distorted image be restored?

We start by creating an empty array of numbers the same size as the distorted image. This array will become the corrected image. Our purpose is to assign grey values to the elements of this array. This can be achieved by performing a two-stage operation: spatial transformation followed by grey level interpolation.

How do we perform the spatial transformation?

Suppose that the true position of a pixel is (x, y) and the distorted position is $(\hat{x}, \hat{y})$ (see *Figure* 6.2). In general there will be a transformation which leads from one set of coordinates to the other, say:

$$\hat{x} \;=\; \mathcal{O}_x(x,y) \tag{6.1}$$

$$\hat{y} \;=\; \mathcal{O}_y(x,y) \tag{6.2}$$

First we must find to which coordinate position in the distorted image each pixel position of the corrected image corresponds. Here we usually make some assumptions. For example, we may say that the above transformation has the following form:

$$\hat{x} \;=\; c_1 x + c_2 y + c_3 xy + c_4 \tag{6.3}$$

$$\hat{y} \;=\; c_5 x + c_6 y + c_7 xy + c_8 \tag{6.4}$$

where $c_1, c_2, \ldots, c_8$ are some parameters. Alternatively, we may assume a more general form, where squares of the coordinates x and y appear on the right hand sides of the above equations. The values of parameters $c_1, \ldots, c_8$ can be determined from the transformation of known points called *tie points*. For example, in aerial photographs of the surface of the Earth, there are certain landmarks with exactly known positions. There are several such points scattered all over the surface of the Earth. We can use, for example, four such points to find the values of the above eight parameters and assume that these transformation equations with the derived parameter values hold inside the whole quadrilateral region defined by these four tie points.

Then, we apply the transformation to find the position A' of point A of the corrected image, in the distorted image.

Why is grey level interpolation needed?

It is likely that point A' will not have integer coordinates even though the coordinates of point A in the (x,y) space are integer. This means that we do not actually know the grey level value at position A'. That is when the grey level interpolation process comes into play. The grey level value at position A' can be estimated from the values at its four nearest neighbouring pixels in the $(\hat{x}, \hat{y})$ space, by some method, for example by bilinear interpolation. We assume that inside each little square the grey level value is a simple function of the positional coordinates:

$$g(\hat{x}, \hat{y}) = \alpha \hat{x} + \beta \hat{y} + \gamma \hat{x}\hat{y} + \delta$$

where $\alpha, \ldots, \delta$ are some parameters. We apply this formula to the four corner pixels to derive values of α, β, γ and δ and then use these values to calculate $g(\hat{x}, \hat{y})$ at the position of A' point.

Figure 6.3 below shows in magnification the neighbourhood of point A' in the distorted image with the four nearest pixels at the neighbouring positions with integer coordinates.

Simpler as well as more sophisticated methods of interpolation may be employed. For example, the simplest method that can be used is the nearest neighbour method where A' gets the grey level value of the pixel which is nearest to it. A more sophisticated method is to fit a higher order surface through a larger patch of pixels around A' and find the value at A' from the equation of that surface.

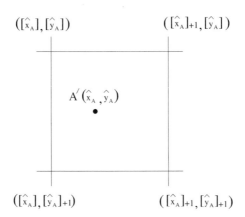

$([\hat{x}_A], [\hat{y}_A])$ $\qquad\qquad$ $([\hat{x}_A]+1, [\hat{y}_A])$

$A'(\hat{x}_A, \hat{y}_A)$

$([\hat{x}_A], [\hat{y}_A]+1)$ $\qquad\qquad$ $([\hat{x}_A]+1, [\hat{y}_A]+1)$

Figure 6.3: The non-integer position of point A' is surrounded by four pixels at integer positions, with known grey values ($[x]$ means integer part of x).

Example 6.1

In the figure below the grid on the right is a geometrically distorted image and has to be registered with the reference image on the left using points A, B, C and D as tie points. The entries in the image on the left indicate coordinate positions. Assuming that the distortion within the rectangle $ABCD$ can be modelled by bilinear interpolation and the grey level value at an interpixel position can be modelled by bilinear interpolation too, find the grey level value at pixel position $(2,2)$ in the reference image.

A			B			A				
$(0,0)$	$(1,0)$	$(2,0)$	$(3,0)$	$(4,0)$		4	3	2	5	1
$(0,1)$	$(1,1)$	$(2,1)$	$(3,1)$	$(4,1)$		3	3	4	2B	4
$(0,2)$	$(1,2)$	$(2,2)$	$(3,2)$	$(4,2)$		0	5	5	3	3
$C(0,3)$	$(1,3)$	$(2,3)$	$(3,3)D$	$(4,3)$		0	4C	2	1	2
$(0,4)$	$(1,4)$	$(2,4)$	$(3,4)$	$(4,4)$		2	1	0	3	5D

Suppose that the position $(\hat{x}, \hat{y})$ of a pixel in the distorted image is given in terms of its position (x, y) in the reference image by:

$$\hat{x} = c_1 x + c_2 y + c_3 xy + c_4$$
$$\hat{y} = c_5 x + c_6 y + c_7 xy + c_8$$

We have the following set of the corresponding coordinates between the two grids using the four tie points:

	Distorted $(\hat{x}, \hat{y})$ coords.	Reference (x, y) coords.
Pixel A	$(0, 0)$	$(0, 0)$
Pixel B	$(3, 1)$	$(3, 0)$
Pixel C	$(1, 3)$	$(0, 3)$
Pixel D	$(4, 4)$	$(3, 3)$

We can use these to calculate the values of the parameters $c_1, \ldots, c_8$:

Pixel A:
$$c_4 = 0$$
$$c_8 = 0$$

Pixel B:
$$\left.\begin{array}{l} 3 = 3c_1 \\ 1 = 3c_5 \end{array}\right\} \Rightarrow \left\{\begin{array}{l} c_1 = 1 \\ c_5 = \frac{1}{3} \end{array}\right.$$

Pixel C
$$\left.\begin{array}{l} 1 = 3c_2 \\ 3 = 3c_6 \end{array}\right\} \Rightarrow \left\{\begin{array}{l} c_2 = \frac{1}{3} \\ c_6 = 1 \end{array}\right.$$

Pixel D
$$\left.\begin{array}{l} 4 = 3 + 3 \times \frac{1}{3} + 9c_3 \\ 4 = 3 \times \frac{1}{3} + 3 + 9c_7 \end{array}\right\} \Rightarrow \left\{\begin{array}{l} c_3 = 0 \\ c_7 = 0 \end{array}\right.$$

The distorted coordinates, therefore, of any pixel within the square ABDC are given by:

$$\hat{x} = x + \frac{y}{3}, \quad \hat{y} = \frac{x}{3} + y$$

For $x = y = 2$ we have $\hat{x} = 2 + \frac{2}{3}$, $\hat{y} = \frac{2}{3} + 2$. So, the coordinates of pixel $(2, 2)$ in the distorted image are $(2\frac{2}{3}, 2\frac{2}{3})$. This position is located between pixels in the distorted image and actually between pixels with the following grey level values:

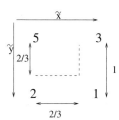

We define a local coordinate system $(\tilde{x}, \tilde{y})$, so that the pixel at the top left corner has coordinate position $(0, 0)$ the pixel at the top right corner has coordinates $(1, 0)$, the one at the bottom left $(0, 1)$ and the one at the bottom right $(1, 1)$. Assuming that the grey level value between four pixels can be computed from the grey level values in the four corner pixels with bilinear interpolation, we have:

$$g(\tilde{x}, \tilde{y}) = \alpha \tilde{x} + \beta \tilde{y} + \gamma \tilde{x} \tilde{y} + \delta$$

Applying this for the four neighbouring pixels we have:

$$5 = \alpha \cdot 0 + \beta \cdot 0 + \gamma \cdot 0 + \delta \quad \Rightarrow \quad \delta = 5$$
$$3 = \alpha \cdot 1 + \beta \cdot 0 + \gamma \cdot 0 + 5 \quad \Rightarrow \quad \alpha = -2$$
$$2 = (-2) \cdot 0 + \beta \cdot 1 + \gamma \cdot 0 + 5 \quad \Rightarrow \quad \beta = -3$$
$$1 = (-2) \cdot 1 + (-3) \cdot 1 + \gamma \cdot 1 \cdot 1 + 5 \quad \Rightarrow \quad \gamma = 0$$

Therefore

$$g(\tilde{x}, \tilde{y}) = -2\tilde{x} - 3\tilde{y} + 5$$

We apply this for $\tilde{x} = \frac{2}{3}$ and $\tilde{y} = \frac{2}{3}$ to obtain:

$$g = -2 \cdot \frac{2}{3} - 3 \cdot \frac{2}{3} + 5 = -\frac{4}{3} + 3 = 1\frac{2}{3}$$

So, the grey level value at position $(2,2)$ should be $1\frac{2}{3}$. Rounding it to the nearest integer we have 2. If the nearest neighbour interpolation were used, we would have assigned to this position grey value 1 as the nearest neighbour is the pixel at the bottom right corner.

How does the degraded image depend on the undegraded image and the point spread function of a linear degradation process?

Under the assumption that the effect which causes the damage is linear, the output image $g(\alpha, \beta)$ can be written in terms of the input image $f(x, y)$ as follows:

$$g(\alpha, \beta) = \int_{-\infty}^{+\infty} \int_{-\infty}^{+\infty} f(x, y) h(x, y, \alpha, \beta) dx dy \tag{6.5}$$

where $h(x, y, \alpha, \beta)$ is the point spread function that expresses the degradation effect.

How does the degraded image depend on the undegraded image and the point spread function of a linear *shift invariant* degradation process?

If we can write:

$$h(x, y, \alpha, \beta) = h(\alpha - x, \beta - y) \tag{6.6}$$

we say that we have a shift invariant point spread function. The meaning is that the output at a point of the image depends only on the **relative** position of the point with respect to other points of the image and not on the **actual** position of the point. Then we have:

$$g(\alpha, \beta) = \int_{-\infty}^{+\infty} \int_{-\infty}^{+\infty} f(x, y) h(\alpha - x, \beta - y) dx dy \tag{6.7}$$

We recognize now that equation (6.7) is the convolution between the undegraded image $f(x, y)$ and the point spread function, and therefore we can write it in terms of their Fourier transforms:

$$\hat{G}(u, v) = \hat{F}(u, v)\hat{H}(u, v) \tag{6.8}$$

where $\hat{G}$, $\hat{F}$ and $\hat{H}$ are the Fourier transforms of functions f, g and h respectively.

What form does equation (6.5) take for the case of discrete images?

$$g(i, j) = \sum_{k=1}^{N} \sum_{l=1}^{N} f(k, l) h(k, l, i, j) \tag{6.9}$$

We have shown that equation (6.9) can be written in matrix form (see equation (1.25)):

$$\mathbf{g} = H\mathbf{f} \tag{6.10}$$

What is the problem of image restoration?

The problem of image restoration is: given the degraded image g, recover the original undegraded image f.

How can the problem of image restoration be solved?

The problem of image restoration can be solved if we have prior knowledge of the *point spread function* or its Fourier transform (the *transfer function*) of the degradation process.

How can we obtain information on the transfer function $\hat{H}(u, v)$ of the degradation process?

1. From the knowledge of the physical process that caused degradation. For example, if the degradation is due to diffraction, $H(u, v)$ can be calculated. Similarly, if the degradation is due to atmospheric turbulence or due to motion, it can be modelled and $H(u, v)$ calculated.

2. We may try to extract information on $\hat{H}(u, v)$ or $h(\alpha - x, \beta - y)$ from the image itself; i.e. from the effect the process has on the images of some known objects, ignoring the actual nature of the underlying physical process that takes place.

Example 6.2

When a certain static scene was being recorded, the camera underwent planar motion parallel to the image plane (x, y). This motion appeared as if the scene moved in the x, y directions by distances $x_0(t)$ and $y_0(t)$ which are functions of time t. The shutter of the camera remained open from $t = 0$ to $t = T$ where T is a constant. Write down the equation that expresses the intensity recorded at pixel position (x, y) in terms of the scene intensity function $f(x, y)$.

The total exposure at any point of the recording medium (say the film) will be T and we shall have for the blurred image:

$$g(x, y) = \int_0^T f(x - x_0(t), y - y_0(t)) dt \qquad (6.11)$$

Example 6.3

In Example 6.2, derive the transfer function with which you can model the degradation suffered by the image due to the camera motion, assuming that the degradation is linear with a shift invariant point spread function.

Consider the Fourier transform of $g(x, y)$ defined in Example 6.2.

$$\hat{G}(u, v) = \int_{-\infty}^{+\infty} \int_{-\infty}^{+\infty} g(x, y) e^{-2\pi j (ux + vy)} dx dy \qquad (6.12)$$

If we substitute (6.11) into (6.12) we have:

$$\hat{G}(u, v) = \int_{-\infty}^{+\infty} \int_{-\infty}^{+\infty} \int_0^T f(x - x_0(t), y - y_0(t)) dt \, e^{-2\pi j (ux + vy)} dx dy \qquad (6.13)$$

We can exchange the order of integrals:

$$\hat{G}(u, v) = \int_0^T \left\{ \underbrace{\int_{-\infty}^{+\infty} \int_{-\infty}^{+\infty} f(x - x_0(t), y - y_0(t)) e^{-2\pi j (ux + vy)} dx dy}_{} \right\} dt \qquad (6.14)$$

This is the Fourier transform of the shifted function by x_0, y_0 in directions x, y respectively

We have shown (see equation (2.67)) that the Fourier transform of a shifted function and the Fourier transform of the unshifted function are related by:

$$(F.T.\ of\ shifted\ function) = (F.T.\ of\ unshifted\ function)e^{-2\pi j(ux_0+vy_0)}$$

Therefore:

$$\hat{G}(u,v) = \int_0^T \hat{F}(u,v)e^{-2\pi j(ux_0+vy_0)}dt$$

where $\hat{F}(u,v)$ is the Fourier transform of the scene intensity function $f(x,y)$, i.e. the unblurred image. $\hat{F}(u,v)$ is independent of time, so it can come out of the integral sign:

$$\hat{G}(u,v) = \hat{F}(u,v)\int_0^T e^{-2\pi j(ux_0(t)+vy_0(t))}dt$$

Comparing this equation with (6.8) we conclude that:

$$\hat{H}(u,v) = \int_0^T e^{-2\pi j(ux_0(t)+vy_0(t))}dt \tag{6.15}$$

Example 6.4

Suppose that the motion in Example 6.2 was in the x direction only and with constant speed $\frac{\alpha}{T}$, so that $y_0(t) = 0$, $x_0(t) = \frac{\alpha t}{T}$. Calculate the transfer function of the motion blurring caused.

In the result of Example 6.3, equation (6.15), substitute $y_0(t)$ and $x_0(t)$ to obtain:

$$\hat{H}(u,v) = \int_0^T e^{-2\pi ju\frac{\alpha t}{T}}dt = \left.\frac{e^{-2\pi ju\frac{\alpha t}{T}}}{-2\pi ju\frac{\alpha}{T}}\right|_0^T = -\frac{T}{2\pi ju\alpha}\left[e^{-2\pi ju\alpha} - 1\right]$$

$$= \frac{T}{2\pi ju\alpha}\left[1 - e^{-2\pi ju\alpha}\right] = \frac{Te^{-\pi ju\alpha}}{2\pi ju\alpha}\left[e^{\pi ju\alpha} - e^{-\pi ju\alpha}\right] \Rightarrow$$

$$\hat{H}(u,v) = \frac{Te^{-\pi ju\alpha}2j\sin(\pi u\alpha)}{2j\pi u\alpha} = T\frac{\sin(\pi u\alpha)}{\pi u\alpha}e^{-j\pi u\alpha} \tag{6.16}$$

Example 6.5 (B)

It was established that during the time interval T when the shutter was open, the camera moved in such a way that it appeared as if the objects in the scene moved along the positive y axis, with constant acceleration 2α and initial velocity s_0, starting from zero displacement. Derive the transfer function of the degradation process for this case.

In this case $x_0(t) = 0$ and

$$\frac{d^2 y_0}{dt^2} = 2\alpha \Rightarrow \frac{dy_0}{dt} = 2\alpha t + b \Rightarrow y_0(t) = \alpha t^2 + bt + c$$

where α is half the constant acceleration and b and c some integration constants. We have the following initial conditions:

$$t = 0 \qquad \text{zero shifting i.e. } c = 0$$
$$t = 0 \quad \text{velocity of shifting} = s_0 \Rightarrow b = s_0$$

Therefore:

$$y_0(t) = \alpha t^2 + s_0 t$$

We substitute $x_0(t)$ and $y_0(t)$ in equation 6.15 for $H(u, v)$:

$$
\begin{aligned}
H(u, v) &= \int_0^T e^{-2\pi j v (\alpha t^2 + s_0 t)} dt \\
&= \int_0^T \cos\left[2\pi v \alpha t^2 + 2\pi v s_0 t\right] dt - j \int_0^T \sin\left[2\pi v \alpha t^2 + 2\pi v s_0 t\right] dt
\end{aligned}
$$

We may use the following formulae:

$$\int \cos(ax^2 + bx + c)dx = \sqrt{\frac{\pi}{2a}}\left\{\cos\frac{ac - b^2}{a}C\left(\frac{ax + b}{\sqrt{a}}\right) - \sin\frac{ac - b^2}{a}S\left(\frac{ax + b}{\sqrt{a}}\right)\right\}$$

$$\int \sin(ax^2 + bx + c)dx = \sqrt{\frac{\pi}{2a}}\left\{\cos\frac{ac - b^2}{a}S\left(\frac{ax + b}{\sqrt{a}}\right) + \sin\frac{ac - b^2}{a}C\left(\frac{ax + b}{\sqrt{a}}\right)\right\}$$

where $S(x)$ and $C(x)$ are

$$S(x) \equiv \sqrt{\frac{2}{\pi}} \int_0^x \sin t^2 dt$$

$$C(x) \equiv \sqrt{\frac{2}{\pi}} \int_0^x \cos t^2 dt$$

and they are called Fresnel integrals.

We shall use the above formulae with

$$a \rightarrow 2\pi v\alpha$$
$$b \rightarrow 2\pi v s_0$$
$$c \rightarrow 0$$

to obtain:

$$H(u,v) = \frac{1}{2\sqrt{v\alpha}} \left\{ \cos\frac{2\pi v s_0^2}{\alpha} C\left(\sqrt{2\pi v}\frac{\alpha t + s_0}{\sqrt{a}}\right) \right.$$

$$- \sin\left(-\frac{2\pi v s_0^2}{\alpha}\right) S\left(\sqrt{2\pi v}\frac{\alpha t + s_0}{\sqrt{a}}\right) - j\cos\frac{2\pi v s_0^2}{\alpha} S\left(\sqrt{2\pi v}\frac{\alpha t + s_0}{\sqrt{a}}\right)$$

$$\left. -j\sin\left(-\frac{2\pi v s_0^2}{\alpha}\right) C\left(\sqrt{2\pi v}\frac{\alpha t + s_0}{\sqrt{a}}\right) \right\}\Bigg|_0^T$$

$$= \frac{1}{2\sqrt{v\alpha}} \left\{ \cos\frac{2\pi v s_0^2}{\alpha} \left[C\left(\sqrt{\frac{2\pi v}{\alpha}}(\alpha T + s_0)\right) - jS\left(\sqrt{2\pi v\alpha}(\alpha T + s_0)\right) \right] \right.$$

$$+ \sin\frac{2\pi v s_0^2}{\alpha} \left[S\left(\sqrt{\frac{2\pi v}{\alpha}}(\alpha T + s_0)\right) + jC\left(\sqrt{2\pi v\alpha}(\alpha T + s_0)\right) \right]$$

$$- \cos\frac{2\pi v s_0^2}{\alpha} \left[C\left(\sqrt{\frac{2\pi v}{\alpha}}s_0\right) - jS\left(\sqrt{2\pi v}s_0\right) \right]$$

$$\left. - \sin\frac{2\pi v s_0^2}{\alpha} \left[S\left(\sqrt{\frac{2\pi v}{\alpha}}s_0\right) + jC\left(\sqrt{2\pi v}s_0\right) \right] \right\} \tag{6.17}$$

Example 6.6 (B)

What is the transfer function for the case of Example 6.5, if the shutter remained open for a very long time and the starting velocity of the shifting was negligible?

It is known that functions $S(x)$ and $C(x)$ that appear in the result of Example 6.5 have the following asymptotic behaviour:

$$\lim_{x \to \infty} S(x) = \frac{1}{2}$$

$$\lim_{x \to \infty} C(x) = \frac{1}{2}$$

$$\lim_{x \to 0} S(x) = 0$$

$$\lim_{x \to 0} C(x) = 0$$

Therefore, for $s_0 \simeq 0$ and $T \to \infty$, we have:

$$C\left(\sqrt{\frac{2\pi v}{\alpha}}\left(\alpha T + s_0\right)\right) \to \frac{1}{2} \qquad C\left(\sqrt{\frac{2\pi v}{\alpha}} s_0\right) \to 0$$

$$S\left(\sqrt{\frac{2\pi v}{\alpha}}\left(\alpha T + s_0\right)\right) \to \frac{1}{2} \qquad S\left(\sqrt{\frac{2\pi v}{\alpha}} s_0\right) \to 0$$

$$\cos\frac{2\pi v s_0^2}{\alpha} \to 1 \qquad \sin\frac{2\pi v s_0^2}{\alpha} \to 0$$

Therefore equation (6.17) becomes:

$$H(u, v) \simeq \frac{1}{2\sqrt{v\alpha}}\left[\frac{1}{2} - j\frac{1}{2}\right] = \frac{1 - j}{4\sqrt{v\alpha}}$$

Example 6.7

How can we infer the point spread function of the degradation process from an astronomical image?

We know that by definition the point spread function is the output of the imaging system when the input is a point source. In an astronomical image, a very distant star can be considered as a point source. By measuring then the brightness profile of a star we immediately have the point spread function of the degradation process this image has been subjected to.

Example 6.8

Suppose that we have an ideal bright straight line in the scene parallel to the image axis x. Use this information to derive the point spread function of the process that degrades the captured image.

Mathematically the undegraded image of a bright line can be represented by:

$$f(x,y) = \delta(y)$$

where we assume that the line actually coincides with the x axis. Then the image of this line will be:

$$h_l(x,y) = \int_{-\infty}^{+\infty} \int_{-\infty}^{+\infty} h(x-x', y-y')\delta(y')dy'dx' = \int_{-\infty}^{+\infty} h(x-x', y)dx'$$

We change variable $\tilde{x} \equiv x - x' \Rightarrow dx' = -d\tilde{x}$. The limits of $\tilde{x}$ are from $+\infty$ to $-\infty$. Then:

$$h_l(x,y) = -\int_{+\infty}^{-\infty} h(\tilde{x}, y)d\tilde{x} = \int_{-\infty}^{+\infty} h(\tilde{x}, y)d\tilde{x} \tag{6.18}$$

The right hand side of this equation does not depend on x, and therefore the left hand side should not depend either; i.e. the image of the line will be parallel to the x axis (or rather coincident with it) and the same all along it:

$$h_l(x,y) = h_l(y) = \underbrace{\int_{-\infty}^{+\infty} h_l(\tilde{x}, y)d\tilde{x}}_{\substack{\tilde{x} \text{ is a dummy variable,} \\ \text{independent of } x}} \tag{6.19}$$

Suppose that we take the Fourier transform of $h_l(y)$:

$$\hat{H}_l(v) \equiv \int_{-\infty}^{+\infty} h_l(y)e^{-2\pi jvy}dy \tag{6.20}$$

The point spread function has as Fourier transform the transfer function given by:

$$\hat{H}(u,v) = \int_{-\infty}^{+\infty} h(x,y)e^{-2\pi j(ux+vy)}dxdy \tag{6.21}$$

If we set $u = 0$ in this expression, we obtain

$$\hat{H}(0,v) = \int_{-\infty}^{+\infty} \underbrace{\left[\int_{-\infty}^{+\infty} h(x,y)dx \right]}_{h_l(y) \; from \; (6.19)} e^{-2\pi jvy} dy \qquad (6.22)$$

By comparing equation (6.20) with (6.22) we get:

$$\hat{H}(0,v) = \hat{H}_l(v) \qquad (6.23)$$

That is, the image of the ideal line gives us the profile of the transfer function along a single direction; i.e. the direction orthogonal to the line. This is understandable, as the cross-section of a line orthogonal to its length is no different from the cross-section of a point. By definition, the cross-section of a point is the point spread function of the blurring process. If now we have lots of ideal lines in various directions in the image, we are going to have information as to how the transfer function looks along the directions orthogonal to the lines in the frequency plane. By interpolation then we can calculate $\hat{H}(u,v)$ at any point in the frequency plane.

Example 6.9

It is known that a certain scene contains a sharp edge. How can the image of the edge be used to infer some information concerning the point spread function of the imaging device?

Let us assume that the ideal edge can be represented by a step function along the x axis, defined by:

$$u(y) = \begin{cases} 1 & for \; y > 0 \\ 0 & for \; y \le 0 \end{cases}$$

The image of this function will be:

$$h_e(x,y) = \int_{-\infty}^{\infty} \int_{-\infty}^{\infty} h(x-x', y-y') u(y') dx' dy'$$

We may define new variables $\tilde{x} \equiv x - x'$, $\tilde{y} \equiv y - y'$. Obviously $dx' = -d\tilde{x}$ and $dy' = -d\tilde{y}$. The limits of both $\tilde{x}$ and $\tilde{y}$ are from $+\infty$ to $-\infty$. Then:

$$h_e(x,y) = \int_{-\infty}^{\infty} \int_{-\infty}^{\infty} h(\tilde{x}, \tilde{y}) u(y - \tilde{y}) d\tilde{x} d\tilde{y}$$

Let us take the partial derivative of both sides of this equation with respect to y:

$$\frac{\partial h_e(x,y)}{\partial y} = \int_{-\infty}^{\infty} \int_{-\infty}^{\infty} h(\tilde{x}, \tilde{y}) \frac{\partial u(y - \tilde{y})}{\partial y} d\tilde{x} d\tilde{y}$$

It is known that the derivative of a step function with respect to its argument is a delta function:

$$\frac{\partial h_e(x,y)}{\partial y} = \int_{-\infty}^{\infty} \int_{-\infty}^{\infty} h(\tilde{x}, \tilde{y}) \delta(y - \tilde{y}) d\tilde{x} d\tilde{y} \Rightarrow$$

$$\frac{\partial h_e(x,y)}{\partial y} = \int_{-\infty}^{\infty} h(\tilde{x}, y) d\tilde{x} \tag{6.24}$$

If we compare (6.24) with equation (6.18) we see that the derivative of the image of the edge is the image of a line parallel to the edge. Therefore, we can derive information concerning the point spread function of the imaging process by obtaining images of ideal step edges at various orientations. Each such image should be differentiated along a direction orthogonal to the direction of the edge. Each resultant derivative image should be treated as the image of an ideal line and used to yield the profile of the point spread function along the direction orthogonal to the line, as described in Example 6.8.

Example 6.10

Use the methodology of Example 6.9 to derive the point spread function of an imaging device.

Using a ruler and black ink we create the chart shown in Figure 6.4.

Figure 6.4: A test chart for the derivation of the point spread function of an imaging device.

*This chart can be used to measure the point spread function of our imaging sys-
tem at orientations 0°, 45°, 90° and 135°. First the test chart is imaged using
our imaging apparatus. Then the partial derivative of the image is computed by
convolution at orientations 0°, 45°, 90° and 135° using the Robinson operators.
These operators are shown in* Figure 6.5.

1	2	1		2	1	0		1	0	-1		0	-1	-2
0	0	0		1	0	-1		2	0	-2		1	0	-1
-1	-2	-1		0	-1	-2		1	0	-1		2	1	0

(a) M0   (b) M1   (c) M2   (d) M3

**Figure 6.5: Filters used to compute the derivative in $0, 45, 90$ and 135
degrees.**

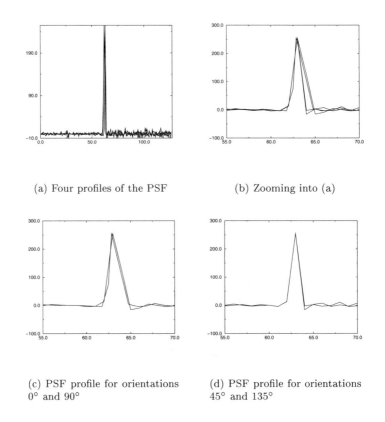

(a) Four profiles of the PSF   (b) Zooming into (a)

(c) PSF profile for orientations   (d) PSF profile for orientations
$0°$ and $90°$ $45°$ and $135°$

**Figure 6.6: The point spread function (PSF) of an imaging system, in
two different scales.**

The profiles of the resultant images along several lines orthogonal to the original edges are computed and averaged to produce the four profiles for 0°, 45°, 90° and 135° plotted in Figure 6.6a. These are the profiles of the point spread function. In Figure 6.6b we zoom into the central part of the plot of Figure 6.6a. Two of the four profiles of the point spread functions plotted there are clearly narrower than the other two. This is because they correspond to orientations 45° and 135° and the distance of the pixels along these orientations is √2 longer than the distance of pixels along 0° and 90°. Thus, the value of the point spread function that is plotted as being 1 pixel away from the peak, in reality is approximately 1.4 pixels away. Indeed, if we take the ratio of the widths of the two pairs of the profiles, we find the value of ∼ 1.4.

In Figures 6.6c and 6.6d we plot separately the two pairs of profiles and see that the system has the same behaviour along the 45°, 135° and 0°, 90° orientations. Taking into account the √2 correction for the 45° and 135°, we conclude that the point spread function of this imaging system is to a high degree circularly symmetric.

In a practical application these four profiles can be averaged to produce a single cross-section of a circularly symmetric point spread function. The Fourier transform of this 2D function is the system transfer function of the imaging device.

If we know the transfer function of the degradation process, isn't the solution to the problem of image restoration trivial?

If we know the transfer function of the degradation and calculate the Fourier transform of the degraded image, it appears that from equation (6.8) we can obtain the Fourier transform of the undegraded image:

$$\hat{F}(u,v) = \frac{\hat{G}(u,v)}{\hat{H}(u,v)} \tag{6.25}$$

Then, by taking the inverse Fourier transform of $\hat{F}(u,v)$, we should be able to recover $f(x,y)$, which is what we want. However, this straightforward approach produces unacceptably poor results.

What happens at points (u,v) where $\hat{H}(u,v) = 0$?

$H(u,v)$ probably becomes 0 at some points in the (u,v) plane and this means that $\hat{G}(u,v)$ will also be zero at the same points as seen from equation (6.8). The ratio $\hat{G}(u,v)/\hat{H}(u,v)$ as appears in (6.25) will be 0/0; i.e. undetermined. All this means is that for the particular frequencies (u,v) the frequency content of the original image cannot be recovered. One can overcome this problem by simply omitting the corresponding points in the frequency plane, provided of course that they are countable.

Will the zeroes of $\hat{H}(u,v)$ and $\hat{G}(u,v)$ always coincide?

No, if there is the slightest amount of noise in equation (6.8), the zeroes of $\hat{H}(u,v)$ will not coincide with the zeroes of $\hat{G}(u,v)$.

How can we take noise into consideration when writing the linear degradation equation?

For additive noise, the complete form of equation (6.8) is:

$$\hat{G}(u,v) = \hat{F}(u,v)\hat{H}(u,v) + \hat{N}(u,v) \tag{6.26}$$

where $\hat{N}(u,v)$ is the Fourier transform of the noise field. $\hat{F}(u,v)$ is then given by:

$$\hat{F}(u,v) = \frac{\hat{G}(u,v)}{\hat{H}(u,v)} - \frac{\hat{N}(u,v)}{\hat{H}(u,v)} \tag{6.27}$$

In places where $\hat{H}(u,v)$ is zero or even just very small, the noise term may be enormously amplified.

How can we avoid the amplification of noise?

In many cases, $|\hat{H}(u,v)|$ drops rapidly away from the origin while $|\hat{N}(u,v)|$ remains more or less constant. To avoid the amplification of noise then when using equation (6.27), we do not use as filter the factor $1/\hat{H}(u,v)$, but a windowed version of it, cutting it off at a frequency before $|\hat{H}(u,v)|$ becomes too small or before its first zero. In other words we use:

$$\hat{F}(u,v) = \hat{M}(u,v)\hat{G}(u,v) - \hat{M}(u,v)\hat{N}(u,v) \tag{6.28}$$

where

$$\hat{M}(u,v) = \begin{cases} 1/\hat{H}(u,v) & \text{for } u^2 + v^2 \leq \omega_0^2 \\ 1 & \text{for } u^2 + v^2 > \omega_0^2 \end{cases} \tag{6.29}$$

where ω_0 is chosen so that all zeroes of $H(u,v)$ are excluded. Of course, one may use other windowing functions instead of the above window with rectangular profile, to make $\hat{M}(u,v)$ go smoothly to zero at ω_0.

Example 6.11

Demonstrate the application of inverse filtering in practice by restoring a motion blurred image.

Let us consider the image of Figure 6.7a. To imitate the way this image would look
if it were blurred by motion, we take every 10 consecutive pixels along the x axis,
find their average value, and assign it to the tenth pixel. This is what would have
happened if, when the image was being recorded, the camera had moved 10 pixels
to the left: the brightness of a line segment in the scene with length equivalent to
10 pixels would have been recorded by a single pixel. The result would look like
Figure 6.7b. The blurred image $g(i,j)$ in terms of the original image $f(i,j)$ is
given by the discrete version of equation (6.11):

$$g(i,j) = \frac{1}{i_T} \sum_{k=0}^{i_T-1} f(i-k,j) \quad i = 0,1,\ldots,N-1 \tag{6.30}$$

where i_T is the total number of pixels with their brightness recorded by the same
cell of the camera, and N is the total number of pixels in a row of the image. In
this example $i_T = 10$ and $N = 128$.

The transfer function of the degradation is given by the discrete version of
the equation derived in Example 6.4. We shall derive it now here. The discrete
Fourier transform of $g(i,j)$ is given by:

$$\hat{G}(m,n) = \frac{1}{N} \sum_{l=0}^{N-1} \sum_{t=0}^{N-1} g(l,t) e^{-j\left(\frac{2\pi ml}{N} + \frac{2\pi nt}{N}\right)} \tag{6.31}$$

If we substitute $g(l,t)$ from equation (6.30) we have:

$$\hat{G}(m,n) = \frac{1}{N}\frac{1}{i_T} \sum_{l=0}^{N-1} \sum_{t=0}^{N-1} \sum_{k=0}^{i_T-1} f(l-k,t) e^{-j\left(\frac{2\pi ml}{N} + \frac{2\pi nt}{N}\right)}$$

We rearrange the order of summations to obtain:

$$\hat{G}(m,n) = \frac{1}{i_T} \sum_{k=0}^{i_T-1} \underbrace{\frac{1}{N} \sum_{l=0}^{N-1} \sum_{t=0}^{N-1} f(l-k,t) e^{-j\left(\frac{2\pi ml}{N} + \frac{2\pi nt}{N}\right)}}_{DFT\ of\ shifted\ f(l,t)}$$

By applying the property of the Fourier transforms concerning shifted functions,
we have:

$$\hat{G}(m,n) = \frac{1}{i_T} \sum_{k=0}^{i_T-1} \hat{F}(m,n) e^{-j\frac{2\pi m}{N}k}$$

where $\hat{F}(m,n)$ is the Fourier transform of the original image.

As $\hat{F}(m,n)$ does not depend on k, it can be taken out of the summation:

$$\hat{G}(m,n) = \hat{F}(m,n)\frac{1}{i_T}\sum_{k=0}^{i_T-1}e^{-j\frac{2\pi m}{N}k}$$

We identify then the Fourier transform of the degradation process as

$$\hat{H}(m,n) = \frac{1}{i_T}\sum_{k=0}^{i_T-1}e^{-j\frac{2\pi m}{N}k} \tag{6.32}$$

The sum on the right hand side of this equation is a geometric progression with ratio between successive terms

$$q \equiv e^{-j\frac{2\pi m}{N}}$$

We apply the formula

$$\sum_{k=0}^{n-1}q^k = \frac{q^n-1}{q-1}\quad where\quad q\neq 1$$

to obtain:

$$\hat{H}(m,n) = \frac{1}{i_T}\frac{e^{-j\frac{2\pi m}{N}i_T}-1}{e^{-j\frac{2\pi m}{N}}-1} = \frac{1}{i_T}\frac{e^{-j\frac{\pi m}{N}i_T}\left(e^{-j\frac{\pi m}{N}i_T}-e^{j\frac{\pi m}{N}i_T}\right)}{e^{-j\frac{\pi m}{N}}\left(e^{-j\frac{\pi m}{N}}-e^{j\frac{\pi m}{N}}\right)}$$

Therefore:

$$\hat{H}(m,n) = \frac{1}{i_T}\frac{\sin\frac{\pi m}{N}i_T}{\sin\frac{\pi m}{N}}e^{-j\frac{\pi m}{N}(i_T-1)} \tag{6.33}$$

Notice that for $m=0$ we have $q=1$ and we cannot apply the formula of the geometric progression. Instead we have a sum of 1's in (6.32) which is equal to i_T and so

$$\hat{H}(0,n) = 1 \quad for \quad 0\leq n\leq N-1$$

It is interesting to compare equation (6.33) with its continuous counterpart, equation (6.16). We can see that there is a fundamental difference between the two equations: in the denominator equation (6.16) has the frequency u along the blurring axis appearing on its own, while in the denominator of equation (6.33) we have the sine of this frequency appearing. This is because discrete images are treated by the discrete Fourier transform as periodic signals, repeated ad infinitum in all directions.

We can analyse the Fourier transform of the blurred image in its real and imaginary parts:

$$\hat{G}(m,n) \equiv G_1(m,n) + jG_2(m,n)$$

We can then write it in magnitude-phase form:

$$\hat{G}(m,n) = \sqrt{G_1^2(m,n) + G_2^2(m,n)}e^{j\phi(m,n)} \tag{6.34}$$

where

$$\cos\phi(m,n) = \frac{G_1(m,n)}{\sqrt{G_1^2(m,n) + G_2^2(m,n)}}$$

$$\sin\phi(m,n) = \frac{G_2(m,n)}{\sqrt{G_1^2(m,n) + G_2^2(m,n)}} \tag{6.35}$$

To obtain the Fourier transform of the original image we divide $\hat{G}(m,n)$ *by* $\hat{H}(m,n)$:

$$\hat{F}(m,n) = \frac{\sqrt{G_1^2(m,n) + G_2^2(m,n)}}{\sin\frac{i_T\pi m}{N}}i_T\sin\frac{\pi m}{N}e^{j\left(\phi(m,n) + \frac{\pi m}{N}(i_T - 1)\right)}$$

Therefore, the real and the imaginary parts of $\hat{F}(m,n)$, $F_1(m,n)$ *and* $F_2(m,n)$ *respectively, are given by:*

$$F_1(m,n) = i_T\sin\frac{\pi m}{N}\frac{\sqrt{G_1^2(m,n) + G_2^2(m,n)}}{\sin\frac{i_T\pi m}{N}}\cos\left(\phi(m,n) + \frac{\pi m}{N}(i_T - 1)\right)$$

$$F_2(m,n) = i_T\sin\frac{\pi m}{N}\frac{\sqrt{G_1^2(m,n) + G_2^2(m,n)}}{\sin\frac{i_T\pi m}{N}}\sin\left(\phi(m,n) + \frac{\pi m}{N}(i_T - 1)\right)$$

If we use the formulae $\cos(a + b) = \cos a\cos b - \sin a\sin b$ *and* $\sin(a + b) = \cos a\sin b + \sin a\cos b$ *and substitute for* $\cos\phi(m,n)$ *and* $\sin\phi(m,n)$ *from equations (6.35) we obtain:*

$$F_1(m,n) = i_T\sin\frac{\pi m}{N}\frac{G_1(m,n)\cos\frac{\pi m(i_T - 1)}{N} - G_2(m,n)\sin\frac{\pi m(i_T - 1)}{N}}{\sin\frac{i_T\pi m}{N}}$$

$$F_2(m,n) = i_T\sin\frac{\pi m}{N}\frac{G_1(m,n)\sin\frac{\pi m(i_T - 1)}{N} + G_2(m,n)\cos\frac{\pi m(i_T - 1)}{N}}{\sin\frac{i_T\pi m}{N}} \tag{6.36}$$

We also remember to set:

$$F_1(0,n) = G_1(0,n) \quad for \quad 0 \leq n \leq N - 1$$
$$F_2(0,n) = G_2(0,n) \quad for \quad 0 \leq n \leq N - 1$$

If we use $F_1(m,n)$ *and* $F_2(m,n)$ *as the real and the imaginary parts of the Fourier transform of the undegraded image and Fourier transform back, we obtain image 6.7d. This is totally wrong because in equations (6.36) we divide by 0 for several values of* m.

Indeed, the denominator $\sin \frac{i_T \pi m}{N}$ *becomes* 0 *every time* $\frac{i_T \pi m}{N}$ *is a multiple of* π:

$$\frac{i_T \pi m}{N} = k\pi \Rightarrow m = \frac{kN}{i_T} \quad where \quad k = 1, 2, \ldots$$

Our image is 128×128, *i.e.* $N = 128$, *and* $i_T = 10$. *Therefore, we divide by* 0 *when* $m = 12.8, 25.6, 38.4$, *etc. As* m *takes only integer values, the denominator becomes very small for* $m = 13, 26, 38$, *etc. It is actually exactly* 0 *only for* $m = 64$. *Let us omit this value for* m, *i.e. let us use:*

$$F_1(64, n) = G_1(64, n) \quad for \quad 0 \le n \le 127$$
$$F_2(64, n) = G_2(64, n) \quad for \quad 0 \le n \le 127$$

The rest of the values of $F_1(m, n)$ *and* $F_2(m, n)$ *are as defined by equations (6.36).*

If we Fourier transform back, we obtain the image in Figure 6.7e. *The image looks now almost acceptable, apart from some horizontal interfering frequency. In practice, instead of trying to identify the values of* m *or* n *for which the denominator of equations (6.36) becomes exactly* 0, *we find the first of those* 0*'s and apply the formula only up to that pair of values. In our case, the first zero is for* $k = 1$, *i.e. for* $m = \frac{N}{i_T} = 12.8$.

We use formulae (6.36), therefore, only for $0 \le m \le 12$ *and* $0 \le n \le 127$. *Otherwise we use:*

$$\left. \begin{array}{l} F_1(m, n) = G_1(m, n) \\ F_2(m, n) = G_2(m, n) \end{array} \right\} \quad for \quad \begin{array}{l} 13 \le m \le 127 \\ 0 \le n \le 127 \end{array}$$

If we Fourier transform back we obtain the image shown in Figure 6.7f. *This image looks more blurred than the previous with the vertical lines (the horizontal interfering frequency) still there, but less prominent. The blurring is understandable: we have effectively done nothing to improve the frequencies above* $m = 12$, *so the high frequencies of the image responsible for any sharp edges will remain degraded. As for the vertical lines, we observe that we have almost* 13 *of them in an image of width* 128, *i.e. they repeat every* 10 *pixels. They are due to the boundary effect: The Fourier transform assumes that the image is repeated ad infinitum in all directions. So it assumes that the pixels on the left of the blurred image carry the true values of the pixels on the right of the image. In reality of course this is not the case, as the blurred pixels on the left carry the true values of some points further left that do not appear in the image. To show that this explanation is correct, we blurred the original image assuming cylindrical boundary conditions, i.e. assuming that the image is repeated on the left. The result is the blurred image of* Figure 6.7c. *The results of restoring this image by the three versions of inverse filtering are shown at the bottom row of* Figure 6.7. *The vertical lines have disappeared entirely and we have a remarkably good restoration in* 6.7h, *obtained by simply omitting the frequency for which the transfer function is exactly* 0.

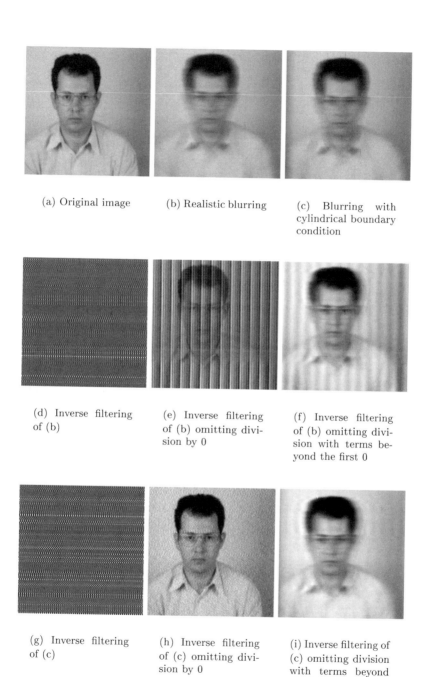

(a) Original image

(b) Realistic blurring

(c) Blurring with cylindrical boundary condition

(d) Inverse filtering of (b)

(e) Inverse filtering of (b) omitting division by 0

(f) Inverse filtering of (b) omitting division with terms beyond the first 0

(g) Inverse filtering of (c)

(h) Inverse filtering of (c) omitting division by 0

(i) Inverse filtering of (c) omitting division with terms beyond the first 0

Figure 6.7: Image restoration with inverse filtering.

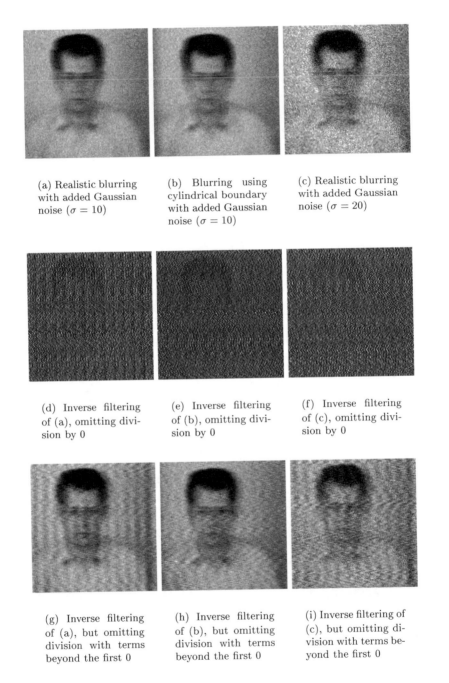

(a) Realistic blurring with added Gaussian noise ($\sigma = 10$)

(b) Blurring using cylindrical boundary with added Gaussian noise ($\sigma = 10$)

(c) Realistic blurring with added Gaussian noise ($\sigma = 20$)

(d) Inverse filtering of (a), omitting division by 0

(e) Inverse filtering of (b), omitting division by 0

(f) Inverse filtering of (c), omitting division by 0

(g) Inverse filtering of (a), but omitting division with terms beyond the first 0

(h) Inverse filtering of (b), but omitting division with terms beyond the first 0

(i) Inverse filtering of (c), but omitting division with terms beyond the first 0

Figure 6.8: Image restoration with inverse filtering in the presence of noise.

> *Unfortunately, in real situations, the blurring is going to be like that of* Figure
> *6.7b and the restoration results are expected to be more like those in* Figures *6.7e*
> *and 6.7f than those in 6.7h and 6.7i.*
>   *To compare how inverse filtering copes with noise, we produced the blurred*
> *and noisy images shown in* Figure *6.8 by adding white Gaussian noise. The noisy*
> *images were subsequently restored using inverse filtering and avoiding the division*
> *by 0. The results, shown in* Figures *6.8d–6.8f are really very bad: High frequencies*
> *dominated by noise are amplified by the filter to the extent that they dominate the*
> *restored image. When the filter is truncated beyond its first 0, the results, shown*
> *in* Figures *6.8g–6.8i are quite reasonable.*

How can we express the problem of image restoration in a formal way?

If $\hat{f}(\mathbf{r})$ is an estimate of the original undegraded image $f(\mathbf{r})$, we wish to calculate
$\hat{f}(\mathbf{r})$ so that the norm of the residual image $f(\mathbf{r}) - \hat{f}(\mathbf{r})$ is minimal over all possible
versions of image $f(\mathbf{r})$. From Chapter 3 we know that this is equivalent to saying
that we wish to identify $\hat{f}(\mathbf{r})$ which minimizes

$$e^2 \equiv E\{[f(\mathbf{r}) - \hat{f}(\mathbf{r})]^2\} \tag{6.37}$$

Equation (6.7) in this section has been assumed to have the form:

$$g(\mathbf{r}) = \int_{-\infty}^{+\infty} \int_{-\infty}^{+\infty} h(\mathbf{r} - \mathbf{r}')f(\mathbf{r}')d\mathbf{r}' + \nu(\mathbf{r}) \tag{6.38}$$

where $g(\mathbf{r})$, $f(\mathbf{r})$ and $\nu(\mathbf{r})$ are considered to be random fields, with $\nu(\mathbf{r})$ being the
noise field.

What is the solution of equation (6.37)?

If no conditions are imposed on the solution, the least squares estimate of $f(\mathbf{r})$ which
minimizes (6.37) turns out to be the *conditional expectation* of $f(\mathbf{r})$ given $g(\mathbf{r})$ which
in general is a non-linear function of $g(\mathbf{r})$ and requires the calculation of the joint
probability density function of the random fields $f(\mathbf{r})$ and $g(\mathbf{r})$. This can be calculated
with the help of non-linear methods like *simulated annealing*. However, such methods
are beyond the scope of this book.

Can we find a linear solution to equation (6.37)?

Yes, by imposing the constraint that the solution $\hat{f}(\mathbf{r})$ is a linear function of $g(\mathbf{r})$.
Clearly, the solution found this way will not give the absolute minimum of e but it
will make e minimum **within the limitations of the constraints** imposed. We

decide that we want the estimated image $\hat{f}(\mathbf{r})$ to be expressed as a linear function of the grey levels of the degraded image, i.e.:

$$\hat{f}(\mathbf{r}) = \int_{-\infty}^{+\infty} \int_{-\infty}^{+\infty} m(\mathbf{r}, \mathbf{r}')g(\mathbf{r}')d\mathbf{r}' \tag{6.39}$$

where $m(\mathbf{r}, \mathbf{r}')$ is the function we want to determine and which gives the weight by which the grey level value of the degraded image g at position $\mathbf{r}'$ affects the value of the estimated image $\hat{f}$ at position $\mathbf{r}$. If the random fields involved are homogeneous, the weighting function $m(\mathbf{r}, \mathbf{r}')$ will depend only on the difference of $\mathbf{r}$ and $\mathbf{r}'$ as opposed to depending on them separately. In that case (6.39) can be written as:

$$\hat{f}(\mathbf{r}) = \int_{-\infty}^{+\infty} \int_{-\infty}^{+\infty} m(\mathbf{r} - \mathbf{r}')g(\mathbf{r}')d\mathbf{r}' \tag{6.40}$$

This equation means that we wish to identify a filter $m(\mathbf{r})$ with which to convolve the degraded image $g(\mathbf{r}')$ in order to obtain an estimate $\hat{f}(\mathbf{r})$ of the undegraded image $f(\mathbf{r})$.

What is the linear least mean square error solution of the image restoration problem?

If $\hat{M}(u, v)$ is the Fourier transform of the filter $m(\mathbf{r})$, it can be shown that the linear solution of equation (6.37) can be obtained if

$$\hat{M}(u, v) = \frac{S_{fg}(u, v)}{S_{gg}(u, v)} \tag{6.41}$$

where $S_{fg}(u, v)$ is the cross-spectral density of the undegraded and the degraded image and $S_{gg}(u, v)$ is the spectral density of the degraded image. $\hat{M}(u, v)$ is the Fourier transform of the *Wiener* filter for image restoration.

Since the original image $f(\mathbf{r})$ is unknown, how can we use equation (6.41) which relies on its cross-spectral density with the degraded image, to derive the filter we need?

In order to proceed we need to make some **extra assumption**: the noise and the true image are uncorrelated and at least one of the two has zero mean. This assumption is a plausible one: we expect the process that gives rise to the image to be entirely different from the process that gives rise to the noise. Further, if the noise has a biasing, i.e. it does not have zero mean, we can always identify and subtract this biasing to make it have zero mean.

Since $f(\mathbf{r})$ and $\nu(\mathbf{r})$ are uncorrelated and since $E\{\nu(\mathbf{r})\} = 0$, we may write:

$$E\{f(\mathbf{r})\nu(\mathbf{r})\} = E\{f(\mathbf{r})\}E\{\nu(\mathbf{r})\} = 0 \tag{6.42}$$

To create the cross-spectral density between the original and the degraded image, we multiply both sides of equation (6.38) with $f(\mathbf{r} - \mathbf{s})$ and take the expectation value:

$$\underbrace{E\{g(\mathbf{r})f(\mathbf{r} - \mathbf{s})\}}_{R_{gf}(\mathbf{s})} = \int_{-\infty}^{+\infty} \int_{-\infty}^{+\infty} h(\mathbf{r} - \mathbf{r}') \underbrace{E\{f(\mathbf{r}')f(\mathbf{r} - \mathbf{s})\}}_{R_{ff}(\mathbf{r}'-\mathbf{r}+\mathbf{s})} d\mathbf{r}' + \underbrace{E\{f(\mathbf{r} - \mathbf{s})\nu(\mathbf{r})\}}_{0 \text{ from (6.42)}}$$

Therefore:

$$R_{gf}(\mathbf{s}) = \int_{-\infty}^{+\infty} \int_{-\infty}^{+\infty} h(\mathbf{r} - \mathbf{r}')R_{ff}(\mathbf{r}' - \mathbf{r} + \mathbf{s})d\mathbf{r}'$$

In terms of Fourier transforms this equation can be written (see Box **B6.4** and Example 6.12) as:

$$S_{gf}(u, v) = \hat{H}^*(u, v)S_{ff}(u, v) \tag{6.43}$$

From equation (6.38) we can also show that (see Box **B6.3**):

$$S_{gg}(u, v) = S_{ff}(u, v)|\hat{H}(u, v)|^2 + S_{\nu\nu}(u, v) \tag{6.44}$$

If we substitute equations (6.43) and (6.44) into (6.41), we obtain:

$$\hat{M}(u, v) = \frac{\hat{H}^*(u, v)S_{ff}(u, v)}{S_{ff}(u, v)|\hat{H}(u, v)|^2 + S_{\nu\nu}(u, v)} \tag{6.45}$$

or

$$\hat{M}(u, v) = \frac{\hat{H}^*(u, v)}{|\hat{H}(u, v)|^2 + \frac{S_{\nu\nu}(u,v)}{S_{ff}(u,v)}} \tag{6.46}$$

If we multiply numerator and denominator by $\hat{H}(u, v)$ we obtain:

$$\hat{M}(u, v) = \frac{1}{\hat{H}(u, v)} \frac{|\hat{H}(u, v)|^2}{|\hat{H}(u, v)|^2 + \frac{S_{\nu\nu}(u,v)}{S_{ff}(u,v)}} \tag{6.47}$$

This equation gives the Fourier transform of the *Wiener filter* for image restoration.

How can we possibly use equation (6.47) if we know nothing about the statistical properties of the unknown image $f(\mathbf{r})$?

If we do not know anything about the statistical properties of the image we want to restore, i.e. we do not know $S_{ff}(u, v)$, we may replace the term $\frac{S_{\nu\nu}(u,v)}{S_{ff}(u,v)}$ in equation (6.47) by a constant Γ and experiment with various values of Γ.

This is clearly rather an oversimplification, as the ratio $\frac{S_{\nu\nu}(u,v)}{S_{ff}(u,v)}$ is a function of (u, v) and not a constant.

What is the relationship of the Wiener filter (6.47) and the inverse filter of equation (6.25)?

In the absence of noise, $S_{\nu\nu}(u,v) = 0$ and the Wiener filter becomes the inverse transfer function filter of equation (6.25). So the linear least square error approach simply determines a correction factor with which the inverse transfer function of the degradation process has to be multiplied before it is used as a filter, so that the effect of noise is taken care of.

Assuming that we know the statistical properties of the unknown image $f(\mathbf{r})$, how can we determine the statistical properties of the noise expressed by $S_{\nu\nu}(\mathbf{r})$?

We usually make the assumption that the noise is white; i.e. that

$$S_{\nu\nu}(u,v) = \text{constant} = S_{\nu\nu}(0,0) = \int_{-\infty}^{+\infty}\int_{-\infty}^{+\infty} R_{\nu\nu}(x,y)dxdy \qquad (6.48)$$

If the noise is assumed to be ergodic, we can obtain $R_{\nu\nu}(x,y)$ from a single pure noise image (the recorded image $g(x,y)$ when there is no original image, i.e. when $f(x,y) = 0$).

B6.1: Show that if $m(\mathbf{r}-\mathbf{r}')$ satisfies

$$E\left\{\left[f(\mathbf{r}) - \int_{-\infty}^{+\infty}\int_{-\infty}^{+\infty} m(\mathbf{r}-\mathbf{r}')g(\mathbf{r}')d\mathbf{r}'\right]g(\mathbf{s})\right\} = 0 \qquad (6.49)$$

then it minimizes the error defined by equation (6.37).

If we substitute equation (6.40) into equation (6.37) we have:

$$e^2 = E\left\{\left[f(\mathbf{r}) - \int_{-\infty}^{+\infty}\int_{-\infty}^{+\infty} m(\mathbf{r}-\mathbf{r}')g(\mathbf{r}')d\mathbf{r}'\right]^2\right\} \qquad (6.50)$$

Consider now another function $m'(\mathbf{r})$ which does not satisfy (6.49). We shall show that $m'(\mathbf{r})$ when used for the restoration of the image, will produce an estimate $\hat{f}'(\mathbf{r})$ with error e'^2, greater than the error of the estimate obtained by $m(\mathbf{r})$ which satisfies (6.49):

$$e'^2 \equiv E\left\{\left[f(\mathbf{r}) - \int_{-\infty}^{+\infty}\int_{-\infty}^{+\infty} m'(\mathbf{r}-\mathbf{r}')g(\mathbf{r}')d\mathbf{r}'\right]^2\right\}$$

Inside the integrand we add to and subtract from $m(\mathbf{r} - \mathbf{r}')$ function $m'(\mathbf{r} - \mathbf{r}')$. We split the integral into two parts and then expand the square:

$$e'^2 = E\left\{\left[f(\mathbf{r}) - \int_{-\infty}^{+\infty}\int_{-\infty}^{+\infty}[m'(\mathbf{r} - \mathbf{r}') + m(\mathbf{r} - \mathbf{r}') - m'(\mathbf{r} - \mathbf{r}')]g(\mathbf{r}')d\mathbf{r}'\right]^2\right\}$$

$$= E\left\{\left[\left(f(\mathbf{r}) - \int_{-\infty}^{+\infty}\int_{-\infty}^{+\infty}m(\mathbf{r} - \mathbf{r}')g(\mathbf{r}')d\mathbf{r}'\right)\right.\right.$$
$$\left.\left. + \left(\int_{-\infty}^{+\infty}\int_{-\infty}^{+\infty}[m(\mathbf{r} - \mathbf{r}') - m'(\mathbf{r} - \mathbf{r}')]g(\mathbf{r}')d\mathbf{r}'\right)\right]^2\right\}$$

$$= \underbrace{E\left\{\left(f(\mathbf{r}) - \int_{-\infty}^{+\infty}\int_{-\infty}^{+\infty}m(\mathbf{r} - \mathbf{r}')g(\mathbf{r}')d\mathbf{r}'\right)^2\right\}}_{e^2}$$

$$+ \underbrace{E\left\{\left(\int_{-\infty}^{+\infty}\int_{-\infty}^{+\infty}[m(\mathbf{r} - \mathbf{r}') - m'(\mathbf{r} - \mathbf{r}')]g(\mathbf{r}')d\mathbf{r}'\right)^2\right\}}_{\text{a non-negative number}}$$

$$+ 2E\left\{\left(f(\mathbf{r}) - \int_{-\infty}^{+\infty}\int_{-\infty}^{+\infty}m(\mathbf{r} - \mathbf{r}')g(\mathbf{r}')d\mathbf{r}'\right)\right.$$
$$\left.\underbrace{\int_{-\infty}^{+\infty}\int_{-\infty}^{+\infty}[m(\mathbf{r} - \mathbf{r}') - m'(\mathbf{r} - \mathbf{r}')]g(\mathbf{r}')d\mathbf{r}'}_{\text{rename } \mathbf{r}' \to \mathbf{s}}\right\} \qquad (6.51)$$

The expectation value of the first term is e^2 and clearly the expectation value of the second term is a non-negative number. In the last term, in the second factor, change the dummy variable of integration from $\mathbf{r}'$ to $\mathbf{s}$. The last term on the right hand side of (6.51) can then be written as:

$$2E\left\{\left(f(\mathbf{r}) - \int_{-\infty}^{+\infty}\int_{-\infty}^{+\infty}m(\mathbf{r}-\mathbf{r}')g(\mathbf{r}')d\mathbf{r}'\right)\int_{-\infty}^{+\infty}\int_{-\infty}^{+\infty}[m(\mathbf{r}-\mathbf{s})-m'(\mathbf{r}-\mathbf{s})]g(\mathbf{s})d\mathbf{s}\right\}$$

The first factor in the above expression does not depend on $\mathbf{s}$ and thus it can be put inside the $\int\int$ sign:

$$2E\left\{\int_{-\infty}^{+\infty}\int_{-\infty}^{+\infty}\left(f(\mathbf{r}) - \int_{-\infty}^{+\infty}\int_{-\infty}^{+\infty}m(\mathbf{r}-\mathbf{r}')g(\mathbf{r}')d\mathbf{r}'\right)[m(\mathbf{r}-\mathbf{s})-m'(\mathbf{r}-\mathbf{s})]g(\mathbf{s})d\mathbf{s}\right\}$$

The difference $[m(\mathbf{r} - \mathbf{s}) - m'(\mathbf{r} - \mathbf{s})]$ is not a random field but the difference of two specific functions. If we change the order of integrating and taking the expectation value, the expectation is not going to affect this factor so this term will become:

$$2 \int_{-\infty}^{+\infty}\int_{-\infty}^{+\infty} E\left\{ \left[f(\mathbf{r}) - \int_{-\infty}^{+\infty}\int_{-\infty}^{+\infty} m(\mathbf{r}-\mathbf{r}')g(\mathbf{r}')d\mathbf{r}' \right] g(\mathbf{s}) \right\} [m(\mathbf{r}-\mathbf{s}) - m'(\mathbf{r}-\mathbf{s})]d\mathbf{s}$$

However, the expectation value in the above term is 0 according to (6.49), so from (6.51) we get that:

$$e'^2 = e^2 + \text{ a non-negative term}$$

We conclude that the error of the restoration created with the $m'(\mathbf{r})$ function is greater than or equal to the error of the restoration created with $m(\mathbf{r})$. So $m(\mathbf{r})$ that satisfies equation (6.49) minimizes the error defined by equation (6.37).

Example 6.12 (B)

If $\hat{F}(u,v)$, $\hat{H}(u,v)$ and $\hat{G}(u,v)$ are the Fourier transforms of $f(\mathbf{r})$, $h(\mathbf{r})$ and $g(\mathbf{r})$ respectively, and

$$g(\mathbf{r}) = \int_{-\infty}^{\infty}\int_{-\infty}^{\infty} h(\mathbf{t}-\mathbf{r})f(\mathbf{t})d\mathbf{t} \tag{6.52}$$

show that

$$\hat{G}(u,v) = \hat{H}^*(u,v)\hat{F}(u,v) \tag{6.53}$$

where $\hat{H}^*(u,v)$ is the complex conjugate of $\hat{H}(u,v)$.

Assume that $\mathbf{r} = (x,y)$ and $\mathbf{t} = (\tilde{x}, \tilde{y})$. The Fourier transforms of the three functions are:

$$\hat{G}(u,v) = \int_{-\infty}^{\infty}\int_{-\infty}^{\infty} g(x,y)e^{-j(ux+vy)}dxdy \tag{6.54}$$

$$\hat{F}(u,v) = \int_{-\infty}^{\infty}\int_{-\infty}^{\infty} f(x,y)e^{-j(ux+vy)}dxdy \tag{6.55}$$

$$\hat{H}(u,v) = \int_{-\infty}^{\infty}\int_{-\infty}^{\infty} h(x,y)e^{-j(ux+vy)}dxdy \tag{6.56}$$

The complex conjugate of $\hat{H}(u,v)$ is

$$\hat{H}^*(u,v) = \int_{-\infty}^{\infty}\int_{-\infty}^{\infty} h(x,y)e^{j(ux+vy)}dxdy \tag{6.57}$$

Let us substitute $g(x, y)$ from (6.52) into the right hand side of (6.54):

$$\hat{G}(u, v) = \int_{-\infty}^{\infty} \int_{-\infty}^{\infty} \int_{-\infty}^{\infty} \int_{-\infty}^{\infty} h(\tilde{x} - x, \tilde{y} - y) f(\tilde{x}, \tilde{y}) d\tilde{x} d\tilde{y} e^{-j(ux+vy)} dx dy$$

We define new variables for integration $s_1 \equiv \tilde{x} - x$ and $s_2 \equiv \tilde{y} - y$ to replace integration over x and y. Since $dx = -ds_1$ and $dy = -ds_2$, $dxdy = ds_1 ds_2$. Also, as the limits of both s_1 and s_2 are from $+\infty$ to $-\infty$, we can change their order without worrying about a change of sign:

$$\hat{G}(u, v) = \int_{-\infty}^{\infty} \int_{-\infty}^{\infty} \int_{-\infty}^{\infty} \int_{-\infty}^{\infty} h(s_1, s_2) f(\tilde{x}, \tilde{y}) d\tilde{x} d\tilde{y} \; e^{-j(u(\tilde{x}-s_1)+v(\tilde{y}-s_2))} ds_1 ds_2$$

The two double integrals are separable:

$$\hat{G}(u, v) = \int_{-\infty}^{\infty} \int_{-\infty}^{\infty} h(s_1, s_2) e^{j(us_1+vs_2)} ds_1 ds_2 \int_{-\infty}^{\infty} \int_{-\infty}^{\infty} f(\tilde{x}, \tilde{y}) e^{-j(u\tilde{x}+v\tilde{y})} d\tilde{x} d\tilde{y}$$

On the right hand side of this equation we recognize the product of $\hat{F}(u, v)$ and $\hat{H}^*(u, v)$ from equations (6.55) and (6.57) respectively. Therefore equation (6.53) is proven.

B6.2: Show that the Fourier transform of the spatial autocorrelation function of a random field $f(x, y)$ is equal to the spectral density $|\hat{F}(u, v)|^2$ of the field (Wiener–Khinchine theorem).

The spatial autocorrelation function of $f(x, y)$ is defined as

$$R_{ff}(\tilde{x}, \tilde{y}) = \int_{-\infty}^{\infty} \int_{-\infty}^{\infty} f(x + \tilde{x}, y + \tilde{y}) f(x, y) dx dy \qquad (6.58)$$

We multiply both sides of equation (6.58) with the kernel of the Fourier transform and integrate to obtain the Fourier transform of $R_{ff}(\tilde{x}, \tilde{y})$, $\hat{R}_{ff}(u, v)$:

$$\hat{R}_{ff}(u, v) = \int_{-\infty}^{\infty} \int_{-\infty}^{\infty} R_{ff}(\tilde{x}, \tilde{y}) e^{-j(\tilde{x}u+\tilde{y}v)} d\tilde{x} d\tilde{y}$$

$$= \int_{-\infty}^{\infty} \int_{-\infty}^{\infty} \int_{-\infty}^{\infty} \int_{-\infty}^{\infty} f(x+\tilde{x}, y+\tilde{y}) f(x, y) e^{-j(\tilde{x}u+\tilde{y}v)} dx dy d\tilde{x} d\tilde{y} \quad (6.59)$$

We define new variables of integration $s_1 \equiv x + \tilde{x}$ and $s_2 \equiv y + \tilde{y}$ to replace the integral over $\tilde{x}$ and $\tilde{y}$. We have $\tilde{x} = s_1 - x$, $\tilde{y} = s_2 - y$, $d\tilde{x}d\tilde{y} = ds_1 ds_2$ and no change in the limits of integration:

$$\hat{R}_{ff}(u,v) = \int_{-\infty}^{\infty} \int_{-\infty}^{\infty} \int_{-\infty}^{\infty} \int_{-\infty}^{\infty} f(s_1, s_2) f(x,y) e^{-j((s_1-x)u+(s_2-y)v)} dx\, dy\, ds_1\, ds_2$$

The two double integrals on the right hand side are separable, so we may write:

$$\hat{R}_{ff}(u,v) = \int_{-\infty}^{\infty} \int_{-\infty}^{\infty} f(s_1, s_2) e^{-j(s_1 u + s_2 v)} ds_1\, ds_2 \int_{-\infty}^{\infty} \int_{-\infty}^{\infty} f(x,y) e^{j(xu+yv)} dx\, dy$$

We recognize the first of the double integrals on the right hand side of this equation to be the Fourier transform $\hat{F}(u,v)$ of $f(s_1, s_2)$ and the second double integral its complex conjugate $\hat{F}^*(u,v)$. Therefore:

$$\hat{R}_{ff}(u,v) \quad = \quad \hat{F}(u,v)\hat{F}^*(u,v) = |\hat{F}(u,v)|^2$$

B6.3: Derive an expression for the Fourier transform of function $m(\mathbf{r})$ that minimizes the error defined by equation (6.37).

Equation (6.49) which is satisfied by $m(\mathbf{r})$ that minimizes (6.37) can be written as:

$$E\{f(\mathbf{r})g(\mathbf{s})\} - E\left\{ \int_{-\infty}^{+\infty} \int_{-\infty}^{+\infty} m(\mathbf{r}-\mathbf{r}')g(\mathbf{r}')g(\mathbf{s})d\mathbf{r}' \right\} = 0$$

where $g(\mathbf{s})$ has gone inside the integral sign because it does not depend on $\mathbf{r}'$. The expectation operator applied to the second term operates really only on the random functions $g(\mathbf{r}')$ and $g(\mathbf{s})$. Therefore, we can write:

$$\underbrace{E\{f(\mathbf{r})g(\mathbf{s})\}}_{R_{fg}(\mathbf{r},\mathbf{s})} = \int_{-\infty}^{+\infty} \int_{-\infty}^{+\infty} m(\mathbf{r}-\mathbf{r}') \underbrace{E\{g(\mathbf{r}')g(\mathbf{s})\}}_{R_{gg}(\mathbf{r}',\mathbf{s})} d\mathbf{r}' \tag{6.60}$$

In this expression we recognize the definitions of the autocorrelation and cross-correlation functions of the random fields, so we can write:

$$\int_{-\infty}^{+\infty} \int_{-\infty}^{+\infty} m(\mathbf{r}-\mathbf{r}')R_{gg}(\mathbf{r}',\mathbf{s})d\mathbf{r}' = R_{fg}(\mathbf{r},\mathbf{s}) \tag{6.61}$$

We have seen that for homogeneous random fields, the correlation function can be written as a function of the difference of its two arguments (see Example 3.7). So:

$$\int_{-\infty}^{+\infty} \int_{-\infty}^{+\infty} m(\mathbf{r} - \mathbf{r}') R_{gg}(\mathbf{r}' - \mathbf{s}) d\mathbf{r}' = R_{fg}(\mathbf{r} - \mathbf{s}) \tag{6.62}$$

We introduce some new variables: $\mathbf{r}' - \mathbf{s} \equiv \mathbf{t}$ and $\mathbf{r} - \mathbf{s} \equiv \boldsymbol{\tau}$. Therefore $d\mathbf{r}' = d\mathbf{t}$ and $\mathbf{r} - \mathbf{r}' = \boldsymbol{\tau} - \mathbf{t}$. Then:

$$\int_{-\infty}^{+\infty} \int_{-\infty}^{+\infty} m(\boldsymbol{\tau} - \mathbf{t}) R_{gg}(\mathbf{t}) d\mathbf{t} = R_{fg}(\boldsymbol{\tau}) \tag{6.63}$$

This is a convolution between the autocorrelation function of the degraded image and the sought filter. According to the convolution theorem the effect is equivalent to the multiplication of the Fourier transforms of the two functions:

$$\hat{M}(u,v) S_{gg}(u,v) = S_{fg}(u,v) \tag{6.64}$$

where S_{gg} and S_{fg} are the spectral density of the degraded image and the cross-spectral density of the degraded and undegraded images respectively; i.e. the Fourier transforms of the autocorrelation function of g and cross-correlation of f and g functions respectively. Therefore:

$$\hat{M}(u,v) = \frac{S_{fg}(u,v)}{S_{gg}(u,v)} \tag{6.65}$$

The Fourier transform of the optimal restoration filter which minimizes the mean square error between the real image and the reconstructed one, is equal to the ratio of the cross-spectral density of the degraded image and the true image, over the spectral density of the degraded image.

B6.4: If $S_{gg}(u,v)$, $S_{ff}(u,v)$ and $S_{\nu\nu}(u,v)$ are the spectral densities of the homogeneous random fields $g(x,y)$, $f(x,y)$ and $\nu(x,y)$ respectively, $\hat{H}(u,v)$ is the Fourier transform of $h(x,y)$ and

$$g(x,y) \quad = \quad \int_{-\infty}^{\infty} \int_{-\infty}^{\infty} h(x - \tilde{x}, y - \tilde{y}) f(\tilde{x}, \tilde{y}) d\tilde{x} d\tilde{y} + \nu(x,y) \tag{6.66}$$

show that

$$S_{gg}(u,v) = S_{ff}(u,v) |H(u,v)|^2 + S_{\nu\nu}(u,v) \tag{6.67}$$

with the additional assumption that $f(x, y)$ and $\nu(x, y)$ are uncorrelated and at least one of the two has zero mean.

If we multiply both sides of equation (6.66) with $g(x + s_1, y + s_2)$ and take the ensemble average over all versions of random field $g(x, y)$, we have:

$$E\{g(x, y)g(x+s_1, y+s_2)\} = E\left\{g(x+s_1, y+s_2)\int_{-\infty}^{\infty}\int_{-\infty}^{\infty} h(x-\tilde{x}, y-\tilde{y})f(\tilde{x}, \tilde{y})d\tilde{x}d\tilde{y}\right\}$$
$$+E\{g(x + s_1, y + s_2)\nu(x, y)\}$$

Since $g(x, y)$ is a homogeneous random field, we recognize on the left hand side the autocorrelation function of g with shifting arguments s_1, s_2, $R_{gg}(s_1, s_2)$. The noise random field $\nu(x, y)$ is also homogeneous, so the last term on the right hand side is the cross-correlation $R_{g\nu}(s_1, s_2)$ between random fields g and ν. Further, $g(x + s_1, y + s_2)$ does not depend on the variables of integration $\tilde{x}$ and $\tilde{y}$, so it may go inside the integral in the first term of the right hand side:

$$R_{gg}(s_1, s_2) = E\left\{\int_{-\infty}^{\infty}\int_{-\infty}^{\infty} h(x-\tilde{x}, y-\tilde{y})f(\tilde{x}, \tilde{y})g(x+s_1, y+s_2)d\tilde{x}d\tilde{y}\right\} + R_{g\nu}(s_1, s_2)$$

Taking the expectation value and integrating are two linear operations that can be interchanged. The expectation operator operates only on random fields f and g, while it leaves unaffected function h. We can write therefore:

$$R_{gg}(s_1, s_2) = \int_{-\infty}^{\infty}\int_{-\infty}^{\infty} h(x-\tilde{x}, y-\tilde{y})E\{f(\tilde{x}, \tilde{y})g(x+s_1, y+s_2)\}\, d\tilde{x}d\tilde{y} + R_{g\nu}(s_1, s_2)$$

We recognize inside the integral the cross correlation R_{gf} between fields f and g calculated for shifting values $x + s_1 - \tilde{x}$ and $y + s_2 - \tilde{y}$:

$$R_{gg}(s_1, s_2) = \int_{-\infty}^{\infty}\int_{-\infty}^{\infty} h(x-\tilde{x}, y-\tilde{y})R_{gf}(x-\tilde{x}+s_1, y-\tilde{y}+s_2)d\tilde{x}d\tilde{y} + R_{g\nu}(s_1, s_2) \quad (6.68)$$

We may define new variables of integration: $x - \tilde{x} \equiv \alpha$, $y - \tilde{y} \equiv \beta$. Then $d\tilde{x}d\tilde{y} = d\alpha d\beta$, and the change of sign of the two sets of limits of integration cancel each other out:

$$R_{gg}(s_1, s_2) = \int_{-\infty}^{\infty}\int_{-\infty}^{\infty} h(\alpha, \beta)R_{gf}(\alpha + s_1, \beta + s_2)d\alpha d\beta + R_{g\nu}(s_1, s_2)$$

We can change variables of integration again, to $w \equiv \alpha + s_1$, $z \equiv \beta + s_2$. Then $\alpha = w - s_1$, $\beta = z - s_2$, $d\alpha d\beta = dwdz$ and the limits of integration are not affected:

$$R_{gg}(s_1, s_2) = \int_{-\infty}^{\infty}\int_{-\infty}^{\infty} h(w - s_1, z - s_2)R_{gf}(w, z)dwdz + R_{g\nu}(s_1, s_2) \quad (6.69)$$

If we take the Fourier transform of both sides of this expression, and make use of the result of Example 6.12, we can write:

$$\hat{R}_{gg}(u, v) = \hat{H}^*(u, v)\hat{R}_{gf}(u, v) + \hat{R}_{g\nu}(u, v)$$

where the quantities with the hat (^) signify the Fourier transforms of the corresponding quantities that appear in equation (6.69).

If the fields were ergodic, the ensemble auto- and cross-correlation functions we calculated here would have been the same as the spatial auto- and cross- correlation functions. Then the Fourier transforms of these functions would have been the spectral or cross-spectral densities of the corresponding random fields (see Box **B6.2**). In general, however, the fields are not ergodic. Then, it is **postulated** that the Fourier transforms of the auto- and cross-correlation functions are the spectral and cross-spectral densities respectively, of the corresponding random fields.

Thus, it must be noted that in the development of the Wiener filter, the ergodicity assumption is tacitly made. With this in mind, we can write:

$$S_{gg}(u, v) = \hat{H}^*(u, v)S_{gf}(u, v) + S_{g\nu}(u, v) \tag{6.70}$$

We notice that we need to calculate the cross-spectral density between random fields f and g. We start again from equation (6.66), but now we multiply both sides with $f(x - s_1, y - s_2)$ and take the expectation value. The reason we multiply with $f(x - s_1, y - s_2)$ and not with $f(x + s_1, y + s_2)$ is because we formed the shifting arguments of R_{gf} in (6.68) by subtracting the arguments of f from the arguments of g. We must follow the same convention again. Proceeding as before, we have:

$$\begin{aligned} R_{gf}(s_1, s_2) &= \int_{-\infty}^{\infty} \int_{-\infty}^{\infty} h(x - \tilde{x}, y - \tilde{y}) E\left\{ f(\tilde{x}, \tilde{y}) f(x - s_1, y - s_2) \right\} d\tilde{x} d\tilde{y} \\ &\quad + E\left\{ f(x - s_1, y - s_2)\nu(x, y) \right\} \end{aligned} \tag{6.71}$$

We remember that random fields f and ν are uncorrelated and at least one of them has 0 mean. Then:

$$E\left\{ f(x - s_1, y - s_2)\nu(x, y) \right\} = E\left\{ f(x - s_1, y - s_2) \right\} E\left\{ \nu(x, y) \right\} = 0 \tag{6.72}$$

Inside the integral on the right hand side of equation (6.71) we recognize the autocorrelation function of random field f computed for shifting argument $(\tilde{x} - x + s_1, \tilde{y} - y + s_2)$. The reason we subtract the arguments of $f(x - s_1, y - s_2)$ from the arguments of $f(\tilde{x}, \tilde{y})$, and not the other way round, is because on the left hand side we subtracted the arguments of the "new" function from the arguments of the existing one (i.e. the arguments of $f(x - s_1, y - s_2)$ from the arguments of $g(x, y)$) to from R_{gf}. So we have:

$$R_{gf}(s_1, s_2) = \int_{-\infty}^{\infty} \int_{-\infty}^{\infty} h(x - \tilde{x}, y - \tilde{y}) R_{ff}(\tilde{x} - x + s_1, \tilde{y} - y + s_2) d\tilde{x} d\tilde{y}$$

We define new variables of integration $\alpha \equiv x - \tilde{x}$, $\beta \equiv y - \tilde{y}$:

$$R_{gf}(s_1, s_2) = \int_{-\infty}^{\infty} \int_{-\infty}^{\infty} h(\alpha, \beta) R_{ff}(s_1 - \alpha, s_2 - \beta) d\alpha d\beta$$

The above equation is a straightforward convolution, and in terms of Fourier transforms it can be written as:

$$\hat{R}_{gf}(u, v) = \hat{H}(u, v) \hat{R}_{ff}(u, v)$$

Invoking again the postulate of the equivalence between Fourier transforms of correlation functions and spectral densities, we can write:

$$\hat{S}_{gf}(u, v) = \hat{H}(u, v) \hat{S}_{ff}(u, v) \tag{6.73}$$

In equation (6.68) we also need the cross-spectral density between random fields g and ν. We start again from equation (6.66), multiply both sides with $\nu(x - s_1, y - s_2)$ and take the expectation value:

$$\hat{R}_{g\nu}(s_1, s_2) = \int_{-\infty}^{\infty} \int_{-\infty}^{\infty} h(x - \tilde{x}, y - \tilde{y}) E\left\{f(\tilde{x}, \tilde{y})\nu(x - s_1, y - s_2)\right\} d\tilde{x} d\tilde{y} + R_{\nu\nu}(s_1, s_2)$$

The integral term vanishes because of equation (6.72). In terms of Fourier transforms we can therefore write:

$$\hat{S}_{g\nu}(u, v) = \hat{S}_{\nu\nu}(u, v) \tag{6.74}$$

If we substitute from equations (6.73) and (6.74) into equation (6.70) we obtain the equation we set out to prove, i.e. equation (6.67).

Example 6.13

Demonstrate how to apply Wiener filtering in practice by restoring a motion blurred image.

Let us consider again the blurred image of Figure 6.7a. From equation (6.33) we have:

$$|\hat{H}(m, n)|^2 = \frac{1}{i_T^2 \sin^2 \frac{\pi m}{N}} \sin^2 \frac{i_T \pi m}{N}$$

The Wiener filter as given by equation (6.46), with the ratio of the spectral densities in the denominator replaced by a constant Γ *is:*

$$\hat{M}(m,n) = \frac{\frac{1}{i_T}\frac{\sin\frac{\pi m}{N}i_T}{\sin\frac{\pi m}{N}}e^{j\frac{\pi m(i_T-1)}{N}}}{\frac{1}{i_T^2}\frac{\sin^2\frac{\pi m}{N}i_T}{\sin^2\frac{\pi m}{N}}+\Gamma}$$

Or

$$\hat{M}(m,n) = \frac{i_T\sin\frac{\pi m}{N}\sin\frac{\pi m}{N}i_T}{\sin^2\frac{\pi m}{N}i_T+\Gamma i_T^2\sin^2\frac{\pi m}{N}}e^{j\frac{\pi m(i_T-1)}{N}}$$

We must be careful for the case $m = 0$, *when we have*

$$\hat{M}(0,n) = \frac{1}{1+\Gamma} \quad \text{for} \quad 0 \le n \le N-1$$

If we multiply with this function the Fourier transform of the blurred image as defined by equation (6.34), we obtain:

$$\hat{F}(m,n) = \frac{i_T\sin\frac{\pi m}{N}\sin\frac{i_T\pi m}{N}\sqrt{G_1^2(m,n)+G_2^2(m,n)}e^{j\left(\phi(m,n)+\frac{(i_T-1)\pi m}{N}\right)}}{\sin^2\frac{i_T\pi m}{N}+\Gamma i_T^2\sin^2\frac{\pi m}{N}}$$

For the case $m = 0$, *we have:*

$$\hat{F}(0,n) = \frac{\sqrt{G_1^2(0,n)+G_2^2(0,n)}}{1+\Gamma}e^{j\phi(0,n)} \quad \text{for} \quad 0 \le n \le N-1$$

The real and the imaginary parts of $\hat{F}(m,n)$ *are given by:*

$$F_1(m,n) = \frac{i_T\sin\frac{\pi m}{N}\sin\frac{i_T\pi m}{N}\sqrt{G_1^2(m,n)+G_2^2(m,n)}}{\sin^2\frac{i_T\pi m}{N}+\Gamma i_T^2\sin^2\frac{\pi m}{N}}\cos\left(\phi(m,n)+\frac{(i_T-1)\pi m}{N}\right)$$

$$F_2(m,n) = \frac{i_T\sin\frac{\pi m}{N}\sin\frac{i_T\pi m}{N}\sqrt{G_1^2(m,n)+G_2^2(m,n)}}{\sin^2\frac{i_T\pi m}{N}+\Gamma i_T^2\sin^2\frac{\pi m}{N}}\sin\left(\phi(m,n)+\frac{(i_T-1)\pi m}{N}\right)$$

If we use formulae $\cos(a+b) = \cos a\cos b - \sin a\sin b$ *and* $\sin(a+b) = \sin a\cos b + \cos a\sin b$ *and substitute* $\cos\phi(m,n)$ *and* $\sin\phi(m,n)$ *from equations (6.35) we obtain:*

$$F_1(m,n) = \frac{i_T\sin\frac{\pi m}{N}\sin\frac{i_T\pi m}{N}}{\sin^2\frac{i_T\pi m}{N}+\Gamma i_T^2\sin^2\frac{\pi m}{N}}\left[G_1(m,n)\cos\frac{(i_T-1)\pi m}{N}-G_2(m,n)\sin\frac{(i_T-1)\pi m}{N}\right]$$

$$F_2(m,n) = \frac{i_T\sin\frac{\pi m}{N}\sin\frac{i_T\pi m}{N}}{\sin^2\frac{i_T\pi m}{N}+\Gamma i_T^2\sin^2\frac{\pi m}{N}}\left[G_1(m,n)\sin\frac{(i_T-1)\pi m}{N}+G_2(m,n)\cos\frac{(i_T-1)\pi m}{N}\right]$$

For $m = 0$ we must remember to use:

$$F_1(0,n) = \frac{G_1(0,n)}{1+\Gamma} \quad for \quad 0 \le n \le N-1$$

$$F_2(0,n) = \frac{G_2(0,n)}{1+\Gamma} \quad for \quad 0 \le n \le N-1$$

If we Fourier transform back using functions $F_1(m,n)$ an $F_2(m,n)$ we obtain the restored image shown in Figure 6.9a. This image should be compared with images 6.7e and 6.7f, which are obtained by inverse filtering.

The restoration of the noisy images of Figures 6.8a and 6.8c by Wiener filtering is shown in Figures 6.9b and 6.9c. These images should be compared with Figures 6.8g and 6.8i respectively. In all cases Wiener filtering produces superior results.

(a) Wiener filtering with $\Gamma = 0.01$ of image 6.7b (realistic blurring)

(b) Wiener filtering with $\Gamma = 0.5$ of image 6.8a (realistic blurring with additive Gaussian noise ($\sigma = 10$))

(c) Wiener filtering with $\Gamma = 1$ of image 6.8c (realistic blurring with additive Gaussian noise ($\sigma = 20$))

Figure 6.9: Image restoration with Wiener filtering.

If the degradation process is assumed linear, why don't we solve a system of linear equations to reverse its effect instead of invoking the convolution theorem?

Indeed, the system of linear equations we must invert is given in matrix form by equation (6.10), $\mathbf{g} = H\mathbf{f}$. However, we saw that it is more realistic to include in this equation an extra term representing noise (see equation (6.26)):

$$\mathbf{g} = H\mathbf{f} + \boldsymbol{\nu} \tag{6.75}$$

where $\boldsymbol{\nu}$ is the noise field written in vector form.

Since we assumed that we have some knowledge about the point spread function of the degradation process, matrix H is assumed to be known. Then:

$$\mathbf{f} = H^{-1}\mathbf{g} - H^{-1}\boldsymbol{\nu} \tag{6.76}$$

where H is an $N^2 \times N^2$ matrix and $\mathbf{f}$, $\mathbf{g}$ and $\boldsymbol{\nu}$ are $N^2 \times 1$ vectors.

Equation (6.76) seems pretty straightforward, why bother with any other approach?

There are two major problems with equation (6.76):

1. It is extremely sensitive to noise. It has been shown that one needs impossibly low levels of noise for the method to work.

2. The solution of equation (6.76) requires the inversion of an $N^2 \times N^2$ square matrix with N typically being 500, which is a formidable task even for modern computers.

Example 6.14

Demonstrate the sensitivity to noise of the inverse matrix restoration.

Let us consider the signal given by:

$$f(x) = 25 \sin \frac{2\pi x}{30} \quad for \quad x = 0, 1, \ldots, 29$$

Suppose that this signal is blurred by a function that averages every three samples after multiplying them with some weights. We can express this by saying that the discrete signal is multiplied with matrix H given below, to produce a blurred signal $g(x)$.

$$H = \begin{pmatrix} 0.4 & 0.3 & 0 & 0 & \ldots & 0 & 0 & 0 & 0.3 \\ 0.3 & 0.4 & 0.3 & 0 & \ldots & 0 & 0 & 0 & 0 \\ 0 & 0.3 & 0.4 & 0.3 & \ldots & 0 & 0 & 0 & 0 \\ \vdots & \vdots & \vdots & \vdots & \vdots & \vdots & \vdots & \vdots & \vdots \\ 0.3 & 0 & 0 & 0 & \ldots & 0 & 0 & 0.4 & 0.3 \end{pmatrix}$$

To introduce some noise we round the elements of $g(x)$ to the nearest integer. To recover the original signal we multiply the blurred signal $g(x)$ with the inverse of matrix H. The original and the restored signal are shown in Figure 6.10.

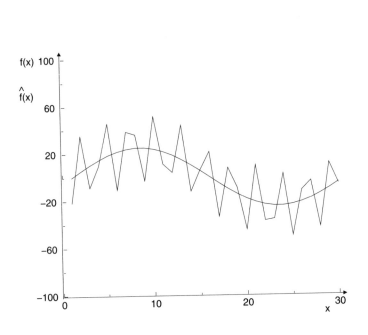

Figure 6.10: An original and a restored signal by direct matrix inversion. The noise in the signal was only rounding error.

Is there any way by which matrix H can be inverted?

Yes, matrix H can easily be inverted because it is a *block circulant matrix*.

When is a matrix block circulant?

A matrix H is *block circulant* if it has the following structure:

$$
H \;=\;
\begin{pmatrix}
H_0 & H_{M-1} & H_{M-2} & \cdots & H_1 \\
H_1 & H_0 & H_{M-1} & \cdots & H_2 \\
H_2 & H_1 & H_0 & \cdots & H_3 \\
\vdots & \vdots & \vdots & & \vdots \\
H_{M-1} & H_{M-2} & H_{M-3} & & H_0
\end{pmatrix}
\tag{6.77}
$$

where $H_0, H_1, \ldots, H_{M-1}$ are partitions of matrix H, and they are themselves *circulant matrices*.

When is a matrix circulant?

A matrix D is circulant if it has the following structure:

$$D = \begin{pmatrix} d(0) & d(M-1) & d(M-2) & \ldots & d(1) \\ d(1) & d(0) & d(M-1) & \ldots & d(2) \\ d(2) & d(1) & d(0) & \ldots & d(3) \\ \vdots & \vdots & \vdots & & \vdots \\ d(M-1) & d(M-2) & d(M-3) & \ldots & d(0) \end{pmatrix} \qquad (6.78)$$

In such a matrix, each column can be obtained from the previous one by shifting all elements one place down and putting the last element at the top.

Why can block circulant matrices be inverted easily?

Circulant and block circulant matrices can easily be inverted because we can find easily their eigenvalues and the eigenvectors.

Which are the eigenvalues and the eigenvectors of a circulant matrix?

We define the following set of scalars:

$$\begin{aligned} \lambda(k) \quad \equiv \quad & d(0) + d(M-1) \exp\left[\frac{2\pi j}{M}k\right] + d(M-2) \exp\left[\frac{2\pi j}{M}2k\right] \\ & + \ldots + d(1) \exp\left[\frac{2\pi j}{M}(M-1)k\right] \end{aligned} \qquad (6.79)$$

and the following set of vectors:

$$\mathbf{w}(k) \quad = \quad \begin{pmatrix} 1 \\ \exp\left[\frac{2\pi j}{M}k\right] \\ \exp\left[\frac{2\pi j}{M}2k\right] \\ \vdots \\ \exp\left[\frac{2\pi j}{M}(M-1)k\right] \end{pmatrix} \qquad (6.80)$$

where k can take up the values $k = 0, 1, 2, \ldots, M-1$. It can be shown then by direct substitution that:

$$D\mathbf{w}(k) = \lambda(k)\mathbf{w}(k) \qquad (6.81)$$

i.e $\lambda(k)$ are the eigenvalues of matrix D (defined by equation (6.78)) and $\mathbf{w}(k)$ are its corresponding eigenvectors.

How does the knowledge of the eigenvalues and the eigenvectors of a matrix help in inverting the matrix?

If we form matrix W which has the eigenvectors of matrix D as columns, we know that we can write

$$D = W \Lambda W^{-1} \tag{6.82}$$

where W^{-1} has elements

$$W^{-1}(k, i) = \frac{1}{M} \exp\left[-j\frac{2\pi}{M}ki\right] \tag{6.83}$$

and Λ is a diagonal matrix with the eigenvalues along its diagonal. Then, the inversion of matrix D is trivial:

$$D^{-1} = (W \Lambda W^{-1})^{-1} = (W^{-1})^{-1} \Lambda^{-1} W^{-1} = W \Lambda^{-1} W^{-1} \tag{6.84}$$

Example 6.15

Consider matrix W the columns of which $\mathbf{w}(0), \mathbf{w}(1), \ldots, \mathbf{w}(M-1)$ are given by equation (6.80). Show that matrix Z with elements:

$$Z(k, i) = \frac{1}{M} \exp\left(-\frac{2\pi j}{M}ki\right)$$

is the inverse of matrix W.

We have :

$$W = \begin{pmatrix} 1 & 1 & 1 & \cdots & 1 \\ 1 & e^{\frac{2\pi j}{M}} & e^{\frac{2\pi j}{M}2} & \cdots & e^{\frac{2\pi j}{M}(M-1)} \\ 1 & e^{\frac{2\pi j}{M}2} & e^{\frac{2\pi j}{M}4} & \cdots & e^{\frac{2\pi j}{M}2(M-1)} \\ \vdots & \vdots & \vdots & & \vdots \\ 1 & e^{\frac{2\pi j}{M}(M-1)} & e^{\frac{2\pi j}{M}2(M-1)} & \cdots & e^{\frac{2\pi j}{M}(M-1)^2} \end{pmatrix}$$

$$Z = \frac{1}{M}\begin{pmatrix} 1 & 1 & 1 & \cdots & 1 \\ 1 & e^{-\frac{2\pi j}{M}} & e^{-\frac{2\pi j}{M}2} & \cdots & e^{-\frac{2\pi j}{M}(M-1)} \\ 1 & e^{-\frac{2\pi j}{M}2} & e^{-\frac{2\pi j}{M}4} & \cdots & e^{-\frac{2\pi j}{M}2(M-1)} \\ \vdots & \vdots & \vdots & & \vdots \\ 1 & e^{-\frac{2\pi j}{M}(M-1)} & e^{-\frac{2\pi j}{M}2(M-1)} & \cdots & e^{-\frac{2\pi j}{M}(M-1)^2} \end{pmatrix}$$

$$ZW = \frac{1}{M}\begin{pmatrix} M & \sum_{k=0}^{M-1} e^{\frac{2\pi j}{M}k} & \cdots & \sum_{k=0}^{M-1} e^{\frac{2\pi j}{M}(M-1)k} \\ \sum_{k=0}^{M-1} e^{-\frac{2\pi j}{M}k} & M & \cdots & \sum_{k=0}^{M-1} e^{-\frac{2\pi j}{M}2(M-2)k} \\ \vdots & \vdots & & \vdots \\ \sum_{k=0}^{M-1} e^{-\frac{2\pi j}{M}(M-1)k} & \sum_{k=0}^{M-1} e^{-\frac{2\pi j}{M}2(M-2)k} & \cdots & M \end{pmatrix}$$

All the off-diagonal elements of this matrix are of the form: $\sum_{k=0}^{M-1} e^{\frac{2\pi jt}{M}k}$ *where t is some positive or negative integer. This sum is a geometric progression with first term 1 and ratio* $q \equiv e^{\frac{2\pi jt}{M}}$. *We apply the formula*

$$\sum_{k=0}^{n-1} q^k = \frac{q^n - 1}{q - 1}$$

and obtain that:

$$\sum_{k=0}^{M-1} e^{\frac{2\pi jt}{M}k} = \left\{ \begin{array}{l} M \text{ for } t = 0 \\ 0 \text{ for } t \neq 0 \end{array} \right.$$

Therefore $ZW = I$, *the identity matrix, i.e.* $Z = W^{-1}$.

Example 6.16

For $M = 3$ **show that** $\lambda(k)$ **defined by equation (6.79) and** $\mathbf{w}(k)$ **defined by (6.80) are the eigenvalues and eigenvectors respectively, of matrix (6.78), for** $k = 0, 1, 2$.

Let us redefine matrix D for M = 3 as:

$$D = \begin{pmatrix} d_0 & d_2 & d_1 \\ d_1 & d_0 & d_2 \\ d_2 & d_1 & d_0 \end{pmatrix}$$

We also have:

$$\mathbf{w}(k) = \begin{pmatrix} 1 \\ e^{\frac{2\pi j}{3}k} \\ e^{\frac{2\pi j}{3}2k} \end{pmatrix} \quad \text{for } k = 0, 1, 2$$

$$\lambda(k) = d_0 + d_2 e^{\frac{2\pi j}{3}k} + d_1 e^{\frac{2\pi j}{3}2k} \quad \text{for } k = 0, 1, 2$$

We must show that

$$D\mathbf{w}(k) = \lambda(k)\mathbf{w}(k) \tag{6.85}$$

We compute first the left hand side of this expression:

$$D\mathbf{w}(k) = \begin{pmatrix} d_0 & d_2 & d_1 \\ d_1 & d_0 & d_2 \\ d_2 & d_1 & d_0 \end{pmatrix} \begin{pmatrix} 1 \\ e^{\frac{2\pi j}{3}k} \\ e^{\frac{2\pi j}{3}2k} \end{pmatrix} = \begin{pmatrix} d_0 + d_2 e^{\frac{2\pi j}{3}k} + d_1 e^{\frac{2\pi j}{3}2k} \\ d_1 + d_0 e^{\frac{2\pi j}{3}k} + d_2 e^{\frac{2\pi j}{3}2k} \\ d_2 + d_1 e^{\frac{2\pi j}{3}k} + d_0 e^{\frac{2\pi j}{3}2k} \end{pmatrix} \quad (6.86)$$

We also compute the right hand side of (6.85):

$$\lambda(k)\mathbf{w}(k) = \begin{pmatrix} d_0 + d_2 e^{\frac{2\pi j}{3}k} + d_1 e^{\frac{2\pi j}{3}2k} \\ d_0 e^{\frac{2\pi j}{3}k} + d_2 e^{\frac{4\pi j}{3}k} + d_1 e^{\frac{2\pi j}{3}3k} \\ d_0 e^{\frac{2\pi j}{3}2k} + d_2 e^{\frac{2\pi j}{3}3k} + d_1 e^{\frac{2\pi j}{3}4k} \end{pmatrix} \quad (6.87)$$

If we compare the elements of the matrices of the right hand sides of expressions (6.86) and (6.87) one by one, and take into consideration the fact that:

$$e^{2\pi jk} = 1 \quad \text{for any integer } k$$

and

$$e^{\frac{2\pi j}{3}4k} = e^{\frac{2\pi j}{3}3k} e^{\frac{2\pi j}{3}k} = e^{2\pi jk} e^{\frac{2\pi j}{3}k} = e^{\frac{2\pi j}{3}k}$$

we see that equation (6.85) is correct.

Example 6.17

Find the inverse of the following matrix:

$$\begin{pmatrix} -1 & 0 & 2 & 3 \\ 3 & -1 & 0 & 2 \\ 2 & 3 & -1 & 0 \\ 0 & 2 & 3 & -1 \end{pmatrix}$$

In this application, $M = 4$, $d(0) = -1$, $d(1) = 3$, $d(2) = 2$, $d(3) = 0$. Then:

$$\lambda(0) = -1 + 2 + 3 = 4 \Rightarrow \lambda(0)^{-1} = \frac{1}{4}$$

$$\lambda(1) = -1 + 2e^{\frac{2\pi j}{4}2} + 3e^{\frac{2\pi j}{4}3} = -1 - 2 - 3j = -3 - v3j \Rightarrow \lambda^{-1}(1) = \frac{-3+3j}{18} = \frac{-1+j}{6}$$

$$\lambda(2) = -1 + 2e^{\frac{2\pi j}{4}4} + 3e^{\frac{2\pi j}{4}6} = -1 + 2 - 3 = -2 \Rightarrow \lambda^{-1}(2) = -\frac{1}{2}$$

$$\lambda(3) = -1 + 2e^{\frac{2\pi j}{4}6} + 3e^{\frac{2\pi j}{4}9} = -1 - 2 + 3j = -3 + 3j \Rightarrow \lambda^{-1}(3) = \frac{-3-3j}{18} = \frac{-1-j}{6}$$

$$
\begin{aligned}
\mathbf{w}^T(0) &= \begin{pmatrix} 1 & 1 & 1 & 1 \end{pmatrix} \\
\mathbf{w}^T(1) &= \begin{pmatrix} 1 & e^{\frac{2\pi j}{4}} & e^{\frac{2\pi j}{4}2} & e^{\frac{2\pi j}{4}3} \end{pmatrix} = \begin{pmatrix} 1 & j & -1 & -j \end{pmatrix} \\
\mathbf{w}^T(2) &= \begin{pmatrix} 1 & e^{\frac{2\pi j}{4}2} & e^{\frac{2\pi j}{4}4} & e^{\frac{2\pi j}{4}6} \end{pmatrix} = \begin{pmatrix} 1 & -1 & 1 & -1 \end{pmatrix} \\
\mathbf{w}^T(3) &= \begin{pmatrix} 1 & e^{\frac{2\pi j}{4}3} & e^{\frac{2\pi j}{4}6} & e^{\frac{2\pi j}{4}9} \end{pmatrix} = \begin{pmatrix} 1 & -j & -1 & j \end{pmatrix}
\end{aligned}
$$

We use these vectors to construct matrices W and W^{-1} and then apply formula (6.84):

$$
\begin{aligned}
D^{-1} &= \frac{1}{4}\begin{pmatrix} 1 & 1 & 1 & 1 \\ 1 & j & -1 & -j \\ 1 & -1 & 1 & -1 \\ 1 & -j & -1 & j \end{pmatrix}\begin{pmatrix} \frac{1}{4} & 0 & 0 & 0 \\ 0 & \frac{-1+j}{6} & 0 & 0 \\ 0 & 0 & -\frac{1}{2} & 0 \\ 0 & 0 & 0 & \frac{-1-j}{6} \end{pmatrix}\begin{pmatrix} 1 & 1 & 1 & 1 \\ 1 & -j & -1 & j \\ 1 & -1 & 1 & -1 \\ 1 & j & -1 & -j \end{pmatrix} \\
&= \frac{1}{4}\begin{pmatrix} 1 & 1 & 1 & 1 \\ 1 & j & -1 & -j \\ 1 & -1 & 1 & -1 \\ 1 & -j & -1 & j \end{pmatrix}\begin{pmatrix} \frac{1}{4} & \frac{1}{4} & \frac{1}{4} & \frac{1}{4} \\ \frac{-1+j}{6} & \frac{1+j}{6} & \frac{1-j}{6} & \frac{-1-j}{6} \\ -\frac{1}{2} & \frac{1}{2} & -\frac{1}{2} & \frac{1}{2} \\ \frac{-1-j}{6} & \frac{1-j}{6} & \frac{1+j}{6} & \frac{-1+j}{6} \end{pmatrix} \\
&= \frac{1}{4}\begin{pmatrix} -\frac{7}{12} & \frac{13}{12} & \frac{1}{12} & \frac{5}{12} \\ \frac{5}{12} & -\frac{7}{12} & \frac{13}{12} & \frac{1}{12} \\ \frac{1}{12} & \frac{5}{12} & -\frac{7}{12} & \frac{13}{12} \\ \frac{13}{12} & \frac{1}{12} & \frac{5}{12} & -\frac{7}{12} \end{pmatrix} = \frac{1}{48}\begin{pmatrix} -7 & 13 & 1 & 5 \\ 5 & -7 & 13 & 1 \\ 1 & 5 & -7 & 13 \\ 13 & 1 & 5 & -7 \end{pmatrix}
\end{aligned}
$$

Example 6.18 (B)

The elements of a matrix W_N are given by:

$$
W_N(k,n) = \frac{1}{\sqrt{N}}\exp\left(\frac{2\pi j}{N}kn\right)
$$

where k and n take values $0,1,2\ldots,N-1$. The elements of the inverse matrix W_N^{-1} (see Example 6.10) are given by:

$$
W_N^{-1}(k,n) = \frac{1}{\sqrt{N}}\exp\left(-\frac{2\pi j}{N}kn\right)
$$

We define matrix W as the Kronecker product of W_N with itself. Show that the inverse of matrix W is formed by the Kronecker product of matrix W_N^{-1} with itself.

Let us consider an element $W(m,l)$ of matrix W. We write integers m and l in terms of their quotients and remainders when divided by N:

$$m \equiv m_1 N + m_2$$
$$l \equiv l_1 N + l_2$$

Since W is $W_N \otimes W_N$ we have:

$$W(m, l) = \frac{1}{N} e^{\left(\frac{2\pi j}{N} m_1 l_1\right)} e^{\left(\frac{2\pi j}{N} m_2 l_2\right)} \tag{6.88}$$

(m_1, l_1) *identifies in which partition of matrix W element $W(m, l)$ belongs. m_2 and l_2 vary inside each partition, taking all their possible values.*

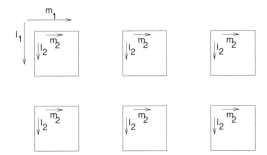

In a similar way we can write an element of matrix $Z \equiv W_N^{-1} \otimes W_N^{-1}$:

$$Z(t, n) = \frac{1}{N} e^{-\frac{2\pi j}{N} t_1 n_1} e^{-\frac{2\pi j}{N} t_2 n_2} \tag{6.89}$$

An element of the product matrix $A \equiv WZ$ is given by:

$$A(k, n) = \sum_{t=0}^{N^2-1} W(k, t) Z(t, n)$$

$$= \frac{1}{N^2} \sum_{t=0}^{N^2-1} e^{\frac{2\pi j}{N} k_1 t_1} e^{\frac{2\pi j}{N} k_2 t_2} e^{-\frac{2\pi j}{N} t_1 n_1} e^{-\frac{2\pi j}{N} t_2 n_2}$$

$$= \frac{1}{N^2} \sum_{t=0}^{N^2-1} e^{\frac{2\pi j}{N} (k_1 - n_1) t_1} e^{\frac{2\pi j}{N} (k_2 - n_2) t_2}$$

If we write again $t \equiv t_1 N + t_2$, we can break the sum over t to two sums, one over t_1 and one over t_2:

$$A(k, n) = \frac{1}{N^2} \sum_{t_1=0}^{N-1} \left\{ e^{\frac{2\pi j}{N} (k_1 - n_1) t_1} \sum_{t_2=0}^{N-1} e^{\frac{2\pi j}{N} (k_2 - n_2) t_2} \right\} \tag{6.90}$$

From Example 6.10 we know that:

$$\sum_{m=0}^{M-1} e^{\frac{2\pi j m}{M} s} = \begin{cases} M & \text{if } s = 0 \\ 0 & \text{if } s \neq 0 \end{cases}$$

We apply this formula to (6.90) for the inner sum first, and the outer sum afterwards:

$$
\begin{aligned}
A(k, n) &= \frac{1}{N^2} \sum_{t_1=0}^{N-1} e^{-\frac{2\pi j}{N}(k_1 - n_1)t_1} \delta(k_2 - n_2) N \\
&= \frac{1}{N} \delta(k_2 - n_2) N \delta(k_1 - n_1) \\
&= \delta(k_2 - n_2) \delta(k_1 - n_1) \\
&= \delta(k - n)
\end{aligned}
$$

Therefore, matrix A has all its elements 0, except the diagonal ones (obtained for $k = n$), which are equal to 1. Therefore A is the unit matrix and this proves that matrix Z with elements given by (6.89) is the inverse of matrix W with elements given by (6.88)

How do we know that matrix H that expresses the linear degradation process is block circulant?

We remember that equation $\mathbf{g} = H\mathbf{f}$ is actually equivalent to:

$$
g(i, j) = \sum_{k=0}^{N-1} \sum_{l=0}^{N-1} h(k, l, i, j) f(k, l) \tag{6.91}
$$

where $h(k, l, i, j)$ is the point spread function of the degradation process. For a shift invariant point spread function we have:

$$
g(i, j) = \sum_{k=0}^{N-1} \sum_{l=0}^{N-1} f(k, l) h(i - k, j - l) \tag{6.92}
$$

Let us consider one of the partitions of the partitioned matrix H (see equation (6.77)). Inside every partition, the values of l and j remain constant; i.e. $j - l$ is constant inside each partition. The value of $i - k$ along each line runs from i to $i - N + 1$, taking all integer values in between. When i is incremented by 1 in the next row, all the values of $i - k$ are shifted by one position to the right (see equations (1.26) and (6.77)). So, each partition submatrix of H is characterized by the value of $j - l \equiv u$ and has a circulant form:

$$
H_u = \begin{pmatrix}
h(0, u) & h(N-1, u) & h(N-2, u) & \cdots & h(1, u) \\
h(1, u) & h(0, u) & h(N-1, u) & \cdots & h(2, u) \\
h(2, u) & h(1, u) & h(0, u) & \cdots & h(3, u) \\
\vdots & \vdots & \vdots & & \vdots \\
h(N-1, u) & h(N-2, u) & h(N-3, u) & \cdots & h(0, u)
\end{pmatrix} \tag{6.93}
$$

Notice that here we assume that $h(v,u)$ is periodic with period N in each of its arguments, and so $h(1-N,u) = h((1-N)+N,u) = h(1,u)$ etc.

The full matrix H can be written in the form:

$$H = \begin{pmatrix} H_0 & H_{-1} & H_{-2} & \cdots & H_{-M+1} \\ H_1 & H_0 & H_{-1} & \cdots & H_{-M+2} \\ H_2 & H_1 & H_0 & \cdots & H_{-M+3} \\ \vdots & \vdots & \vdots & & \vdots \\ H_{M-1} & H_{M-2} & H_{M-3} & \cdots & H_0 \end{pmatrix} \tag{6.94}$$

where again owing to the periodicity of $h(v,u)$, $H_{-1} = H_{M-1}$, $H_{-M+1} = H_1$ etc.

How can we diagonalize a block circulant matrix?

Define a matrix with elements

$$W_N(k,n) = \frac{1}{\sqrt{N}} \exp\left[\frac{2\pi j}{N} kn\right] \tag{6.95}$$

and matrix

$$W = W_N \otimes W_N \tag{6.96}$$

where $\otimes$ is the Kronecker product of the two matrices. The inverse of $W_N(k,n)$ is a matrix with elements:

$$W_N^{-1}(k,n) = \frac{1}{\sqrt{N}} \exp\left[-\frac{2\pi j}{N} kn\right] \tag{6.97}$$

The inverse of W is given by (see Example 6.18):

$$W_N^{-1}(k,n) = W_N^{-1} \otimes W_N^{-1} \tag{6.98}$$

We also define a diagonal matrix Λ as follows:

$$\Lambda(k,i) = \begin{cases} N\hat{H}\left(k_{mod\,N},\left[\frac{k}{N}\right]\right) & \text{if } i = k \\ 0 & \text{if } i \neq k \end{cases} \tag{6.99}$$

where $\hat{H}$ is the discrete Fourier transform of the point spread function h:

$$\hat{H}(u,v) = \frac{1}{N} \sum_{x=0}^{N-1} \sum_{y=0}^{N-1} h(x,y) e^{-2\pi j\left(\frac{ux}{N} + \frac{vy}{N}\right)} \tag{6.100}$$

It can be shown then by direct matrix multiplication that:

$$H = W\Lambda W^{-1} \tag{6.101}$$

Thus H can be inverted easily since it has been written as the product of matrices the inversion of which is trivial.

B6.5: What is the proof of equation (6.101)?

First we have to find how an element $H(f, g)$ of matrix H is related to the point spread function $h(x, y)$. Let us write indices f, g as multiples of the dimension N of one of the partitions, plus a remainder:

$$f \equiv f_1 N + f_2$$
$$g \equiv g_1 N + g_2$$

As f and g scan all possible values from 0 to $N^2 - 1$ each, we can visualize the $N \times N$ partitions of matrix H, indexed by subscript u, as follows:

$f_1 = 0$	$f_1 = 0$	$f_1 = 0$
$g_1 = 0$	$g_1 = 1$	$g_1 = 2$
$u = 0$	$u = -1$	$u = -2$

...

$f_1 = 1$	$f_1 = 1$	$f_1 = 1$
$g_1 = 0$	$g_1 = 1$	$g_1 = 2$
$u = 1$	$u = 0$	$u = -1$

...

We observe that each partition is characterized by index $f_1 - g_1$ and inside each partition the elements computed from $h(x, y)$ are computed for various values of $f_2 - g_2$. We conclude that:

$$H(f, g) = h(f_2 - g_2, f_1 - g_1) \tag{6.102}$$

Let us consider next an element of matrix $W \Lambda W^{-1}$:

$$A(m, n) \equiv \sum_{l=0}^{N^2-1} \sum_{t=0}^{N^2-1} W_{ml} \Lambda_{lt} W_{tn}^{-1}$$

Since matrix Λ is diagonal, the sum over t collapses only to values of $t = l$. Then:

$$A(m, n) = \sum_{l=0}^{N^2-1} W_{ml} \Lambda_{ll} W_{ln}^{-1}$$

Λ_{ll} is a scalar and therefore it may change position inside the summand:

$$A(m, n) = \sum_{l=0}^{N^2-1} W_{ml} W_{ln}^{-1} \Lambda_{ll} \tag{6.103}$$

In Box **B6.2** we showed how the elements of matrices W_{ml} and W_{ln}^{-1} can be written if we write their indices in terms of their quotients and remainders when divided by N:

$$m \equiv Nm_1 + m_2$$
$$l \equiv Nl_1 + l_2$$
$$n \equiv Nn_1 + n_2$$

Using these expressions, and the definition of Λ_{ll} as $N\hat{H}(l_2, l_1)$ from equation (6.99), we obtain:

$$A(m,n) = \sum_{l=0}^{N^2-1} e^{\frac{2\pi j}{N}m_1 l_1} e^{\frac{2\pi j}{N}m_2 l_2} \frac{1}{N^2} e^{-\frac{2\pi j}{N}l_1 n_1} e^{-\frac{2\pi j}{N}l_2 n_2} N\hat{H}(l_2, l_1)$$

On rearranging, we have:

$$A(m,n) = \frac{1}{N} \sum_{l_1=0}^{N-1} \sum_{l_2=0}^{N-1} \hat{H}(l_2, l_1) e^{\frac{2\pi j}{N}(m_1 - n_1)l_1} e^{\frac{2\pi j}{N}(m_2 - n_2)l_2} \tag{6.104}$$

We recognize this expression as the inverse Fourier transform of $h(m_2 - n_2, m_1 - n_1)$. Therefore:

$$A(m,n) = h(m_2 - n_2, m_1 - n_1) \tag{6.105}$$

By comparing equations (6.102) and (6.105) we can see that the elements of matrix H and $W\Lambda W^{-1}$ have been shown to be identical.

B6.6: What is the transpose of matrix H?

We shall show that $H^T = W\Lambda^*W^{-1}$, where Λ^* means the complex conjugate of matrix Λ.

According to equation (6.102) of Box **B6.5**, an element of the transpose of matrix H will be given by:

$$H^T(f,g) = h(g_2 - f_2, g_1 - f_1) \tag{6.106}$$

An element $A(m,n)$ of matrix $W\Lambda^*W^{-1}$ will be given by an equation similar to (6.104) of Box **B6.5**, but instead of having factor $\hat{H}(l_2, l_1)$ it will have factor $\hat{H}(-l_2, -l_1)$, coming from the element of Λ_{ll}^* being defined in terms of the complex conjugate of the Fourier transform $\hat{H}(u,v)$ given by equation (6.100):

$$A(m,n) = \frac{1}{N} \sum_{l_1=0}^{N-1} \sum_{l_2=0}^{N-1} \hat{H}(-l_2, -l_1) e^{\frac{2\pi j}{N}(m_1 - n_1)l_1} e^{\frac{2\pi j}{N}(m_2 - n_2)l_2}$$

We change the dummy variables of summation to:

$$\tilde{l}_1 \equiv -l_1 \text{ and } \tilde{l}_2 \equiv -l_2$$

Then:

$$A(m,n) = \frac{1}{N} \sum_{\tilde{l}_1=0}^{-N+1} \sum_{\tilde{l}_2=0}^{-N+1} \hat{H}(\tilde{l}_2, \tilde{l}_1) e^{\frac{2\pi j}{N}(-m_1+n_1)\tilde{l}_1} e^{\frac{2\pi j}{N}(-m_2+n_2)\tilde{l}_2}$$

Since we are dealing with periodic functions summed over a period, the range over which we sum does not really matter, as long as N consecutive values are considered. Then we can write:

$$A(m,n) = \frac{1}{N} \sum_{\tilde{l}_1=0}^{N-1} \sum_{\tilde{l}_2=0}^{N-1} \hat{H}(\tilde{l}_2, \tilde{l}_1) e^{\frac{2\pi j}{N}(-m_1+n_1)\tilde{l}_1} e^{\frac{2\pi j}{N}(-m_2+n_2)\tilde{l}_2}$$

We recognize on the right hand side of the above expression the inverse Fourier transform of $\hat{H}(\tilde{l}_2, \tilde{l}_1)$ computed at $(n_2 - m_2, n_1 - m_1)$:

$$A(m,n) = h(n_2 - m_2, n_1 - m_1) \tag{6.107}$$

By direct comparison with equation (6.106), we prove that matrices H^T and $W\Lambda^*W^{-1}$ are equal, element by element.

Example 6.19

Show that the Laplacian, i.e. the sum of the second derivatives, of a discrete image at a pixel position (i,j) can be estimated by:

$$\Delta^2 f(i,j) = f(i-1,j) + f(i,j-1) + f(i+1,j) + f(i,j+1) - 4f(i,j)$$

At inter-pixel position $(i+\frac{1}{2},j)$, the first derivate of the image function along the i axis is approximated by the first difference:

$$\Delta_i f(i+\frac{1}{2},j) = f(i+1,j) - f(i,j)$$

Similarly, the first difference at $(i-\frac{1}{2},j)$ along the i axis is:

$$\Delta_i f(i - \frac{1}{2}, j) \;=\; f(i, j) - f(i-1, j)$$

The second derivative at (i, j) along the i axis can be approximated by the first difference of the first differences, computed at positions $(i + \frac{1}{2}, j)$ and $(i - \frac{1}{2}, j)$, that is:

$$
\begin{aligned}
\Delta_i^2 f(i, j) &= \Delta_i f(i + \frac{1}{2}, j) - \Delta_i f(i - \frac{1}{2}, j) \\
&= f(i+1, j) - 2f(i, j) + f(i-1, j) \qquad (6.108)
\end{aligned}
$$

Similarly, the second derivative at (i, j) along the j axis can be approximated by:

$$
\begin{aligned}
\Delta_j^2 f(i, j) &= \Delta_j f(i, j + \frac{1}{2}) - \Delta_j f(i, j - \frac{1}{2}) \\
&= f(i, j+1) - 2f(i, j) + f(i, j-1) \qquad (6.109)
\end{aligned}
$$

Adding equations (6.108) and (6.109) part by part we obtain the result.

Example 6.20

Consider a 3×3 image represented by a column vector f. Identify a 9×9 matrix L such that if we multiply vector f by it, the output will be a vector with the estimate of the value of the Laplacian at each position. Assume that image f is periodic in each direction with period 3. What type of matrix is L?

From Example 6.19 we know that the matrix operator that returns the estimate of the Laplacian at each position is:

$$
\begin{array}{ccc}
0 & 1 & 0 \\
1 & -4 & 1 \\
0 & 1 & 0
\end{array}
$$

To avoid boundary effects, we first extend the image in all directions periodically:

$$
\begin{array}{ccc}
 & \begin{array}{ccc} f_{31} & f_{32} & f_{33} \end{array} & \\
\begin{array}{c} f_{13} \\ f_{21} \\ f_{33} \end{array} & \left(\begin{array}{ccc} f_{11} & f_{12} & f_{13} \\ f_{21} & f_{22} & f_{23} \\ f_{31} & f_{32} & f_{33} \end{array} \right) & \begin{array}{c} f_{11} \\ f_{21} \\ f_{31} \end{array} \\
 & \begin{array}{ccc} f_{11} & f_{12} & f_{13} \end{array} &
\end{array}
$$

By observing which values will contribute to the value of the Laplacian at a pixel, and with what weight, we construct the 9×9 matrix below with which we must multiply the column vector $\mathbf{f}$:

$$
\begin{pmatrix}
-4 & 1 & 1 & 1 & 0 & 0 & 1 & 0 & 0 \\
1 & -4 & 1 & 0 & 1 & 0 & 0 & 1 & 0 \\
1 & 1 & -4 & 0 & 0 & 1 & 0 & 0 & 1 \\
1 & 0 & 0 & -4 & 1 & 1 & 1 & 0 & 0 \\
0 & 1 & 0 & 1 & -4 & 1 & 0 & 1 & 0 \\
0 & 0 & 1 & 1 & 1 & -4 & 0 & 0 & 1 \\
1 & 0 & 0 & 1 & 0 & 0 & -4 & 1 & 1 \\
0 & 1 & 0 & 0 & 1 & 0 & 1 & -4 & 1 \\
0 & 0 & 1 & 0 & 0 & 1 & 1 & 1 & -4
\end{pmatrix}
\begin{pmatrix}
f_{11} \\ f_{21} \\ f_{31} \\ f_{12} \\ f_{22} \\ f_{32} \\ f_{13} \\ f_{23} \\ f_{33}
\end{pmatrix}
$$

This matrix is a block circulant matrix with easily identifiable partitions of size 3×3.

Example 6.21 (B)

Using the matrix defined in Example 6.20, estimate the Laplacian for the following image:

$$
\begin{pmatrix}
3 & 2 & 1 \\
2 & 0 & 1 \\
0 & 0 & 1
\end{pmatrix}
$$

Then re-estimate the Laplacian of the above image using the formula of Example 6.19.

$$
\begin{pmatrix}
-4 & 1 & 1 & 1 & 0 & 0 & 1 & 0 & 0 \\
1 & -4 & 1 & 0 & 1 & 0 & 0 & 1 & 0 \\
1 & 1 & -4 & 0 & 0 & 1 & 0 & 0 & 1 \\
1 & 0 & 0 & -4 & 1 & 1 & 1 & 0 & 0 \\
0 & 1 & 0 & 1 & -4 & 1 & 0 & 1 & 0 \\
0 & 0 & 1 & 1 & 1 & -4 & 0 & 0 & 1 \\
1 & 0 & 0 & 1 & 0 & 0 & -4 & 1 & 1 \\
0 & 1 & 0 & 0 & 1 & 0 & 1 & -4 & 1 \\
0 & 0 & 1 & 0 & 0 & 1 & 1 & 1 & -4
\end{pmatrix}
\begin{pmatrix}
3 \\ 2 \\ 0 \\ 2 \\ 0 \\ 0 \\ 1 \\ 1 \\ 1
\end{pmatrix}
=
\begin{pmatrix}
-7 \\ -4 \\ 6 \\ -4 \\ 5 \\ 3 \\ 3 \\ 0 \\ -2
\end{pmatrix}
$$

If we use the formula, we need to augment first the image by writing explicitly the boundary pixels:

$$
\begin{array}{ccc}
 & 0 \ \ 0 \ \ 1 & \\
1 & \begin{pmatrix} 3 & 2 & 1 \\ 2 & 0 & 1 \\ 0 & 0 & 1 \end{pmatrix} & 3 \\
1 & & 2 \\
1 & & 0 \\
 & 3 \ \ 2 \ \ 1 &
\end{array}
$$

The Laplacian is:

$$
\begin{pmatrix}
1+2+2-4\times3 & 3+1-4\times2 & 1+2+1+3-4\times1 \\
3+1-4\times2 & 2+2+1 & 1+1+2-4\times1 \\
2+1+3 & 2+1 & 1+1-4\times1
\end{pmatrix}
=
\begin{pmatrix}
-7 & -4 & 3 \\
-4 & 5 & 0 \\
6 & 3 & -2
\end{pmatrix}
$$

Note that we obtain the same answer, whether we use the local formula or the matrix multiplication.

Example 6.22 (B)

Find the eigenvectors and eigenvalues of the matrix derived in Example 6.20.

Matrix L derived in Example 6.20 is:

$$
\begin{pmatrix}
-4 & 1 & 1 & 1 & 0 & 0 & 1 & 0 & 0 \\
1 & -4 & 1 & 0 & 1 & 0 & 0 & 1 & 0 \\
1 & 1 & -4 & 0 & 0 & 1 & 0 & 0 & 1 \\
1 & 0 & 0 & -4 & 1 & 1 & 1 & 0 & 0 \\
0 & 1 & 0 & 1 & -4 & 1 & 0 & 1 & 0 \\
0 & 0 & 1 & 1 & 1 & -4 & 0 & 0 & 1 \\
1 & 0 & 0 & 1 & 0 & 0 & -4 & 1 & 1 \\
0 & 1 & 0 & 0 & 1 & 0 & 1 & -4 & 1 \\
0 & 0 & 1 & 0 & 0 & 1 & 1 & 1 & -4
\end{pmatrix}
$$

We can clearly see that this matrix is a block circulant matrix with easily identifiable 3×3 partitions. To find its eigenvectors we first use equation (6.80) for $M = 3$ to define vectors $\mathbf{w}$:

$$
\mathbf{w}(0) = \begin{pmatrix} 1 \\ 1 \\ 1 \end{pmatrix} \quad
\mathbf{w}(1) = \begin{pmatrix} 1 \\ e^{\frac{2\pi j}{3}} \\ e^{\frac{4\pi j}{3}} \end{pmatrix} \quad
\mathbf{w}(2) = \begin{pmatrix} 1 \\ e^{\frac{4\pi j}{3}} \\ e^{\frac{8\pi j}{3}} \end{pmatrix}
$$

These vectors are used as columns to create a matrix. We normalize this matrix by dividing by $\sqrt{M} = \sqrt{3}$, and thus construct the matrix defined by equation (6.95):

$$W_3 = \frac{1}{\sqrt{3}} \begin{pmatrix} 1 & 1 & 1 \\ 1 & e^{\frac{2\pi j}{3}} & e^{\frac{4\pi j}{3}} \\ 1 & e^{\frac{4\pi j}{3}} & e^{\frac{8\pi j}{3}} \end{pmatrix}$$

We take the Kronecker product of this matrix with itself to create matrix W as defined by equation (6.96):

$$W = \frac{1}{3} \begin{pmatrix} 1 & 1 & 1 & 1 & 1 & 1 & 1 & 1 & 1 \\ 1 & e^{\frac{2\pi j}{3}} & e^{\frac{4\pi j}{3}} & 1 & e^{\frac{2\pi j}{3}} & e^{\frac{4\pi j}{3}} & 1 & e^{\frac{2\pi j}{3}} & e^{\frac{4\pi j}{3}} \\ 1 & e^{\frac{4\pi j}{3}} & e^{\frac{8\pi j}{3}} & 1 & e^{\frac{4\pi j}{3}} & e^{\frac{8\pi j}{3}} & 1 & e^{\frac{4\pi j}{3}} & e^{\frac{8\pi j}{3}} \\ 1 & 1 & 1 & e^{\frac{2\pi j}{3}} & e^{\frac{2\pi j}{3}} & e^{\frac{2\pi j}{3}} & e^{\frac{4\pi j}{3}} & e^{\frac{4\pi j}{3}} & e^{\frac{4\pi j}{3}} \\ 1 & e^{\frac{2\pi j}{3}} & e^{\frac{4\pi j}{3}} & e^{\frac{2\pi j}{3}} & e^{\frac{4\pi j}{3}} & e^{\frac{6\pi j}{3}} & e^{\frac{4\pi j}{3}} & e^{\frac{6\pi j}{3}} & e^{\frac{8\pi j}{3}} \\ 1 & e^{\frac{4\pi j}{3}} & e^{\frac{8\pi j}{3}} & e^{\frac{2\pi j}{3}} & e^{\frac{6\pi j}{3}} & e^{\frac{10\pi j}{3}} & e^{\frac{4\pi j}{3}} & e^{\frac{8\pi j}{3}} & e^{\frac{12\pi j}{3}} \\ 1 & 1 & 1 & e^{\frac{4\pi j}{3}} & e^{\frac{4\pi j}{3}} & e^{\frac{4\pi j}{3}} & e^{\frac{8\pi j}{3}} & e^{\frac{8\pi j}{3}} & e^{\frac{8\pi j}{3}} \\ 1 & e^{\frac{2\pi j}{3}} & e^{\frac{4\pi j}{3}} & e^{\frac{4\pi j}{3}} & e^{\frac{6\pi j}{3}} & e^{\frac{8\pi j}{3}} & e^{\frac{8\pi j}{3}} & e^{\frac{10\pi j}{3}} & e^{\frac{12\pi j}{3}} \\ 1 & e^{\frac{4\pi j}{3}} & e^{\frac{8\pi j}{3}} & e^{\frac{4\pi j}{3}} & e^{\frac{8\pi j}{3}} & e^{\frac{12\pi j}{3}} & e^{\frac{8\pi j}{3}} & e^{\frac{12\pi j}{3}} & e^{\frac{16\pi j}{3}} \end{pmatrix}$$

The columns of this matrix are the eigenvectors of matrix L. Note that these eigenvectors are the same for all block circulant matrices with the same structure, independent of what the exact values of the elements are. The inverse of matrix W can be constructed using equation (6.97), i.e. by taking the complex conjugate of matrix W. (Note that for a general unitary matrix we must take the complex conjugate of its transpose in order to construct its inverse. This is not necessary here as W is a symmetric matrix and therefore it is equal to its transpose.)

$$W^{-1} = \frac{1}{3} \begin{pmatrix} 1 & 1 & 1 & 1 & 1 & 1 & 1 & 1 & 1 \\ 1 & e^{-\frac{2\pi j}{3}} & e^{-\frac{4\pi j}{3}} & 1 & e^{-\frac{2\pi j}{3}} & e^{-\frac{4\pi j}{3}} & 1 & e^{-\frac{2\pi j}{3}} & e^{-\frac{4\pi j}{3}} \\ 1 & e^{-\frac{4\pi j}{3}} & e^{-\frac{8\pi j}{3}} & 1 & e^{-\frac{4\pi j}{3}} & e^{-\frac{8\pi j}{3}} & 1 & e^{-\frac{4\pi j}{3}} & e^{-\frac{8\pi j}{3}} \\ 1 & 1 & 1 & e^{-\frac{2\pi j}{3}} & e^{-\frac{2\pi j}{3}} & e^{-\frac{2\pi j}{3}} & e^{-\frac{4\pi j}{3}} & e^{-\frac{4\pi j}{3}} & e^{-\frac{4\pi j}{3}} \\ 1 & e^{-\frac{2\pi j}{3}} & e^{-\frac{4\pi j}{3}} & e^{-\frac{2\pi j}{3}} & e^{-\frac{4\pi j}{3}} & e^{-\frac{6\pi j}{3}} & e^{-\frac{4\pi j}{3}} & e^{-\frac{6\pi j}{3}} & e^{-\frac{8\pi j}{3}} \\ 1 & e^{-\frac{4\pi j}{3}} & e^{-\frac{8\pi j}{3}} & e^{-\frac{2\pi j}{3}} & e^{-\frac{6\pi j}{3}} & e^{-\frac{10\pi j}{3}} & e^{-\frac{4\pi j}{3}} & e^{-\frac{8\pi j}{3}} & e^{-\frac{12\pi j}{3}} \\ 1 & 1 & 1 & e^{-\frac{4\pi j}{3}} & e^{-\frac{4\pi j}{3}} & e^{-\frac{4\pi j}{3}} & e^{-\frac{8\pi j}{3}} & e^{-\frac{8\pi j}{3}} & e^{-\frac{8\pi j}{3}} \\ 1 & e^{-\frac{2\pi j}{3}} & e^{-\frac{4\pi j}{3}} & e^{-\frac{4\pi j}{3}} & e^{-\frac{6\pi j}{3}} & e^{-\frac{8\pi j}{3}} & e^{-\frac{8\pi j}{3}} & e^{-\frac{10\pi j}{3}} & e^{-\frac{12\pi j}{3}} \\ 1 & e^{-\frac{4\pi j}{3}} & e^{-\frac{8\pi j}{3}} & e^{-\frac{4\pi j}{3}} & e^{-\frac{8\pi j}{3}} & e^{-\frac{12\pi j}{3}} & e^{-\frac{8\pi j}{3}} & e^{-\frac{12\pi j}{3}} & e^{-\frac{16\pi j}{3}} \end{pmatrix}$$

The eigenvalues of matrix L can be computed from its Fourier transform, using equation (6.99). First, however, we need to identify the kernel $l(x, y)$ of the operator represented by matrix L and take its Fourier transform $\hat{L}(u, v)$ using equation (6.100).

   From Example 6.20 we know that the kernel function is:

$$
\begin{array}{ccc}
0 & 1 & 0 \\
1 & -4 & 1 \\
0 & 1 & 0
\end{array}
$$

We can identify then the following values for the discrete function $l(x,y)$:

$$l(0,0) = -4,\ l(-1,-1) = 0,\ l(-1,0) = 1,\ l(-1,1) = 0$$
$$l(0,-1) = 1,\ l(0,1) = 1,\ l(1,-1) = 0,\ l(1,0) = 1,\ l(1,1) = 0$$

Note that these values cannot be directly used in equation (6.100), which assumes a function $h(x,y)$ defined with positive values of its arguments only. We therefore need a shifted version of our kernel, one that puts the value -4 at the top left corner of the matrix representation of the kernel. We can obtain such a version by reading the first column of matrix L and wrapping it around to form a 3×3 matrix:

$$
\begin{array}{ccc}
-4 & 1 & 1 \\
1 & 0 & 0 \\
1 & 0 & 0
\end{array}
$$

Then we have:

$$l(0,0) = -4,\ l(0,1) = 1,\ l(0,2) = 1,\ l(1,0) = 1$$
$$l(2,0) = 1,\ l(1,1) = 0,\ l(1,2) = 0,\ l(2,1) = 0,\ l(2,2) = 0$$

We can use these values in equation (6.100) to derive:

$$\hat{L}(u,v) = \frac{1}{3}\left[-4 + e^{-\frac{2\pi j}{3}v} + e^{-\frac{2\pi j}{3}2v} + e^{-\frac{2\pi j}{3}u} + e^{-\frac{2\pi j}{3}2u} \right] \tag{6.110}$$

Formula (6.99) says that the eigenvalues of matrix L, which appear along the diagonal of matrix $\Lambda(k,i)$, are the values of the Fourier transform $\hat{L}(u,v)$, computed for $u = k_{mod\,3}$ and $v = \left[\frac{k}{3}\right]$ where $k = 0,1,\ldots,8$. These values can be computed using formula (6.110):

$$\hat{L}(0,0) = 0$$
$$\hat{L}(0,1) = \frac{1}{3}\left[-4 + e^{-\frac{2\pi j}{3}} + e^{-\frac{4\pi j}{3}} + 1 + 1 \right] = \frac{1}{3}[-2 - 2\cos 60°] = -1$$
$$\hat{L}(0,2) = \frac{1}{3}\left[-4 + e^{-\frac{4\pi j}{3}} + e^{-\frac{8\pi j}{3}} + 2 \right] = \frac{1}{3}[-2 + e^{-\frac{4\pi j}{3}} + e^{-\frac{2\pi j}{3}}] = -1$$
$$\hat{L}(1,0) = \hat{L}(0,1) = -1$$
$$\hat{L}(1,1) = \frac{1}{3}\left[-4 + 2e^{-\frac{2\pi j}{3}} + 2e^{-\frac{4\pi j}{3}} \right] = \frac{1}{3}[-4 - 4\cos 60°] = -2$$
$$\hat{L}(1,2) = \frac{1}{3}\left[-4 + e^{-\frac{4\pi j}{3}} + e^{-\frac{8\pi j}{3}} + e^{-\frac{2\pi j}{3}} + e^{-\frac{4\pi j}{3}} \right] = -2$$

$$\hat{L}(2,0) = \hat{L}(0,2) = -1$$

$$\hat{L}(2,1) = \hat{L}(1,2) = -2$$

$$\hat{L}(2,2) = \frac{1}{3}\left[-4 + 2e^{-\frac{4\pi j}{3}} + 2e^{-\frac{8\pi j}{3}}\right] = -2$$

where we have made use of the following:

$$e^{-\frac{2\pi j}{3}} = -\cos 60° - j\sin 60° = -\frac{1}{2} - j\frac{\sqrt{3}}{2}$$

$$e^{-\frac{4\pi j}{3}} = -\cos 60° + j\sin 60° = -\frac{1}{2} + j\frac{\sqrt{3}}{2}$$

$$e^{-\frac{6\pi j}{3}} = 1$$

$$e^{-\frac{8\pi j}{3}} = e^{-\frac{2\pi j}{3}} = -\cos 60° - j\sin 60° = -\frac{1}{2} - j\frac{\sqrt{3}}{2}$$

Note that the first eigenvalue of matrix L is 0. This means that matrix L is singular, and even though we can diagonalize it using equation (6.101), we cannot invert it by taking the inverse of this equation. This should not be surprising as matrix L expresses the Laplacian operator for an image, and we know that from the knowledge of the Laplacian alone we can never recover the original image. Applying equation (6.99) we define matrix Λ for L to be:

$$\Lambda = \begin{bmatrix} 0 & 0 & 0 & 0 & 0 & 0 & 0 & 0 & 0 \\ 0 & -3 & 0 & 0 & 0 & 0 & 0 & 0 & 0 \\ 0 & 0 & -3 & 0 & 0 & 0 & 0 & 0 & 0 \\ 0 & 0 & 0 & -3 & 0 & 0 & 0 & 0 & 0 \\ 0 & 0 & 0 & 0 & -6 & 0 & 0 & 0 & 0 \\ 0 & 0 & 0 & 0 & 0 & -6 & 0 & 0 & 0 \\ 0 & 0 & 0 & 0 & 0 & 0 & -3 & 0 & 0 \\ 0 & 0 & 0 & 0 & 0 & 0 & 0 & -6 & 0 \\ 0 & 0 & 0 & 0 & 0 & 0 & 0 & 0 & -6 \end{bmatrix}$$

Having defined matrices W, W^{-1} and Λ we can write:

$$L = W\Lambda W^{-1}$$

This equation can be confirmed by direct substitution. First we compute matrix ΛW^{-1}:

$$\begin{bmatrix}
0 & 0 & 0 & 0 & 0 & 0 & 0 & 0 & 0 \\
-1 & -e^{-\frac{2\pi j}{3}} & -e^{-\frac{4\pi j}{3}} & -1 & -e^{-\frac{2\pi j}{3}} & -e^{-\frac{4\pi j}{3}} & -1 & -e^{-\frac{2\pi j}{3}} & -e^{-\frac{4\pi j}{3}} \\
-1 & -e^{-\frac{4\pi j}{3}} & -e^{-\frac{8\pi j}{3}} & -1 & -e^{-\frac{4\pi j}{3}} & -e^{-\frac{8\pi j}{3}} & -1 & -e^{-\frac{4\pi j}{3}} & -e^{-\frac{8\pi j}{3}} \\
-1 & -1 & -1 & -e^{-\frac{2\pi j}{3}} & -e^{-\frac{2\pi j}{3}} & -e^{-\frac{2\pi j}{3}} & -e^{-\frac{4\pi j}{3}} & -e^{-\frac{4\pi j}{3}} & -e^{-\frac{4\pi j}{3}} \\
-2 & -2e^{-\frac{2\pi j}{3}} & -2e^{-\frac{4\pi j}{3}} & -2e^{-\frac{2\pi j}{3}} & -2e^{-\frac{4\pi j}{3}} & -2e^{-\frac{6\pi j}{3}} & -2e^{-\frac{4\pi j}{3}} & -2e^{-\frac{6\pi j}{3}} & -2e^{-\frac{8\pi j}{3}} \\
-2 & -2e^{-\frac{4\pi j}{3}} & -2e^{-\frac{8\pi j}{3}} & -2e^{-\frac{2\pi j}{3}} & -2e^{-\frac{6\pi j}{3}} & -2e^{-\frac{10\pi j}{3}} & -2e^{-\frac{4\pi j}{3}} & -2e^{-\frac{8\pi j}{3}} & -2e^{-\frac{12\pi j}{3}} \\
-1 & -1 & -1 & -e^{-\frac{4\pi j}{3}} & -e^{-\frac{4\pi j}{3}} & -e^{-\frac{4\pi j}{3}} & -e^{-\frac{8\pi j}{3}} & -e^{-\frac{8\pi j}{3}} & -e^{-\frac{8\pi j}{3}} \\
-2 & -2e^{-\frac{2\pi j}{3}} & -2e^{-\frac{4\pi j}{3}} & -2e^{-\frac{4\pi j}{3}} & -2e^{-\frac{6\pi j}{3}} & -2e^{-\frac{8\pi j}{3}} & -2e^{-\frac{8\pi j}{3}} & -2e^{-\frac{10\pi j}{3}} & -2e^{-\frac{12\pi j}{3}} \\
-2 & -2e^{-\frac{4\pi j}{3}} & -2e^{-\frac{8\pi j}{3}} & -2e^{-\frac{4\pi j}{3}} & -2e^{-\frac{8\pi j}{3}} & -2e^{-\frac{12\pi j}{3}} & -2e^{-\frac{8\pi j}{3}} & -2e^{-\frac{12\pi j}{3}} & -2e^{-\frac{16\pi j}{3}}
\end{bmatrix}$$

If we take into consideration that:

$$e^{-\frac{10\pi j}{3}} = e^{-\frac{4\pi j}{3}} = -\cos 60° + j \sin 60° = -\frac{1}{2} + j\frac{\sqrt{3}}{2}$$

$$e^{-\frac{12\pi j}{3}} = 1$$

$$e^{-\frac{16\pi j}{3}} = e^{-\frac{4\pi j}{3}} = -\cos 60° + j \sin 60° = -\frac{1}{2} + j\frac{\sqrt{3}}{2}$$

and multiply the above matrix with W from the left, we recover matrix L.

OK, now we know how to overcome the problem of inverting H; however, how can we overcome the extreme sensitivity of equation (6.76) to noise?

We can do it by imposing a smoothness constraint to the solution, so that it does not fluctuate too much. Let us say that we would like the second derivative of the reconstructed image to be small overall. At each pixel, the sum of the second derivatives of the image along each axis can be approximated by:

$$\Delta^2 f(i,k) = f(i-1,k) + f(i,k-1) + f(i+1,k) + f(i,k+1) - 4f(i,k) \quad (6.111)$$

This is the value of the *Laplacian* at position (i,k). The constraint we choose to impose then is for the sum of the squares of the Laplacian values at each pixel position to be minimal:

$$\sum_{k=1}^{N}\sum_{i=1}^{N} \left[\Delta^2 f(i,k)\right]^2 = \text{minimal} \quad (6.112)$$

The value of the Laplacian at each pixel position can be computed by the Laplacian operator which has the form of an $N^2 \times N^2$ matrix acting on the column vector $\mathbf{f}$ (of size $N^2 \times 1$), $L\mathbf{f}$. $L\mathbf{f}$ is a vector. The sum of the squares of its elements are given by $(L\mathbf{f})^T L\mathbf{f}$. The constraint then is:

$$(L\mathbf{f})^T L\mathbf{f} = \text{minimal} \quad (6.113)$$

How can we incorporate the constraint in the inversion of the matrix?

Let us write again in matrix form the equation we want to solve for $\mathbf{f}$:

$$\mathbf{g} = H\mathbf{f} + \boldsymbol{\nu} \qquad (6.114)$$

We assume that the noise vector $\boldsymbol{\nu}$ is not known but some of its statistical properties are known; say we know that:

$$\boldsymbol{\nu}^T \boldsymbol{\nu} = \varepsilon \qquad (6.115)$$

This quantity ε is related to the variance of noise and could be estimated from the image itself using areas of uniform brightness only. If we substitute $\boldsymbol{\nu}$ from (6.114) into (6.115) we have:

$$(\mathbf{g} - H\mathbf{f})^T (\mathbf{g} - H\mathbf{f}) = \varepsilon \qquad (6.116)$$

The problem then is to minimize (6.113) under the constraint (6.116). The solution of this problem is a filter with Fourier transform:

$$\hat{M}(u, v) = \frac{1}{N} \frac{\hat{H}^*(u, v)}{|\hat{H}(u, v)|^2 + \gamma |\hat{L}(u, v)|^2} \qquad (6.117)$$

By multiplying numerator and denominator by $\hat{H}(u, v)$, we can extract the transfer function of the restoration filter as being:

$$\hat{M}(u, v) = \frac{1}{N} \frac{1}{\hat{H}(u, v)} \frac{|\hat{H}(u, v)|^2}{|\hat{H}(u, v)|^2 + \gamma |\hat{L}(u, v)|^2} \qquad (6.118)$$

where γ is a constant and $\hat{L}(u, v)$ is the Fourier transform of an $N \times N$ matrix L, with the following property: If we use it to multiply the image from the left, the output will be an array the same size as the image, with an estimate of the value of the Laplacian at each pixel position. The role of parameter γ is to strike the balance between smoothing the output and paying attention to the data.

B6.7: Find the solution of the problem: minimize $(L\mathbf{f})^T L\mathbf{f}$ with the constraint

$$[\mathbf{g} - H\mathbf{f}]^T [\mathbf{g} - H\mathbf{f}] = \varepsilon \qquad (6.119)$$

According to the method of *Lagrange multipliers* (see Box **B6.9**) the solution must satisfy:

$$\frac{\partial}{\partial \mathbf{f}} [\mathbf{f}^T L^T L\mathbf{f} + \lambda(\mathbf{g} - H\mathbf{f})^T (\mathbf{g} - H\mathbf{f})] = 0 \qquad (6.120)$$

where λ is a constant. This differentiation is with respect to a vector and it will yield a system of N^2 equations (one for each component of vector $\mathbf{f}$) which with equation (6.116) form a system of $N^2 + 1$ equations, for the $N^2 + 1$ unknowns: N^2 the components of $\mathbf{f}$ plus λ.

If $\mathbf{a}$ is a vector and $\mathbf{b}$ another one, then it can be shown (see Example 6.23) that:

$$\frac{\partial \mathbf{f}^T \mathbf{a}}{\partial \mathbf{f}} = \mathbf{a} \tag{6.121}$$

$$\frac{\partial \mathbf{b}^T \mathbf{f}}{\partial \mathbf{f}} = \mathbf{b} \tag{6.122}$$

Also, if A is an $N^2 \times N^2$ square matrix, then (see Example 6.24):

$$\frac{\partial \mathbf{f}^T A \mathbf{f}}{\partial \mathbf{f}} = (A + A^T)\mathbf{f} \tag{6.123}$$

We apply equations (6.121), (6.122) and (6.123) to (6.120) to perform the differentiation:

$$\frac{\partial}{\partial \mathbf{f}} \left[\underbrace{\mathbf{f}^T (L^T L)\mathbf{f}}_{\substack{\text{eqn(6.123)} \\ \text{with } A = L^T L}} + \lambda(\mathbf{g}^T\mathbf{g} - \underbrace{\mathbf{g}^T H \mathbf{f}}_{\substack{\text{eqn(6.122)} \\ \text{with } \mathbf{b} = H^T\mathbf{g}}} - \underbrace{\mathbf{f}^T H^T \mathbf{g}}_{\substack{\text{eqn(6.121)} \\ \text{with } \mathbf{a} = H^T\mathbf{g}}} + \underbrace{\mathbf{f}^T H^T H \mathbf{f}}_{\substack{\text{eqn(6.123) with} \\ A = H^T H}}) \right] = 0$$

$$\Rightarrow (2L^T L)\mathbf{f} + \lambda(-H^T\mathbf{g} - H^T\mathbf{g} + 2H^T H\mathbf{f}) = 0$$
$$\Rightarrow (H^T H + \gamma L^T L)\mathbf{f} = H^T\mathbf{g}$$

$$\tag{6.124}$$

where $\gamma \equiv \frac{1}{\lambda}$. Equation (6.124) can easily be solved in terms of block circulant matrices. Then:

$$\mathbf{f} = [H^T H + \gamma L^T L]^{-1} H^T \mathbf{g} \tag{6.125}$$

γ can be specified by substitution in equation (6.119).

B6.8 Solve equation (6.124).

Since H and L are block circulant matrices (see Examples 6.20, 6.22 and Box B6.6), they can be written as:

$$H = W\Lambda_h W^{-1} \qquad H^T = W\Lambda_h^* W^{-1}$$
$$L = W\Lambda_l W^{-1} \qquad L^T = W\Lambda_l^* W^{-1} \qquad (6.126)$$

Then:

$$
\begin{aligned}
H^T H + \gamma L^T L &= W\Lambda_h^* W^{-1} W \Lambda_h W^{-1} + \gamma W \Lambda_l^* W^{-1} W \Lambda_l W^{-1} \\
&= W\Lambda_h^* \Lambda_h W^{-1} + \gamma W \Lambda_l^* \Lambda_l W^{-1} \\
&= W(\Lambda_h^* \Lambda_h + \gamma \Lambda_l^* \Lambda_l) W^{-1} \qquad (6.127)
\end{aligned}
$$

We substitute from (6.126) and (6.127) into (6.124) to obtain:

$$W(\Lambda_h^* \Lambda_h + \gamma \Lambda_l^* \Lambda_l) W^{-1} \mathbf{f} = W \Lambda_h^* W^{-1} \mathbf{g} \qquad (6.128)$$

First we multiply both sides of the equation from the left with W^{-1}, to get:

$$(\Lambda_h^* \Lambda_h + \gamma \Lambda_l^* \Lambda_l) W^{-1} \mathbf{f} = \Lambda_h^* W^{-1} \mathbf{g} \qquad (6.129)$$

Notice that as Λ_h^*, $\Lambda_h^* \Lambda_h$ and $\Lambda_l^* \Lambda_l$ are diagonal matrices, this equation expresses a relationship between the corresponding elements of vectors $W^{-1}\mathbf{f}$ and $W^{-1}\mathbf{g}$ one by one.

Applying the result of Example 6.26, we can write:

$$\Lambda_h^* \Lambda_h = N^2 |\hat{H}(u,v)|^2 \quad \text{and} \quad \Lambda_l^* \Lambda_l = N^2 |\hat{L}(u,v)|^2$$

where $|\hat{L}(u,v)|^2$ is the Fourier transform of matrix L. Also, by applying the results of Example 6.25, we can write:

$$W^{-1}\mathbf{f} = \hat{F}(u,v) \quad \text{and} \quad W^{-1}\mathbf{g} = \hat{G}(u,v)$$

Finally, we replace Λ_h^* by its definition, equation (6.99), so that (6.129) becomes:

$$N^2 \left[|\hat{H}(u,v)|^2 + \gamma |\hat{L}(u,v)|^2 \right] \hat{F}(u,v) = N \hat{H}^*(u,v)\hat{G}(u,v) \Rightarrow$$

$$\hat{F}(u,v) = \frac{1}{N} \frac{\hat{H}^*(u,v)\hat{G}(u,v)}{|\hat{H}(u,v)|^2 + \gamma |\hat{L}(u,v)|^2} \qquad (6.130)$$

What is the relationship between the Wiener filter and the constrained matrix inversion filter?

Both filters look similar (see equations (6.47) and (6.118)), but they differ in many ways:

1. The Wiener filter is designed to optimize the restoration in an average statistical sense over a large ensemble of similar images. The constrained matrix inversion deals with one image only and imposes constraints on the solution sought.

2. The Wiener filter is based on the assumption that the random fields involved are homogeneous with known spectral densities. In the constrained matrix inversion it is assumed that we know only some statistical property of the noise.

In the constraint matrix restoration approach, various filters may be constructed using the same formulation by simply changing the smoothing criterion. For example, one may try to minimize the sum of the squares of the first derivatives at all positions as opposed to the second derivatives. The only difference from the formula (6.117) will be in matrix L.

Example 6.23 (B)

Differentiation by a vector is defined as differentiation by each of the elements of the vector. For vectors a, b and f show that:

$$\frac{\partial \mathbf{f}^T \mathbf{a}}{\partial \mathbf{f}} = \mathbf{a} \quad \text{and} \quad \frac{\partial \mathbf{f}^T \mathbf{b}}{\partial \mathbf{f}} = \mathbf{b}$$

Assume that vectors **a**, **b** *and* **f** *are* $N \times 1$. *Then we have:*

$$\mathbf{f}^T \mathbf{a} = \begin{pmatrix} f_1 & f_2 & \cdots & f_N \end{pmatrix} \begin{pmatrix} a_1 \\ a_2 \\ \vdots \\ a_N \end{pmatrix} = f_1 a_1 + f_2 a_2 + \ldots + f_N a_N$$

Use this in:

$$\frac{\partial \mathbf{f}^T \mathbf{a}}{\partial \mathbf{f}} \equiv \begin{pmatrix} \frac{\partial \mathbf{f}^T \mathbf{a}}{\partial f_1} \\ \frac{\partial \mathbf{f}^T \mathbf{a}}{\partial f_2} \\ \vdots \\ \frac{\partial \mathbf{f}^T \mathbf{a}}{\partial f_N} \end{pmatrix} = \begin{pmatrix} a_1 \\ a_2 \\ \vdots \\ a_N \end{pmatrix} \Rightarrow \frac{\partial \mathbf{f}^T \mathbf{a}}{\partial \mathbf{f}} = \mathbf{a}$$

Similarly:

$$\mathbf{b}^T \mathbf{f} = b_1 f_1 + b_2 f_2 + \ldots + b_N f_N$$

Then

$$\frac{\partial \mathbf{b}^T \mathbf{f}}{\partial \mathbf{f}} = \begin{pmatrix} \frac{\partial \mathbf{b}^T \mathbf{f}}{\partial f_1} \\ \frac{\partial \mathbf{b}^T \mathbf{f}}{\partial f_2} \\ \vdots \\ \frac{\partial \mathbf{b}^T \mathbf{f}}{\partial f_N} \end{pmatrix} = \begin{pmatrix} b_1 \\ b_2 \\ \vdots \\ b_N \end{pmatrix} \Rightarrow \frac{\partial \mathbf{b}^T \mathbf{f}}{\partial \mathbf{f}} = \mathbf{b}$$

Example 6.24 (B)

If **f** is an $N \times 1$ vector and A is an $N \times N$ matrix, show that

$$\frac{\partial \mathbf{f}^T A \mathbf{f}}{\partial \mathbf{f}} = (A + A^T)\mathbf{f}$$

Using the results of Example 6.23 we can easily see that:

$$\begin{aligned} \frac{\partial \mathbf{f}^T A \mathbf{f}}{\partial \mathbf{f}} &= \frac{\partial \mathbf{f}^T (A\mathbf{f})}{\partial \mathbf{f}} + \frac{\partial (\mathbf{f}^T A)\mathbf{f}}{\partial \mathbf{f}} \\ &= A\mathbf{f} + \frac{\partial (A^T \mathbf{f})^T \mathbf{f}}{\partial \mathbf{f}} \\ &= A\mathbf{f} + A^T \mathbf{f} = (A + A^T)\mathbf{f} \end{aligned}$$

(We make use of the fact that $A\mathbf{f}$ and $A^T\mathbf{f}$ are vectors.)

Example 6.25 (B)

If **g** is the column vector that corresponds to a 3×3 image G and matrix W^{-1} is defined as in Example 6.18 for $N = 3$, show that vector $W^{-1}\mathbf{g}$ is the discrete Fourier transform $\hat{G}$ of G.

Assume that:

$$G = \begin{pmatrix} g_{11} & g_{12} & g_{13} \\ g_{21} & g_{22} & g_{23} \\ g_{31} & g_{32} & g_{33} \end{pmatrix} \quad and \quad W_3^{-1} = \frac{1}{\sqrt{3}} \begin{pmatrix} 1 & 1 & 1 \\ 1 & e^{-\frac{2\pi j}{3}} & e^{-\frac{2\pi j}{3}2} \\ 1 & e^{-\frac{2\pi j}{3}2} & e^{-\frac{2\pi j}{3}} \end{pmatrix}$$

Then

$$W^{-1} = W_3^{-1} \otimes W_3^{-1}$$

$$= \frac{1}{3} \begin{pmatrix} 1 & 1 & 1 & 1 & 1 & 1 & 1 & 1 & 1 \\ 1 & e^{-\frac{2\pi j}{3}} & e^{-\frac{2\pi j}{3}2} & 1 & e^{-\frac{2\pi j}{3}} & e^{-\frac{2\pi j}{3}2} & 1 & e^{-\frac{2\pi j}{3}} & e^{-\frac{2\pi j}{3}2} \\ 1 & e^{-\frac{2\pi j}{3}2} & e^{-\frac{2\pi j}{3}} & 1 & e^{-\frac{2\pi j}{3}2} & e^{-\frac{2\pi j}{3}} & 1 & e^{-\frac{2\pi j}{3}2} & e^{-\frac{2\pi j}{3}} \\ 1 & 1 & 1 & e^{-\frac{2\pi j}{3}} & e^{-\frac{2\pi j}{3}} & e^{-\frac{2\pi j}{3}} & e^{-\frac{2\pi j}{3}2} & e^{-\frac{2\pi j}{3}2} & e^{-\frac{2\pi j}{3}2} \\ 1 & e^{-\frac{2\pi j}{3}} & e^{-\frac{2\pi j}{3}2} & e^{-\frac{2\pi j}{3}} & e^{-\frac{2\pi j}{3}2} & e^{-\frac{2\pi j}{3}3} & e^{-\frac{2\pi j}{3}2} & e^{-\frac{2\pi j}{3}3} & e^{-\frac{2\pi j}{3}4} \\ 1 & e^{-\frac{2\pi j}{3}2} & e^{-\frac{2\pi j}{3}} & e^{-\frac{2\pi j}{3}} & e^{-\frac{2\pi j}{3}3} & e^{-\frac{2\pi j}{3}2} & e^{-\frac{2\pi j}{3}2} & e^{-\frac{2\pi j}{3}4} & e^{-\frac{2\pi j}{3}3} \\ 1 & 1 & 1 & e^{-\frac{2\pi j}{3}2} & e^{-\frac{2\pi j}{3}2} & e^{-\frac{2\pi j}{3}2} & e^{-\frac{2\pi j}{3}} & e^{-\frac{2\pi j}{3}} & e^{-\frac{2\pi j}{3}} \\ 1 & e^{-\frac{2\pi j}{3}} & e^{-\frac{2\pi j}{3}2} & e^{-\frac{2\pi j}{3}2} & e^{-\frac{2\pi j}{3}3} & e^{-\frac{2\pi j}{3}4} & e^{-\frac{2\pi j}{3}} & e^{-\frac{2\pi j}{3}2} & e^{-\frac{2\pi j}{3}3} \\ 1 & e^{-\frac{2\pi j}{3}2} & e^{-\frac{2\pi j}{3}} & e^{-\frac{2\pi j}{3}2} & e^{-\frac{2\pi j}{3}4} & e^{-\frac{2\pi j}{3}3} & e^{-\frac{2\pi j}{3}} & e^{-\frac{2\pi j}{3}3} & e^{-\frac{2\pi j}{3}2} \end{pmatrix}$$

If we use $e^{-\frac{2\pi j}{3}3} = e^{-2\pi j} = 1$ and $e^{-\frac{2\pi j}{3}4} = e^{-\frac{2\pi j}{3}3}e^{-\frac{2\pi j}{3}} = e^{-\frac{2\pi j}{3}}$, this matrix simplifies somehow. So we get:

$$W^{-1}g =$$

$$\frac{1}{3} \begin{pmatrix} 1 & 1 & 1 & 1 & 1 & 1 & 1 & 1 & 1 \\ 1 & e^{-\frac{2\pi j}{3}} & e^{-\frac{2\pi j}{3}2} & 1 & e^{-\frac{2\pi j}{3}} & e^{-\frac{2\pi j}{3}2} & 1 & e^{-\frac{2\pi j}{3}} & e^{-\frac{2\pi j}{3}2} \\ 1 & e^{-\frac{2\pi j}{3}2} & e^{-\frac{2\pi j}{3}} & 1 & e^{-\frac{2\pi j}{3}2} & e^{-\frac{2\pi j}{3}} & 1 & e^{-\frac{2\pi j}{3}2} & e^{-\frac{2\pi j}{3}} \\ 1 & 1 & 1 & e^{-\frac{2\pi j}{3}} & e^{-\frac{2\pi j}{3}} & e^{-\frac{2\pi j}{3}} & e^{-\frac{2\pi j}{3}2} & e^{-\frac{2\pi j}{3}2} & e^{-\frac{2\pi j}{3}2} \\ 1 & e^{-\frac{2\pi j}{3}} & e^{-\frac{2\pi j}{3}2} & e^{-\frac{2\pi j}{3}} & e^{-\frac{2\pi j}{3}2} & 1 & e^{-\frac{2\pi j}{3}2} & 1 & e^{-\frac{2\pi j}{3}} \\ 1 & e^{-\frac{2\pi j}{3}2} & e^{-\frac{2\pi j}{3}} & e^{-\frac{2\pi j}{3}} & 1 & e^{-\frac{2\pi j}{3}2} & e^{-\frac{2\pi j}{3}2} & e^{-\frac{2\pi j}{3}} & 1 \\ 1 & 1 & 1 & e^{-\frac{2\pi j}{3}2} & e^{-\frac{2\pi j}{3}2} & e^{-\frac{2\pi j}{3}2} & e^{-\frac{2\pi j}{3}} & e^{-\frac{2\pi j}{3}} & e^{-\frac{2\pi j}{3}} \\ 1 & e^{-\frac{2\pi j}{3}} & e^{-\frac{2\pi j}{3}2} & e^{-\frac{2\pi j}{3}2} & 1 & e^{-\frac{2\pi j}{3}} & e^{-\frac{2\pi j}{3}} & e^{-\frac{2\pi j}{3}2} & 1 \\ 1 & e^{-\frac{2\pi j}{3}2} & e^{-\frac{2\pi j}{3}} & e^{-\frac{2\pi j}{3}2} & e^{-\frac{2\pi j}{3}} & 1 & e^{-\frac{2\pi j}{3}} & 1 & e^{-\frac{2\pi j}{3}2} \end{pmatrix} \begin{pmatrix} g_{11} \\ g_{21} \\ g_{31} \\ g_{12} \\ g_{22} \\ g_{32} \\ g_{13} \\ g_{23} \\ g_{33} \end{pmatrix}$$

$$= \frac{1}{3} \begin{pmatrix} g_{11}+g_{21}+g_{31}+g_{12}+g_{22}+g_{32}+g_{13}+g_{23}+g_{33} \\ g_{11}+g_{21}e^{-\frac{2\pi j}{3}}+g_{31}e^{-\frac{2\pi j}{3}2}+g_{12}+g_{22}e^{-\frac{2\pi j}{3}}+g_{32}e^{-\frac{2\pi j}{3}2}+g_{13}+g_{23}e^{-\frac{2\pi j}{3}}+g_{33}e^{-\frac{2\pi j}{3}2} \\ \vdots \\ g_{11}+g_{21}+g_{31}+g_{12}e^{-\frac{2\pi j}{3}2}+g_{22}e^{-\frac{2\pi j}{3}2}+g_{32}e^{-\frac{2\pi j}{3}2}+g_{13}e^{-\frac{2\pi j}{3}}+g_{23}e^{-\frac{2\pi j}{3}}+g_{33}e^{-\frac{2\pi j}{3}} \\ \vdots \end{pmatrix}$$

Careful examination of the elements of this vector shows that they are the Fourier components of G computed at various combinations of frequencies (u, v), for $u = 0, 1, 2$ and $v = 0, 1, 2$, arranged as follows:

$$\begin{pmatrix} \hat{G}(0,0) \\ \hat{G}(1,0) \\ \hat{G}(2,0) \\ \hat{G}(0,1) \\ \hat{G}(1,1) \\ \hat{G}(2,1) \\ \hat{G}(0,2) \\ \hat{G}(1,2) \\ \hat{G}(2,2) \end{pmatrix}$$

This shows that $W^{-1}\mathbf{g}$ yields the Fourier transform of G, as a column vector.

Example 6.26 (B)

Show that if matrix Λ is defined by equation (6.99) then $\Lambda^*\Lambda$ is a diagonal matrix with its kth element along the diagonal being $N^2|\hat{H}(k_2,k_1)|^2$ where $k_2 \equiv k_{mod\ N}$ and $k_1 \equiv \left[\frac{k}{N}\right]$.

From the definition of Λ, equation (6.99), we can write:

$$\Lambda = \begin{pmatrix} N\hat{H}(0,0) & 0 & 0 & \cdots & 0 \\ 0 & N\hat{H}(1,0) & 0 & \cdots & 0 \\ 0 & 0 & N\hat{H}(2,0) & \cdots & 0 \\ \vdots & \vdots & \vdots & & \vdots \\ 0 & 0 & 0 & \cdots & N\hat{H}(N-1,N-1) \end{pmatrix}$$

Then:

$$\Lambda^* = \begin{pmatrix} N\hat{H}^*(0,0) & 0 & 0 & \cdots & 0 \\ 0 & N\hat{H}^*(1,0) & 0 & \cdots & 0 \\ 0 & 0 & N\hat{H}^*(2,0) & \cdots & 0 \\ \vdots & \vdots & \vdots & & \vdots \\ 0 & 0 & 0 & \cdots & N\hat{H}^*(N-1,N-1) \end{pmatrix}$$

Obviously

$$\Lambda^*\Lambda = \begin{pmatrix} N|\hat{H}(0,0)|^2 & 0 & 0 & \dots & 0 \\ 0 & N|\hat{H}(1,0)|^2 & 0 & \dots & 0 \\ 0 & 0 & N|\hat{H}(2,0)|^2 & \dots & 0 \\ \vdots & \vdots & \vdots & & \vdots \\ 0 & 0 & 0 & \dots & N|\hat{H}(N-1,N-1)|^2 \end{pmatrix}$$

B6.9 Why does the method of Lagrange multipliers work?

Suppose that we wish to satisfy two equations simultaneously:

$$f(x,y) = 0$$
$$g(x,y) = 0 \qquad\qquad (6.131)$$

Let us assume that in the (x,y) plane the first of these equations is satisfied at point A and the other at point B, so that it is impossible to satisfy both equations exactly for the same value of (x,y).

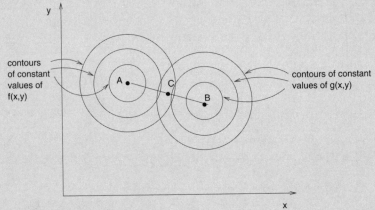

We wish to find a point C on the plane where we make the least compromise in violating these two equations. The location of this point will depend on how fast the values of $f(x,y)$ and $g(x,y)$ change from 0 as we move away from points A and B respectively.

Let us consider the isocontours of f and g around each of the points A and B respectively. As the contours grow away from point A, function $|f(x,y)|$ takes larger and larger values, while as contours grow away from point B, the values function $|g(x,y)|$ takes become larger as well. Point C where the values of $|f(x,y)|$ and $g|(x,y)|$ are as small as possible (minimum violation of the constraints which demand that $|f(x,y)| = |g(x,y)| = 0$), must be the point where an isocontour around A just touches an isocontour around B, without crossing each other. When two curves just touch each other, their tangents become parallel. The tangent vector to a curve along which $f =$ constant is ∇f, and the tangent vector to a curve along which $g =$ constant is ∇g. The two tangent vectors do not need to have the same magnitude for the minimum violation of the constraints. It is enough for them to have the same orientation. Therefore, we say that point C is determined by the solution of equation $\nabla f = \mu \nabla g$ where μ is some constant that takes care of the (possibly) different magnitudes of the two vectors. In other words, the solution to the problem of simultaneous satisfaction of the two incompatible equations (6.131) is the solution of the differential set of equations:

$$\nabla f + \lambda \nabla g = 0$$

where λ is the Lagrange multiplier, an arbitrary constant.

Example 6.27

Demonstrate how to apply the constraint matrix inversion in practice, by restoring a motion blurred image.

To apply the filter given by equation (6.117) to the blurred image of Figure 6.7a, we must first define matrix $\hat{L}(u,v)$ which expresses the constraint.

Following the steps of Example 6.19, we can see that matrix $L(i,j)$ with which we have to multiply an $N \times N$ image in order to obtain the value of the Laplacian at each position is given by an $N^2 \times N^2$ matrix of the following structure:

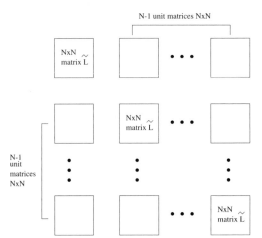

Matrix $\tilde{L}$ has the following form:

$$\tilde{L} = \begin{pmatrix} -4 & 1 & \overbrace{0 & 0 & \dots & 0}^{N-3 \ zeroes} & 1 \\ 1 & -4 & 1 & 0 & \dots & 0 & 0 \\ 0 & 1 & -4 & 1 & \dots & 0 & 0 \\ 0 & 0 & 1 & -4 & \dots & 0 & 0 \\ 0 & 0 & 0 & 1 & \dots & 0 & 0 \\ \vdots & \vdots & \vdots & \vdots & \dots & \vdots & \vdots \\ 0 & 0 & 0 & 0 & \dots & -4 & 1 \\ 1 & 0 & 0 & 0 & \dots & 1 & -4 \end{pmatrix}$$

To form the kernel we require, we must take the first column of matrix L and wrap it to form an $N \times N$ matrix. The first column of matrix L consists of the first column of matrix $\tilde{L}$ (N elements) plus the first columns of $N-1$ unit matrices of size $N \times N$. These N^2 elements have to be written as N columns of size N next to each other, to form an $N \times N$ matrix L', say

$$L' = \begin{pmatrix} -4 & 1 & 1 & \dots & 1 \\ 1 & 0 & 0 & \dots & 0 \\ 0 & 0 & 0 & \dots & 0 \\ \vdots & \vdots & \vdots & \dots & \vdots \\ 0 & 0 & 0 & \dots & 0 \\ 1 & 0 & 0 & \dots & 0 \end{pmatrix}$$

It is the Fourier transform of this matrix that appears in the constraint matrix inversion filter.

We calculate the Fourier transform using formula (6.100) with $N = 128$. Let us say that the real and the imaginary parts of this transform are $L_1(m, n)$ and $L_2(m, n)$ respectively. Then

$$\left|\hat{L}(m, n)\right|^2 = L_1^2(m, n) + L_2^2(m, n)$$

The transfer function of the filter we must use then is given by substituting the transfer function (6.33) into equation (6.117):

$$\hat{M}(m, n) = \frac{1}{N} \frac{\frac{1}{i_T \sin \frac{\pi m}{N}} \sin \frac{i_T \pi m}{N}}{\frac{\sin^2 \frac{i_T \pi m}{N}}{i_T^2 \sin^2 \frac{\pi m}{N}} + \gamma \left(L_1^2(m, n) + L_2^2(m, n)\right)} e^{j \frac{\pi m}{N}(i_T - 1)}$$

Note that for $m = 0$ we must use:

$$\hat{M}(0, n) = \frac{1}{N} \frac{1}{1 + \gamma \left(L_1^2(0, n) + L_2^2(0, n)\right)} \quad \text{for} \quad 0 \le n \le N - 1$$

Working as for the case of the Wiener filtering we can find that the real and imaginary parts of the Fourier transform of the original image are given by:

$$F_1(m, n) = \frac{1}{N} \frac{i_T \sin \frac{\pi m}{N} \sin \frac{i_T \pi m}{N} \left[G_1(m, n) \cos \frac{(i_T - 1)\pi m}{N} - G_2(m, n) \sin \frac{(i_T - 1)\pi m}{N}\right]}{\sin^2 \frac{i_T \pi m}{N} + \gamma \left(L_1^2(m, n) + L_2^2(m, n)\right) i_T^2 \sin^2 \frac{\pi m}{N}}$$

$$F_2(m, n) = \frac{1}{N} \frac{i_T \sin \frac{\pi m}{N} \sin \frac{i_T \pi m}{N} \left[G_1(m, n) \sin \frac{(i_T - 1)\pi m}{N} + G_2(m, n) \cos \frac{(i_T - 1)\pi m}{N}\right]}{\sin^2 \frac{i_T \pi m}{N} + \gamma \left(L_1^2(m, n) + L_2^2(m, n)\right) i_T^2 \sin^2 \frac{\pi m}{N}}$$

These formulae are valid for $0 < m \le N - 1$ and $0 \le n \le N - 1$. For $m = 0$ we must use the formulae:

$$F_1(0, n) = \frac{1}{N} \frac{G_1(0, n)}{1 + \gamma \left(L_1^2(0, n) + L_2^2(0, n)\right)}$$

$$F_2(0, n) = \frac{1}{N} \frac{G_2(0, n)}{1 + \gamma \left(L_1^2(0, n) + L_2^2(0, n)\right)}$$

If we take the inverse Fourier transform using functions $F_1(m, n)$ and $F_2(m, n)$ as the real and the imaginary parts, we obtain the restored image. The results of restoring images 6.7b, 6.8a and 6.8c are shown in Figure 6.11. Note that different values of γ, i.e. different levels of smoothing, have to be used for different levels of noise in the image.

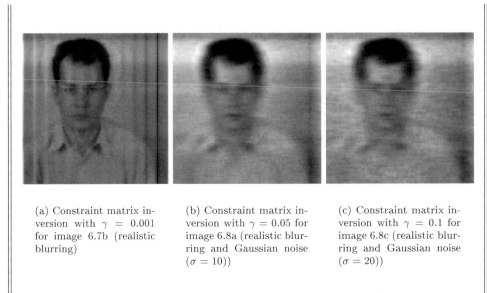

(a) Constraint matrix inversion with $\gamma = 0.001$ for image 6.7b (realistic blurring)

(b) Constraint matrix inversion with $\gamma = 0.05$ for image 6.8a (realistic blurring and Gaussian noise $(\sigma = 10)$)

(c) Constraint matrix inversion with $\gamma = 0.1$ for image 6.8c (realistic blurring and Gaussian noise $(\sigma = 20)$)

Figure 6.11: Restoration with constraint matrix inversion.

What is the "take home" message of this chapter?

This chapter explored some techniques used to correct (i.e. restore) the damaged values of an image. The problem of restoration requires some prior knowledge concerning the original uncorrupted signal or the imaged scene, and in that way differs from the image enhancement problem. Geometric restoration of an image requires knowledge of the correct location of some reference points.

Grey level restoration of an image requires knowledge of some statistical properties of the corrupting noise, the blurring process and the original image itself. Often, we bypass the requirement for knowing the statistical properties of the original image by imposing some spatial smoothness constraints on the solution, based on the heuristic that "the world is largely smooth". Having chosen the correct model for the degradation process and the uncorrupted image, we have then to solve the problem of recovering the original image values.

The full problem of image restoration is a very difficult one as it is non-linear. It can be solved with the help of local or global optimization approaches. However, simpler solutions can be found, in the form of convolution filters, if we make the assumption that the degradation process is shift invariant and if we restrict the domain of the sought solution to that of linear solutions only. *Figure 6.12* summarizes the results obtained by the various restoration methods discussed, for an image blurred by motion along the horizontal axis.

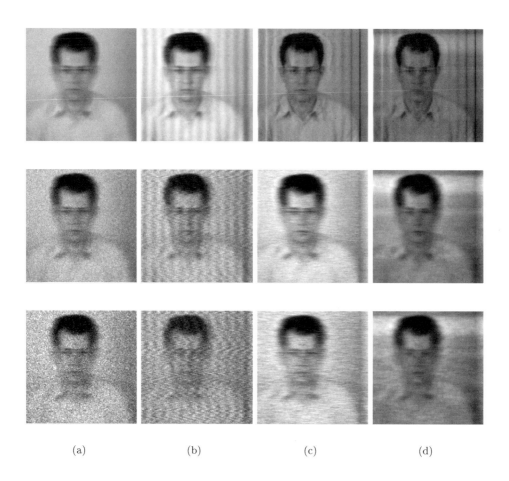

(a)   (b)   (c)   (d)

Figure 6.12: (a) Original images. (b) Restoration by inverse filtering with omission of all terms beyond the first zero of the filter transfer function. (c) Restoration by Wiener filtering. (d) Restoration by constraint matrix inversion.

Chapter 7

Image Segmentation and Edge Detection

What is this chapter about?

This chapter is about those Image Processing techniques that are used in order to prepare an image as an input to an automatic vision system. These techniques perform *image segmentation* and *edge detection*, and their purpose is to **extract** information from an image in such a way that the output image contains much **less** information than the original one, but the little information it contains is much more relevant to the other modules of an automatic vision system than the discarded information.

What exactly is the purpose of image segmentation and edge detection?

The purpose of image segmentation and edge detection is to extract the outlines of different regions in the image; i.e. to divide the image into regions which are made up of pixels which have something in common. For example, they may have similar brightness, or colour, which may indicate that they belong to the same object or facet of an object.

How can we divide an image into uniform regions?

One of the simplest methods is that of histogramming and *thresholding*. If we plot the number of pixels which have a specific grey level value, versus that value, we create the histogram of the image. Properly normalized, the histogram is essentially the probability density function for a certain grey level value to occur. Suppose that we have images consisting of bright objects on a dark background and suppose that we want to extract the objects. For such an image, the histogram will have two peaks and a valley between them.

We can choose as the threshold then the grey level value which corresponds to the valley of the histogram, indicated by t_0 in *Figure* 7.1, and *label* all pixels with grey

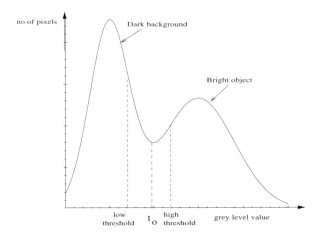

Figure 7.1: The histogram of an image with a bright object on a dark background.

level values greater than t_0 as object pixels and pixels with grey level values smaller than t_0 as background pixels.

What do we mean by "labelling" an image?

When we say we "extract" an object in an image, we mean that we identify the pixels that make it up. To express this information, we create an array of the same size as the original image and we give to each pixel a *label*. All pixels that make up the object are given the same label and all pixels that make up the background are given a different label. The label is usually a number, but it could be anything: a letter or a colour. It is essentially a name and it has symbolic meaning only. Labels, therefore, cannot be treated as numbers. Label images cannot be processed in the same way as grey level images. Often label images are also referred to as *classified images* as they indicate the *class* to which each pixel belongs.

What can we do if the valley in the histogram is not very sharply defined?

If there is no clear valley in the histogram of an image, it means that there are several pixels in the background which have the same grey level value as pixels in the object and vice versa. Such pixels are particularly encountered near the boundaries of the objects which may be fuzzy and not sharply defined. One can use then what is called *hysteresis thresholding*: instead of one, two threshold values (see *Figure* 7.1) are chosen on either side of the valley.

The highest of the two thresholds is used to define the "hard core" of the object. The lowest is used in conjunction with spatial proximity of the pixels: a pixel with intensity value greater than the smaller threshold but less than the larger threshold is labelled as object pixel only if it is adjacent to a pixel which is a core object pixel.

Figure 7.2 shows an image depicting a dark object on a bright background and its histogram. In 7.2c the image is segmented with a single threshold, marked with a t in the histogram, while in 7.2d it has been segmented using two thresholds marked t_1 and t_2 in the histogram.

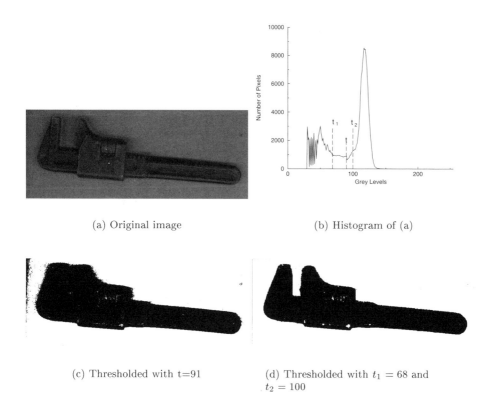

(a) Original image

(b) Histogram of (a)

(c) Thresholded with t=91

(d) Thresholded with $t_1 = 68$ and $t_2 = 100$

Figure 7.2: Simple thresholding versus hysteresis thresholding.

Alternatively, we may try to choose the global threshold value in an optimal way. Since we know we are bound to misclassify some pixels, we may try to minimize the number of misclassified pixels.

How can we minimize the number of misclassified pixels?

We can minimize the number of misclassified pixels if we have some prior knowledge about the distributions of the grey values that make up the object and the background.

For example, if we know that the objects occupy a certain fraction θ of the area of the picture then this θ is the prior probability for a pixel to be an object pixel. Clearly the background pixels occupy $1 - \theta$ of the area and a pixel has $1 - \theta$ prior probability to be a background pixel. We may choose the threshold then so that the pixels we classify as object pixels are a θ fraction of the total number of pixels. This method is called *p-tile* method. Further, if we also happen to know the probability density functions of the grey values of the object pixels and the background pixels, then we may choose the threshold that exactly minimizes the error.

B7.1 Differentiation of an integral with respect to a parameter.

Suppose that the definite integral $I(\lambda)$ depends on a parameter λ as follows:

$$I(\lambda) = \int_{a(\lambda)}^{b(\lambda)} f(x; \lambda)dx$$

Its derivative with respect to λ is given by the following formula, known as the *Leibnitz rule*:

$$\frac{dI(\lambda)}{d\lambda} = \frac{db(\lambda)}{d\lambda}f(b(\lambda); \lambda) - \frac{da(\lambda)}{d\lambda}f(a(\lambda); \lambda) + \int_{a(\lambda)}^{b(\lambda)} \frac{\partial f(x; \lambda)}{\partial \lambda}dx \qquad (7.1)$$

How can we choose the minimum error threshold?

Let us assume that the pixels which make up the object are distributed according to the probability density function $p_o(x)$ and the pixels which make up the background are distributed according to function $p_b(x)$.

Suppose that we choose a threshold value t (see *Figure 7.3*). Then the error committed by misclassifying object pixels as background pixels will be given by:

$$\int_{-\infty}^{t} p_o(x)dx$$

and the error committed by misclassifying background pixels as object pixels is:

$$\int_{t}^{+\infty} p_b(x)dx$$

In other words, the error that we commit arises from misclassifying the two tails of the two probability density functions on either side of threshold t. Let us also assume that the fraction of the pixels that make up the object is θ, and by inference, the

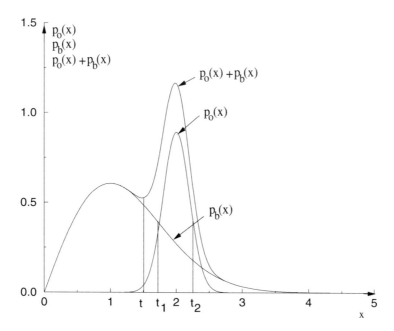

Figure 7.3: The probability density functions of the grey values of the pixels that make up the object ($p_o(x)$) and the background ($p_b(x)$). Their sum, normalized to integrate to 1, is what we obtain if we take the histogram of an image and normalize it.

fraction of the pixels that make up the background is $1 - \theta$. Then the total error is:

$$E(t) = \theta \int_{-\infty}^{t} p_o(x)dx + (1 - \theta) \int_{t}^{+\infty} p_b(x)dx \qquad (7.2)$$

We would like to choose t so that $E(t)$ is minimum. We take the first derivative of $E(t)$ with respect to t (see Box **B7.1**) and set it to zero:

$$\frac{\partial E}{\partial t} = \theta p_o(t) - (1 - \theta)p_b(t) = 0 \Rightarrow$$

$$\theta p_o(t) = (1 - \theta)p_b(t) \qquad (7.3)$$

The solution of this equation gives the minimum error threshold, for any type of distributions the two pixel populations have.

Example 7.1 (B)

Derive equation (7.3) from (7.2).

We apply the Leibnitz rule given by equation (7.1) to perform the differentiation of $E(t)$ given by equation (7.2). We have the following correspondences:
Parameter λ corresponds to t.
For the first integral:

$$a(\lambda) \rightarrow -\infty \qquad \text{(a constant, with zero derivative)}$$
$$b(\lambda) \rightarrow t$$
$$f(x; \lambda) \rightarrow p_0(x) \qquad \text{(independent from the parameter with respect to which we differentiate)}$$

For the second integral:

$$a(\lambda) \rightarrow t$$
$$b(\lambda) \rightarrow -\infty \qquad \text{(a constant, with zero derivative)}$$
$$f(x; \lambda) \rightarrow p_b(x) \qquad \text{(independent from t)}$$

Equation (7.3) then follows.

Example 7.2

The grey level values of the object and the background pixels are distributed according to the probability density function:

$$p(x) = \begin{cases} \frac{3}{4a^3}\left[a^2 - (x-b)^2\right] & \text{for } b - a \le x \le b + a \\ 0 & \text{otherwise} \end{cases}$$

with $a = 1$, $b = 5$ for the background and $a = 2$, $b = 7$ for the object. Sketch the two distributions and determine the range of possible thresholds.

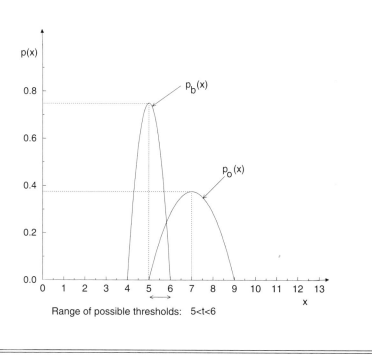

Range of possible thresholds: 5<t<6

Example 7.3

If the object pixels are eight-ninths $\left(\frac{8}{9}\right)$ of the total number of pixels, determine the threshold that minimizes the fraction of misclassified pixels for the problem of Example 7.2.

We substitute into equation (7.3) the following:

$$\theta = \frac{8}{9} \Rightarrow 1 - \theta = \frac{1}{9}$$

$$p_b(t) = \frac{3}{4}(-t^2 - 24 + 10t) \qquad p_0(t) = \frac{3}{32}(-t^2 - 45 + 14t)$$

Then:

$$\frac{1}{9}\frac{3}{4}(-t^2 - 24 + 10t) = \frac{8}{9}\frac{3}{32}(-t^2 - 45 + 14t) \Rightarrow$$

$$-24 + 10t = -45 + 14t \Rightarrow 4t = 21 \Rightarrow t = \frac{21}{4} = 5.25 \qquad (7.4)$$

Example 7.4

The grey level values of the object and the background pixels are distributed according to the probability density function:

$$p(x) = \begin{cases} \frac{\pi}{4a} \cos \frac{(x - x_0)\pi}{2a} & \text{for } x_0 - a \leq x \leq x_0 + a \\ 0 & \text{otherwise} \end{cases}$$

with $x_0 = 1$, $a = 1$ for the objects, and $x_0 = 3$, $a = 2$ for the background. Sketch the two probability density functions. If one-third of the total number of pixels are object pixels, determine the fraction of misclassified object pixels by optimal thresholding.

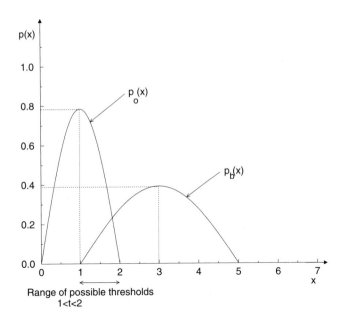

Apply formula (7.3) with:

$$\theta = \frac{1}{3} \Rightarrow 1 - \theta = \frac{2}{3}$$

$$p_o(x) = \frac{\pi}{4} \cos \frac{(x - 1)\pi}{2} \qquad p_b(x) = \frac{\pi}{8} \cos \frac{(x - 3)\pi}{4}$$

Equation (7.3) becomes:

$$\frac{1}{3} \frac{\pi}{4} \cos \frac{(t - 1)\pi}{2} = \frac{2}{3} \frac{\pi}{8} \cos \frac{(t - 3)\pi}{4} \Rightarrow$$

$$\Rightarrow \cos \frac{(t-1)\pi}{2} = \cos \frac{(t-3)\pi}{4} \Rightarrow$$

$$\Rightarrow \frac{(t-1)\pi}{2} \stackrel{+}{=} \frac{(t-3)\pi}{4}$$

Consider first $\frac{t-1}{2}\pi = \frac{t-3}{4}\pi \Rightarrow 2t - 2 = t - 3 \Rightarrow t = -1$.
This value is outside the acceptable range, so it is a meaningless solution.
Then:

$$\frac{(t-1)\pi}{2} = -\frac{(t-3)\pi}{4} \Rightarrow 2t - 2 = -t + 3 \Rightarrow 3t = 5 \Rightarrow t = \frac{5}{3}$$

This is the threshold for minimum error. The fraction of misclassified object pixels will be given by all those object pixels that have grey value greater than $\frac{5}{3}$. We define a new variable of integration $y \equiv x - 1$, to obtain:

$$\int_{\frac{5}{3}}^{2} \frac{\pi}{4} \cos \frac{(x-1)\pi}{2} dx = \frac{\pi}{4} \int_{\frac{2}{3}}^{1} \cos \frac{y\pi}{2} dy = \frac{\pi}{4} \frac{\sin \frac{y\pi}{2}}{\frac{\pi}{2}} \Big|_{\frac{2}{3}}^{1}$$

$$= \frac{1}{2} \left(\sin \frac{\pi}{2} - \sin \frac{\pi}{3} \right) = \frac{1}{2}(1 - \sin 60^\circ) = \frac{1}{2} \left(1 - \frac{\sqrt{3}}{2} \right)$$

$$= \frac{2 - 1.7}{4} = \frac{0.3}{4} = 0.075 = 7.5\%$$

What is the minimum error threshold when object and background pixels are normally distributed?

Let us assume that the pixels that make up the object are normally distributed with mean μ_o and standard deviation σ_o and the pixels that make up the background are normally distributed with mean μ_b and the standard deviation σ_b:

$$p_o(x) = \frac{1}{\sqrt{2\pi}\sigma_o} \exp\left[-\frac{(x-\mu_o)^2}{2\sigma_o^2} \right]$$

$$p_b(x) = \frac{1}{\sqrt{2\pi}\sigma_b} \exp\left[-\frac{(x-\mu_b)^2}{2\sigma_b^2} \right] \tag{7.5}$$

Upon substitution into equation (7.3) we obtain:

$$\theta \frac{1}{\sqrt{2\pi}\sigma_o} \exp\left[-\frac{(t-\mu_o)^2}{2\sigma_o^2} \right] = (1-\theta) \frac{1}{\sqrt{2\pi}\sigma_b} \exp\left[-\frac{(t-\mu_b)^2}{2\sigma_b^2} \right] \Rightarrow$$

$$\exp\left[-\frac{(t-\mu_o)^2}{2\sigma_o^2} + \frac{(t-\mu_b)^2}{2\sigma_b^2} \right] = \frac{1-\theta}{\theta} \frac{\sigma_o}{\sigma_b} \Rightarrow$$

$$-\frac{(t-\mu_o)^2}{2\sigma_o^2} + \frac{(t-\mu_b)^2}{2\sigma_b^2} = \ln\left[\frac{\sigma_o}{\sigma_b}\frac{1-\theta}{\theta}\right] \Rightarrow$$

$$(t^2 + \mu_b^2 - 2t\mu_b)\sigma_o^2 - (t^2 + \mu_o^2 - 2\mu_o t)\sigma_b^2 = 2\sigma_o^2\sigma_b^2\ln\left[\frac{\sigma_o}{\sigma_b}\frac{1-\theta}{\theta}\right] \Rightarrow$$

$$(\sigma_o^2 - \sigma_b^2)t^2 + 2(-\mu_b\sigma_o^2 + \mu_o\sigma_b^2)t + \mu_b^2\sigma_o^2 - \mu_o^2\sigma_b^2 - 2\sigma_o^2\sigma_b^2\ln\left[\frac{\sigma_o}{\sigma_b}\frac{1-\theta}{\theta}\right] = 0 \quad (7.6)$$

This is a quadratic equation in t. It has two solutions in general, except when the two populations have the same standard deviation. If $\sigma_o = \sigma_b$, the above expression takes the form:

$$2(\mu_o - \mu_b)\sigma_o^2 t + (\mu_b^2 - \mu_o^2)\sigma_o^2 - 2\sigma_o^4\ln\left(\frac{1-\theta}{\theta}\right) = 0 \Rightarrow$$

$$t = \frac{\sigma_o^2}{\mu_o - \mu_b}\ln\left(\frac{1-\theta}{\theta}\right) - \frac{\mu_o + \mu_b}{2} \quad (7.7)$$

This is the minimum error threshold.

What is the meaning of the two solutions of (7.6)?

When $\sigma_o \neq \sigma_b$, the quadratic term in (7.6) does not vanish and we have two thresholds, t_1 and t_2. These turn out to be one on either side of the sharpest distribution. Let us assume that the sharpest distribution is that of the object pixels (see *Figure* 7.3). Then the correct thresholding will be to label as object pixels only those pixels with grey value x such that $t_1 < x < t_2$.

The meaning of the second threshold is that the flatter distribution has such a long tail that the pixels with grey values $x \geq t_2$ are more likely to belong to the long tail of the flat distribution, than to the sharper distribution.

Example 7.5

Derive the optimal threshold for image 7.4a and use it to threshold it.

Figure 7.4d shows the image of Figure 7.2a thresholded with the optimal threshold method. First the two main peaks in its histogram were identified. Then a Gaussian was fitted to the peak on the left and its standard deviation was chosen by trial and error so that the best fitting was obtained. The reason we fit first the peak on the left is because it is flatter, so it is expected to have the longest tails which contribute to the value of the peak of the other distribution. Once the first peak has been fitted, the values of the fitting Gaussian are subtracted from the histogram. If the result of this subtraction is negative, it is simply set to zero. Figure 7.4a shows the full histogram. Figure 7.4b shows the histogram with the Gaussian with which the first peak has been fitted superimposed. The mean and the standard deviation of this Gaussian are $\mu_o = 50$ and $\sigma_o = 7.5$. Figure 7.4c shows the histogram that is left after we subtract this Gaussian, with the negative numbers set to zero, and a second fitting Gaussian superimposed. The second Gaussian has $\mu_b = 117$ and $\sigma_b = 7$. The amplitude of the first Gaussian was $A_o = 20477$, and of the second $A_b = 56597$. We can estimate θ, i.e. the fraction of object pixels, by integrating the two fitting functions:

$$\frac{\theta}{1-\theta} = \frac{A_o \int_{-\infty}^{\infty} e^{-\frac{(x-\mu_o)^2}{2\sigma_o^2}} dx}{A_b \int_{-\infty}^{\infty} e^{-\frac{(x-\mu_b)^2}{2\sigma_b^2}} dx} = \frac{A_o \sqrt{2\pi}\sigma_o}{A_b \sqrt{2\pi}\sigma_b} = \frac{A_o\sigma_o}{A_b\sigma_b}$$

Therefore

$$\theta = \frac{A_o\sigma_o}{A_o\sigma_o + A_b\sigma_b}$$

In our case we estimate $\theta = 0.272$. Substituting these values into equation (7.6) we obtain two solutions $t_1 = -1213$ and $t_2 = 74$. The original image 7.2a thresholded with t_2 is shown in Figure 7.4d. After thresholding, we may wish to check how the distributions of the pixels of each class agree with the assumed distributions. Figures 7.4e and 7.4f show the histograms of the pixels of the object and the background respectively, with the assumed Gaussians superimposed. One can envisage an iterative scheme according to which these two histograms are used to estimate new improved parameters for each class which are then used to define a new threshold, and so on. However, it is not certain that such a scheme will converge. What is more, this result is worse than that obtained by hysteresis thresholding with two heuristically chosen thresholds. This demonstrates how powerful the combination of using criteria of spatial proximity and attribute similarity is.

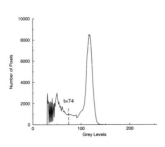

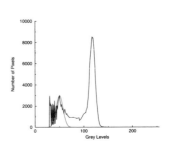

(a) Histogram with the optimal threshold marked

(b) Histogram with the Gaussian model for the object pixels superimposed

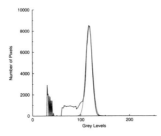

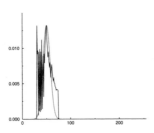

(c) Histogram after subtraction of the Gaussian used to model the object pixels with the Gaussian model for the background pixels superimposed.

(d) Image thresholded with the optimal threshold

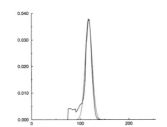

(e) Gaussian model used for the object pixels, and their real histogram

(f) Gaussian model used for the background pixels, and their real histogram

Figure 7.4: Optimal thresholding ($t = 74$).

Example 7.6

The grey level values of the object pixels are distributed according to the probability density function:

$$p_o = \frac{1}{2\sigma_o} \exp\left(-\frac{|x - \mu_o|}{\sigma_o}\right)$$

while the grey level values of the background pixels are distributed according to the probability density function:

$$p_b = \frac{1}{2\sigma_b} \exp\left(-\frac{|x - \mu_b|}{\sigma_b}\right)$$

If $\mu_o = 60$, $\mu_b = 40$ and $\sigma_o = 10$ and $\sigma_b = 5$, **find the thresholds that minimize the fraction of misclassified pixels when we know that the object occupies two-thirds of the area of the image.**

We substitute in equation (7.3):

$$\frac{\theta}{2\sigma_o}e^{-\frac{|t-\mu_o|}{\sigma_o}} = \frac{1-\theta}{2\sigma_b}e^{-\frac{|t-\mu_b|}{\sigma_b}}$$

$$\Rightarrow \exp\left(-\frac{|t-\mu_o|}{\sigma_o} + \frac{|t-\mu_b|}{\sigma_b}\right) = \frac{\sigma_o(1-\theta)}{\sigma_b\theta}$$

We have $\theta = \frac{2}{3}$ (therefore $1 - \theta = \frac{1}{3}$) and $\sigma_o = 10$, $\sigma_b = 5$. Then:

$$-\frac{|t-\mu_o|}{\sigma_o} + \frac{|t-\mu_b|}{\sigma_b} = \ln\frac{10 \times \frac{1}{3}}{5 \times \frac{2}{3}} = \ln 1 = 0$$

We have the following cases:

1. $t < \mu_b < \mu_o \Rightarrow |t - \mu_o| = -t + \mu_o$ and $|t - \mu_b| = -t + \mu_b \Rightarrow$ $\frac{t-\mu_o}{\sigma_o} + \frac{-t+\mu_b}{\sigma_b} = 0 \Rightarrow (\sigma_b - \sigma_o)t = \mu_o\sigma_b - \sigma_o\mu_b \Rightarrow t = \frac{\mu_o\sigma_b - \sigma_o\mu_b}{\sigma_b - \sigma_o} \Rightarrow$ $t = \frac{60 \times 5 - 10 \times 40}{-5} \Rightarrow t_1 = 20$

2. $\mu_b < t < \mu_o \Rightarrow |t - \mu_o| = -t + \mu_o$ and $|t - \mu_b| = t - \mu_b \Rightarrow$ $\frac{t-\mu_o}{\sigma_o} + \frac{t-\mu_b}{\sigma_b} = 0 \Rightarrow (\sigma_o + \sigma_b)t = \mu_o\sigma_b + \sigma_o\mu_b \Rightarrow t = \frac{\mu_o\sigma_b + \sigma_o\mu_b}{\sigma_b + \sigma_o}$ $= \frac{60 \times 5 + 10 \times 40}{15} = \frac{700}{15} = 47 \Rightarrow t_2 = 47$

3. $\mu_b < \mu_o < t \Rightarrow |t - \mu_o| = t - \mu_o$ and $|t - \mu_b| = t - \mu_b \Rightarrow$ $-\frac{t-\mu_o}{\sigma_o} + \frac{t-\mu_b}{\sigma_b} = 0 \Rightarrow (\sigma_o - \sigma_b)t = \sigma_o\mu_b - \mu_o\sigma_b \Rightarrow t = \frac{-\mu_o\sigma_b + \sigma_o\mu_b}{\sigma_o - \sigma_b}$ $= \frac{-60 \times 5 + 10 \times 40}{5} = 20 < \mu_o$, *rejected because t was assumed* $> \mu_o$.

So, there are two thresholds, $t_1 = 20$ and $t_2 = 47$.

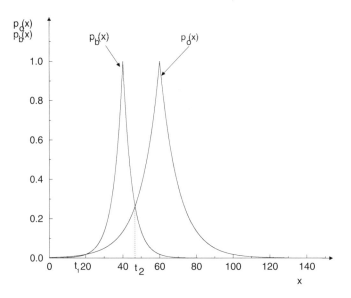

Only pixels with the grey level values between t_1 and t_2 should be classified as background pixels in order to minimize the error.

What are the drawbacks of the minimum error threshold method?

The method has various drawbacks. For a start, we must know the prior probabilities for the pixels to belong to the object or the background; i.e. we must know θ. Next, we must know the distributions of the two populations. Often it is possible to approximate these distributions by normal distributions, but even in that case one would have to estimate the parameters σ and μ of each distribution.

Is there any method that does not depend on the availability of models for the distributions of the object and the background pixels?

A method which does not depend on modelling the probability density functions has been developed by Otsu. Unlike the previous analysis, this method has been developed directly in the discrete domain.

Consider that we have an image with L grey levels in total and its normalized histogram, so that for each grey level value x, p_x represents the frequency with which the particular value arises. Then suppose that we set the threshold at t. Let us assume that we are dealing with the case of a bright object on a dark background.

The fraction of pixels that will be classified as background ones will be:

$$\theta(t) = \sum_{x=1}^{t} p_x \tag{7.8}$$

The fraction of pixels that will be classified as object pixels will be:

$$1 - \theta(t) = \sum_{x=t+1}^{L} p_x \tag{7.9}$$

The mean grey level value of the background pixels and the object pixels respectively will be:

$$\mu_b = \frac{\sum_{x=t+1}^{t} x p_x}{\sum_{x=1}^{t} p_x} \equiv \frac{\mu(t)}{\theta(t)} \tag{7.10}$$

$$\mu_o = \frac{\sum_{x=t+1}^{L} x p_x}{\sum_{x=t+1}^{L} p_x} = \frac{\sum_{x=1}^{L} x p_x - \sum_{x=1}^{t} x p_x}{1 - \theta(t)} = \frac{\mu - \mu(t)}{1 - \theta(t)} \tag{7.11}$$

where we defined $\mu(t) \equiv \sum_{x=1}^{t} x p_x$, and μ is the mean grey level value over the whole image, defined by:

$$\mu \equiv \frac{\sum_{x=1}^{L} x p_x}{\sum_{x=1}^{L} p_x} \tag{7.12}$$

Similarly, we may define the variance of each of the two populations created by the choice of a threshold t as:

$$\sigma_b^2 \equiv \frac{\sum_{x=1}^{t} (x - \mu_b)^2 p_x}{\sum_{x=1}^{t} p_x} = \frac{1}{\theta(t)} \sum_{x=1}^{t} (x - \mu_b)^2 p_x$$

$$\sigma_o^2 \equiv \frac{\sum_{x=t+1}^{L} (x - \mu_o)^2 p_x}{\sum_{x=t+1}^{L} p_x} = \frac{1}{1 - \theta(t)} \sum_{x=t+1}^{L} (x - \mu_o)^2 p_x \tag{7.13}$$

Let us consider next the total variance of the distribution of the pixels in the image:

$$\sigma_T^2 = \sum_{x=1}^{L} (x - \mu)^2 p_x$$

We may split this sum into two:

$$\sigma_T^2 = \sum_{x=1}^{t} (x - \mu)^2 p_x + \sum_{x=t+1}^{L} (x - \mu)^2 p_x$$

As we would like eventually to involve the statistics defined for the two populations, we add and subtract inside each sum the corresponding mean:

$$
\begin{aligned}
\sigma_T^2 &= \sum_{x=1}^{t} (x - \mu_b + \mu_b - \mu)^2 p_x + \sum_{x=t+1}^{L} (x - \mu_o + \mu_o - \mu)^2 p_x \\
&= \sum_{x=1}^{t} (x - \mu_b)^2 p_x + \sum_{x=1}^{t} (\mu_b - \mu)^2 p_x + 2 \sum_{x=1}^{t} (x - \mu_b)(\mu_b - \mu) p_x \\
&\quad + \sum_{x=t+1}^{L} (x - \mu_o)^2 p_x + \sum_{x=t+1}^{L} (\mu_o - \mu)^2 p_x + 2 \sum_{x=t+1}^{L} (x - \mu_o)(\mu_o - \mu) p_x
\end{aligned}
$$

Next we substitute the two sums on the left of each line in terms of σ_b^2 and σ_o^2 using equations (7.13). We also notice that the two sums in the middle of each line can be expressed in terms of equations (7.8) and (7.9), since μ, μ_b and μ_o are constants and do not depend on the summing variable x:

$$
\begin{aligned}
\sigma_T^2 &= \theta(t)\sigma_b^2 + (\mu_b - \mu)^2 \theta(t) + 2(\mu_b - \mu) \sum_{x=1}^{t} (x - \mu_b) p_x \\
&\quad + (1 - \theta(t))\sigma_o^2 + (\mu_o - \mu)^2 \theta(t) + 2(\mu_o - \mu) \sum_{x=t+1}^{L} (x - \mu_b) p_x
\end{aligned}
$$

The two terms with the sums are zero, since, for example:

$$
\sum_{x=1}^{t} (x - \mu_b) p_x = \sum_{x=1}^{t} x p_x - \sum_{x=1}^{t} \mu_b p_x = \mu_b \theta(t) - \mu_b \theta(t) = 0
$$

Then by rearranging the remaining terms:

$$
\sigma_T^2 = \underbrace{\theta(t)\sigma_b^2 + (1 - \theta(t))\sigma_o^2}_{\substack{\text{terms depending on the variance} \\ \textbf{within } \text{each class}}} + \underbrace{(\mu_b - \mu)^2 \theta(t) + (\mu_o - \mu)^2 (1 - \theta(t))}_{\substack{\text{terms depending on the variance} \\ \textbf{between } \text{the two classes}}}
$$

$$
\equiv \sigma_W^2(t) + \sigma_B^2(t) \tag{7.14}
$$

where $\sigma_W^2(t)$ is defined to be the within-class variance and $\sigma_B^2(t)$ is defined to be the between-class variance. Clearly σ_T^2 is a constant. We want to specify t so that $\sigma_W^2(t)$ is as small as possible, i.e. the classes that are created are as compact as possible, and $\sigma_B^2(t)$ is as large as possible. Suppose that we choose to work with $\sigma_B^2(t)$, i.e. choose t so that it maximizes $\sigma_B^2(t)$. We substitute in the definition of $\sigma_B^2(t)$ the expression for μ_b and μ_o as given by equations (7.10) and (7.11) respectively:

$$
\begin{aligned}
\sigma_B^2(t) &= (\mu_b - \mu)^2 \theta(t) + (\mu_o - \mu)^2 (1 - \theta(t)) \\
&= \left[\frac{\mu(t)}{\theta(t)} - \mu\right]^2 \theta(t) + \left[\frac{\mu - \mu(t)}{1 - \theta(t)} - \mu\right]^2 (1 - \theta(t))
\end{aligned}
$$

$$= \frac{[\mu(t) - \mu\theta(t)]^2}{\theta(t)} + \frac{[\mu - \mu(t) - \mu + \mu\theta(t)]^2}{1 - \theta(t)}$$

$$= \frac{[\mu(t) - \mu\theta(t)]^2[1 - \theta(t)] + \theta(t)[-\mu(t) + \mu\theta(t)]^2}{\theta(t)[1 - \theta(t)]}$$

$$= \frac{[\mu(t) - \mu\theta(t)]^2}{\theta(t)[1 - \theta(t)]} \tag{7.15}$$

This function expresses the **interclass** variance $\sigma_B^2(t)$ in terms of the mean grey value of the image μ, and quantities that can be computed once we know the values of the image histogram up to the chosen threshold t.

The idea is then to start from the beginning of the histogram and test each grey level value for the possibility of being the threshold that maximizes $\sigma_B^2(t)$, by calculating the values of $\mu(t) = \sum_{x=1}^{t} x p_x$ and $\theta(t) = \sum_{x=1}^{t} p_x$ and substituting into equation (7.15). We stop testing once the value of σ_B^2 starts decreasing. This way we identify t for which $\sigma_B^2(t)$ becomes maximal. This method tacitly assumes that function $\sigma_B^2(t)$ is well-behaved; i.e. that it has only one maximum.

Example 7.7

Calculate Otsu's threshold for the image of Figure 7.2a and use it to threshold the image.

Figure 7.5a shows how $\sigma_B^2(t)$ varies as t scans all possible grey values. The first maximum of this function is at $t = 84$ and we choose this threshold to produce the result shown in Figure 7.5b. We can see that the result is not noticeably different from the result obtained with the empirical threshold (Figure 7.5c) and a little worse than the optimal threshold result (Figure 7.4d). It is much worse than the result obtained by hysteresis thresholding, reinforcing again the conclusion that spatial and grey level characteristics used in thresholding is a powerful combination.

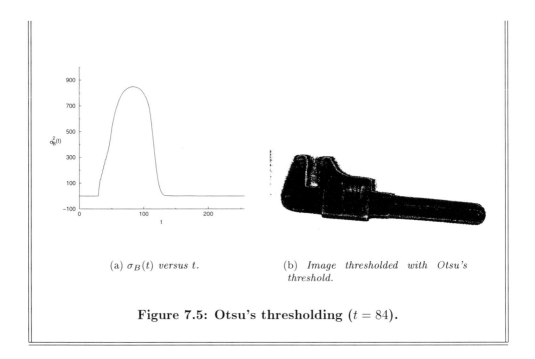

(a) $\sigma_B(t)$ *versus* t.

(b) *Image thresholded with Otsu's threshold.*

Figure 7.5: Otsu's thresholding ($t = 84$).

Are there any drawbacks to Otsu's method?

Yes, a few:

1. Although the method does not make any assumption about the probability density functions $p_o(x)$ and $p_b(x)$, it describes them by using only their means and variances. Thus it tacitly assumes that these two statistics are sufficient in representing them. This may not be true.

2. The method breaks down when the two populations are very unequal. When the two populations become very different in size from each other, $\sigma_B^2(t)$ may have two maxima and actually the correct maximum is not necessarily the global maximum. That is why in practice the correct maximum is selected from among all maxima of $\sigma_B^2(t)$ by checking that the value of the histogram at the selected threshold, p_t, is actually a valley (i.e. $p_t < p_{\mu_o}$, $p_t < p_{\mu_b}$) and only if this is true should t be accepted as the best threshold.

3. The method, as presented above, assumes that the histogram of the image is bimodal; i.e. the image contains two classes. For more than two classes present in the image, the method has to be modified so that multiple thresholds are defined which maximize the *interclass* variance and minimize the *intraclass* variance.

4. The method will divide the image into two classes even if this division does not make sense. A case when the method should not be directly applied is that of variable illumination.

How can we threshold images obtained under variable illumination?

In the chapter on image enhancement, we saw that an image is essentially the **product** of a reflectance function $r(x,y)$ which is intrinsic to the viewed surfaces, and an illumination function $i(x,y)$:

$$f(x,y) = r(x,y)i(x,y)$$

Thus any spatial variation of the illumination results in a multiplicative interference to the reflectance function that is recorded during the imaging process. We can convert the **multiplicative** interference into **additive**, if we take the logarithm of the image:

$$\ln f(x,y) = \ln r(x,y) + \ln i(x,y) \tag{7.16}$$

Then instead of forming the histogram of $f(x,y)$, we can form the histogram of $\ln f(x,y)$.

If we threshold the image according to the histogram of $\ln f(x,y)$, are we thresholding it according to the reflectance properties of the imaged surfaces?

The question really is, what the histogram of $\ln f(x,y)$ is in terms of the histograms of $\ln r(x,y)$ and $\ln i(x,y)$. For example, if $\ln f(x,y)$ is the sum of $\ln r(x,y)$ and $\ln i(x,y)$ which may be reasonably separated functions apart from some overlap, then by thresholding $\ln f(x,y)$ we may be able to identify the $\ln r(x,y)$ component; i.e. the component of interest.

Let us define some new variables:

$$z(x,y) \equiv \ln f(x,y)$$
$$\tilde{r}(x,y) \equiv \ln r(x,y)$$
$$\tilde{i}(x,y) \equiv \ln i(x,y)$$

Therefore, equation (7.16) can be written as:

$$z(x,y) = \tilde{r}(x,y) + \tilde{i}(x,y) \tag{7.17}$$

If $f(x,y)$, $r(x,y)$ and $i(x,y)$ are thought of as random variables, then $z(x,y)$, $\tilde{r}(x,y)$ and $\tilde{i}(x,y)$ are also random variables. So, the question can be rephrased into: **What is the histogram of the sum of two random variables in terms of the histograms of the two variables?** A histogram can be thought of as a probability density function. Rephrasing the question again, we have: **What is the probability density function of the sum of two random variables in terms of the probability density functions of the two variables?** We have seen that the probability density function of a random variable is the derivative of the distribution function of the variable. So, we can rephrase the question again: **What is the distribution**

function of the sum of two random variables in terms of the probability density functions or the distribution functions of the two variables? In the $(\tilde{i}, \tilde{r})$ space, equation (7.17) represents a line for a given value of z. By definition, we know that:

Distribution function of $z = P_z(u) =$ Probability of $z \le u \equiv \mathcal{P}(z \le u)$

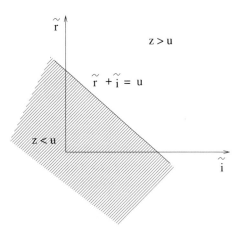

Figure 7.6: z is less than u in the shadowed half plane.

The line $\tilde{r} + \tilde{i} = u$ divides the $(\tilde{i}, \tilde{r})$ plane into two half planes, one in which $z > u$ and one where $z < u$. The probability of $z < u$ is equal to the integral of the probability density function of pairs $(\tilde{i}, \tilde{r})$ over the area of the half plane in which $z < u$ (see *Figure* 7.6):

$$P_z(u) = \int_{\tilde{i}=-\infty}^{+\infty} \int_{\tilde{r}=-\infty}^{u-\tilde{i}} p_{\tilde{r}\tilde{i}}(\tilde{r}, \tilde{i}) d\tilde{r} d\tilde{i}$$

where $p_{\tilde{r}\tilde{i}}(\tilde{r}, \tilde{i})$ is the joint probability density function of the two random variables $\tilde{r}$ and $\tilde{i}$.

To find the probability density function of z, we differentiate $P_z(u)$ with respect to u, using Leibnitz's rule (see Box **B7.1**), applied twice, once with

$$f(x; \lambda) \rightarrow \int_{\tilde{r}=-\infty}^{u-\tilde{i}} p_{\tilde{u}\tilde{i}}(\tilde{r}, \tilde{i}) d\tilde{r}$$
$$b(\lambda) \rightarrow +\infty$$
$$a(\lambda) \rightarrow -\infty$$

and once more when we need to differentiate $f(x; \lambda)$ which itself is an integral that depends on parameter u with respect to which we differentiate:

$$
\begin{aligned}
p_z(u) &= \frac{dP_z u}{du} \int_{\tilde{i}=-\infty}^{\infty} \frac{d}{du} \left[\int_{\tilde{r}=-\infty}^{u-\tilde{i}} p_{\tilde{r}\tilde{i}}(\tilde{r}, \tilde{i}) d\tilde{r} \right] d\tilde{i} \\
&= \int_{\tilde{i}=-\infty}^{\infty} p_{\tilde{r}\tilde{i}}(u - \tilde{i}, \tilde{i}) d\tilde{i}
\end{aligned}
\tag{7.18}
$$

The two random variables $\tilde{r}$ and $\tilde{i}$, one associated with the imaged surface and one with the source of illumination, are independent, and therefore their joint probability density function can be written as the product of their two probability density functions:

$$
p_{\tilde{r}\tilde{i}}(\tilde{r}, \tilde{i}) = p_{\tilde{r}}(\tilde{r}) p_{\tilde{i}}(\tilde{i})
$$

Upon substitution in (7.18) we obtain:

$$
p_z(u) = \int_{-\infty}^{\infty} p_{\tilde{r}}(u - \tilde{i}) p_{\tilde{i}}(\tilde{i}) d\tilde{i}
\tag{7.19}
$$

which shows that the histogram (= probability density function) of z is equal to the **convolution** of the two histograms of the two random variables $\tilde{r}$ and $\tilde{i}$.

If the illumination is uniform, then:

$$
i(x, y) = \text{constant} \quad \Rightarrow \quad \tilde{i} = \ln i(x, y) = \tilde{i}_o = \text{constant}
$$

Then $p_{\tilde{i}}(\tilde{i}) = \delta(\tilde{i} - \tilde{i}_o)$, and after substitution in (7.19) and integration we obtain $p_z(u) = p_{\tilde{r}}(u)$.

That is, under uniform illumination, the histogram of the reflectance function (intrinsic to the object) is essentially unaffected. If, however, the illumination is not uniform then, even if we had a perfectly distinguishable object, the histogram is badly distorted and the various thresholding methods break down.

Since straightforward thresholding methods break down under variable illumination, how can we cope with it?

There are two ways in which we can circumvent the problem of variable illumination:

1. Divide the image into more or less uniformly illuminated patches and histogram and threshold each patch as if it were a separate image. Some adjustment may be needed when the patches are put together as the threshold essentially will jump from one value in one patch to another value in a neighbouring patch.

2. Obtain an image of just the illumination field, using the image of a surface with uniform reflectance and divide the image $f(x, y)$ by $i(x, y)$; i.e. essentially subtract the illumination component $\tilde{i}(x, y)$ from $z(x, y)$. Then multiply $\frac{f(x,y)}{i(x,y)}$ with a reference value, say $i(0, 0)$, to bring the whole image under the same illumination and proceed using the corrected image.

Example 7.8

Threshold the image of Figure 7.7a.

This image exhibits an illumination variation from left to right. Figure 7.7b shows the histogram of the image. Using Otsu's method, we identify threshold $t = 75$. The result of thresholding the image with this threshold is shown in Figure 7.7c. The result of dividing the image into four subimages from left to right and applying Otsu's method to each subimage separately is shown in Figure 7.7d.

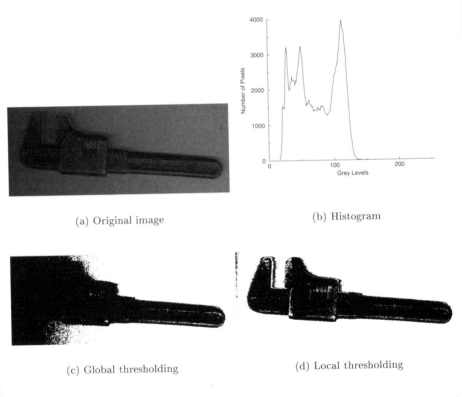

(a) Original image

(b) Histogram

(c) Global thresholding

(d) Local thresholding

Figure 7.7: Global versus local thresholding for an image with variable illumination.

Are there any shortcomings of the thresholding methods?

With the exception of hysteresis thresholding which is of limited use, the spatial proximity of the pixels in the image is not considered at all in the segmentation process. Instead, only the grey level values of the pixels are used.

For example, consider the two images in *Figure* 7.8.

Figure 7.8: Two very different images with identical histograms.

Clearly, the first image is the image of a uniform region, while the second image contains two quite distinct regions. Even so, both images have identical histograms, shown in *Figure* 7.9. Their histograms are bimodal and we can easily choose a threshold. However, if we use it to segment the first image, we shall get nonsense.

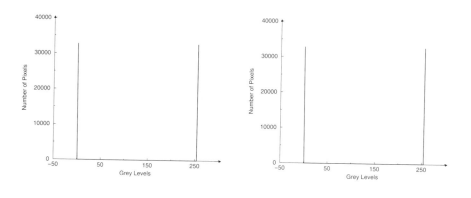

(a) Histogram of image in Figure 7.8a   (b) Histogram of image in Figure 7.8b

Figure 7.9: The two identical histograms of the very different images shown in *Figure* 7.8.

How can we cope with images that contain regions that are not uniform but they are *perceived* as uniform?

Regions that are not uniform in terms of the grey values of their pixels but are perceived as uniform, are called *textured regions*. For segmentation purposes then, each pixel cannot only be characterized by its grey level value but also by another number or numbers which quantify the variation of the grey values in a small patch around that pixel. The point is that the problem posed by the segmentation of textured images can be solved by using more than one attribute to segment the image. We can envisage that each pixel is characterized not by one number but by a vector of numbers, each component of the vector measuring something at the pixel position. Then each pixel is represented by a point in a multidimensional space, where we measure one such number, *a feature*, along each axis. Pixels belonging to the same region will have similar or identical values in their attributes and thus will cluster together. The problem then becomes one of identifying clusters of pixels in a multidimensional space. Essentially it is similar to histogramming only now we deal with multidimensional histograms. There are several *clustering* methods that may be used but they are in the realm of *Pattern Recognition* and thus beyond the scope of this book.

Are there any segmentation methods that take into consideration the spatial proximity of pixels?

Yes, they are called *region growing* methods. In general, one starts from some seed pixels and attaches neighbouring pixels to them provided the attributes of the pixels in the region created in this way vary within a predefined range. So, each seed grows gradually by accumulating more and more neighbouring pixels until all pixels in the image have been assigned to a region.

How can one choose the seed pixels?

There is no clear answer to this question, and this is the most important drawback of this type of method. In some applications the choice of seeds is easy. For example, in target tracking in infrared images, the target will appear bright, and one can use as seeds the few brightest pixels. A method which does not need a predetermined number of regions or seeds is that of split and merge.

How does the split and merge method work?

Initially the whole image is considered as one region. If the range of attributes within this region is greater than a predetermined value, then the region is split into four quadrants and each quadrant is tested in the same way until every square region created in this way contains pixels with range of attributes within the given value. At the end all adjacent regions with attributes within the same range may be merged.

An example is shown in *Figure* 7.10 where for simplicity a binary 8×8 image is considered. The tree structure shows the successive splitting of the image into quadrants. Such a tree is called *quadtree*.

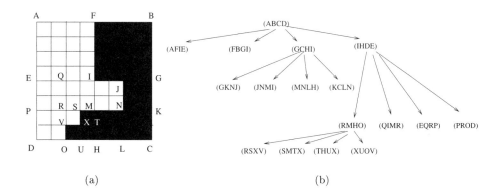

(a) (b)

Figure 7.10: Image segmentation by splitting.

We end up having the following regions:

$$(AFIE)(FBGI)(GKNJ)(JNMI)(MNLH)(KCLN)$$
$$(RSXV)(SMTX)(THUX)(XUOV)(QIMR)(EQRP)(PROD)$$

i.e all the children of the quadtree. Any two adjacent regions then are checked for merging and eventually only the two main regions of irregular shape emerge. The above quadtree structure is clearly favoured when the image is square with $N = 2^n$ pixels in each side.

Split and merge algorithms often start at some intermediate level of the quadtree (i.e some blocks of size $2^l \times 2^l$ where $l < n$) and check each block for further splitting into four square sub-blocks and any two adjacent blocks for merging. At the end again we check for merging any two adjacent regions.

Is it possible to segment an image by considering the *dissimilarities* between regions, as opposed to considering the similarities between pixels?

Yes, in such an approach we examine the differences between neighbouring pixels and say that pixels with different attribute values belong to different regions and therefore we postulate a boundary separating them. Such a boundary is called an *edge* and the process is called *edge detection*.

How do we measure the dissimilarity between neighbouring pixels?

We may slide a window across the image and at each position calculate the statistical properties of the pixels within each half of the window and compare the two results.

The places where these statistical properties differ most are where the boundaries of the regions are.

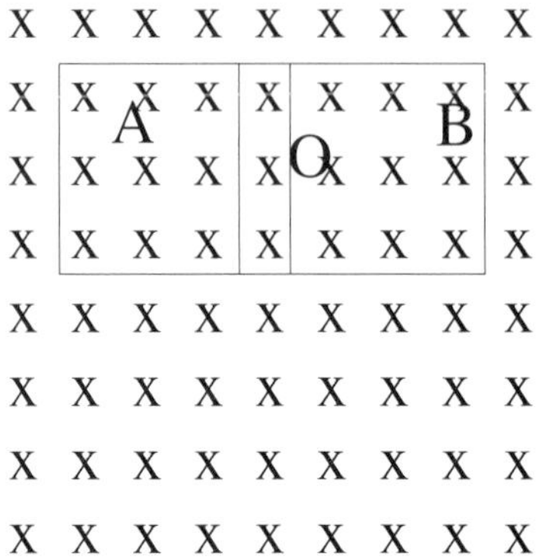

```
X  X  X  X  X  X  X  X  X

X  X  X  X  X  X  X  X  X
    A              B
X  X  X  X  X  X  X  X  X
              O
X  X  X  X  X  X  X  X  X

X  X  X  X  X  X  X  X  X

X  X  X  X  X  X  X  X  X

X  X  X  X  X  X  X  X  X

X  X  X  X  X  X  X  X  X
```

Figure 7.11: Measuring the dissimilarity between two image regions using a sliding widow.

For example, consider the 8×8 image in *Figure* 7.11. Each X represents a pixel. The rectangle drawn is a 3×7 window which could be placed so that its centre O coincides with every pixel in the image, apart from those too close to the edge of the image. We can calculate the statistical properties of the nine pixels on the left of the window (part A) and those of the nine pixels on the right of the window (part B) and assign their difference to pixel O. For example, we may calculate the standard deviation of the grey values of the pixels within each half of the window, say σ_A and σ_B, calculate the standard deviation of the pixels inside the whole window, say σ, and assign the value $E \equiv 2\sigma - \sigma_A - \sigma_B$ to the central pixel. We can slide this window horizontally to scan the whole image. Local maxima of the assigned values are candidate positions for vertical boundaries. Local maxima where the value of E is greater than a certain threshold are accepted as vertical boundaries between adjacent regions. We can repeat the process by rotating the window by $90°$ and sliding it vertically to scan the whole image again. Clearly the size of the window here plays a crucial role as we need a large enough window to calculate the statistics properly and a small enough window to include within each half only part of a single region and avoid contamination from neighbouring regions.

What is the smallest possible window we can choose?

The smallest possible window we can choose consists of two adjacent pixels. The only "statistic" we can calculate from such a window is the difference of the grey values of the two pixels. When this difference is high we say we have an *edge* passing between the two pixels. Of course, the difference of the grey values of the two pixels is not a statistic but is rather an estimate of the first derivative of the intensity function with respect to the spatial variable along the direction of which we take the difference.

This is because first derivatives are approximated by first differences in the discrete case:

$$\Delta f_x = f(i+1, j) - f(i, j)$$
$$\Delta f_y = f(i, j+1) - f(i, j)$$

Calculating Δf_x at each pixel position is equivalent to convolving the image with a mask (filter) of the form $\boxed{-1}\,\boxed{+1}$ in the x direction, and calculating Δf_y is equivalent to convolving the image with the filter $\begin{array}{|c|}\hline -1 \\ \hline +1 \\ \hline\end{array}$ in the y direction.

The first and the simplest edge detection scheme then is to convolve the image with these two masks and produce two outputs. Note that these small masks have even lengths so their centres are not associated with any particular pixel in the image as they slide across the image. So, the output of each calculation should be assigned to the position in between the two adjacent pixels. These positions are said to constitute the *dual* grid of the image grid. In practice, we seldom invoke the dual grid. We usually adopt a convention and try to be consistent. For example, we may always assign the output value to the first pixel of the mask. If necessary, we later may remember that this value actually measures the difference between the two adjacent pixels at the position half an interpixel distance to the left or the bottom of the pixel to which it is assigned. So, with this understanding, and for simplicity from now on, we shall be talking about *edge pixels*.

In the first output, produced by convolution with mask $\boxed{-1}\,\boxed{+1}$, any pixel that has an absolute value larger than values of its left and right neighbours is a candidate pixel to be a vertical edge pixel. In the second output, produced by convolution with mask $\begin{array}{|c|}\hline -1 \\ \hline +1 \\ \hline\end{array}$, any pixel that has an absolute value larger than the values of its top and bottom neighbours is a candidate pixel to be a horizontal edge pixel. The process of identifying the local maxima as candidate edge pixels (=*edgels*) is called *non-maxima suppression*.

In the case of zero noise this scheme will clearly pick up the discontinuities in intensity.

What happens when the image has noise?

In the presence of noise every small and irrelevant fluctuation in the intensity value will be amplified by differentiating the image. It is common sense then that one should smooth the image first with a lowpass filter and then find the local differences. *Figure* 7.12 shows an original image and the output obtained if the $\boxed{-1}\,\boxed{+1}$ and $\begin{array}{|c|}\hline -1 \\ \hline +1 \\ \hline\end{array}$ convolution filters are applied along the horizontal and vertical directions respectively. The outputs of the two convolutions are squared, added and square rooted to produce the gradient magnitude associated with each pixel. *Figure* 7.13 shows the results obtained using these minimal convolution filters and some more sophisticated filters that take consideration the noise present in the image.

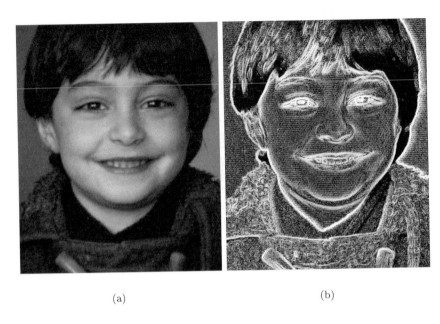

(a)   (b)

Figure 7.12: (a) Original image. (b) The image of the gradient magnitudes computed by simple differencing without applying any smoothing. For displaying purposes the gradient image has been subjected to histogram equalization.

Let us consider for example a 1-dimensional signal. Suppose that one uses as lowpass filter a simple averaging procedure. We smooth the signal by replacing each intensity value with the average of three successive intensity values:

$$A_i \equiv \frac{I_{i-1} + I_i + I_{i+1}}{3} \qquad (7.20)$$

Then we estimate the derivative at position i by averaging the two differences between the value at position i and its left and right neighbours:

$$F_i \equiv \frac{(A_{i+1} - A_i) + (A_i - A_{i-1})}{2} = \frac{A_{i+1} - A_{i-1}}{2} \qquad (7.21)$$

If we substitute from (7.20) into (7.21), we obtain:

$$F_i = \frac{1}{6}[I_{i+2} + I_{i+1} - I_{i-1} - I_{i-2}] \qquad (7.22)$$

It is obvious from this example that one can combine the two linear operations of smoothing and finding differences into one operation if one uses large enough masks. In this case, the first difference at each position could be estimated by using a mask like $\boxed{-\frac{1}{6} \mid -\frac{1}{6} \mid 0 \mid \frac{1}{6} \mid \frac{1}{6}}$. It is clear that the larger the mask used, the better

(a)

(b)

(c)

(d)

Figure 7.13: (a) Result obtained by simply thresholding the gradient values obtained without any smoothing. (b) The same as in (a) but using a higher threshold, so some noise is removed. However, useful image information has also been removed. (c) Result obtained by smoothing first along one direction and differentiating afterwards along the orthogonal direction, using a Sobel mask. (d) Result obtained using the optimal filter of size 7×7. In all cases the same value of threshold was used.

is the smoothing. But it is also clear that the more blurred the edge becomes so the more inaccurately its position will be specified (see *Figure 7.14*).

For an image which is a 2D signal one should use 2-dimensional masks. The smallest mask that one should use to combine minimum smoothing with differencing is a 3×3 mask. In this case we also have the option to smooth in one direction and take the difference along the other. This implies that the 2D mask may be the result of applying in a cascaded way first a 3×1 smoothing mask and then a 1×3 differencing masks, or vice versa. In general, however, a 2D 3×3 mask will have the form:

a_{11}	a_{12}	a_{13}
a_{21}	a_{22}	a_{23}
a_{31}	a_{32}	a_{33}

How can we choose the weights of a 3×3 mask for edge detection?

We are going to use one such mask to calculate Δf_x and another to calculate Δf_y. Such masks must obey the following conditions:

1. The mask which calculates Δf_x must be produced from the mask that calculates Δf_y by rotation by $90°$. Let us consider from now on the mask which will produce Δf_y only. The calculated value will be assigned to the central pixel.

2. We do not want to give any extra weight to the left or the right neighbours of the central pixel, so we must have identical weights in the left and right columns. The 3×3 mask therefore must have the form:

a_{11}	a_{12}	a_{11}
a_{21}	a_{22}	a_{21}
a_{31}	a_{32}	a_{31}

3. Let us say that we want to subtract the signal "in front" of the central pixel from the signal "behind" it, in order to find local differences, and we want these two subtracted signals to have equal weights. The 3×3 mask therefore must have the form:

a_{11}	a_{12}	a_{11}
a_{21}	a_{22}	a_{21}
$-a_{11}$	$-a_{12}$	$-a_{11}$

4. If the image is absolutely smooth we want to have zero response. So the sum of all the weights must be zero. Therefore, $a_{22} = -2a_{21}$:

a_{11}	a_{12}	a_{11}
a_{21}	$-2a_{21}$	a_{21}
$-a_{11}$	$-a_{12}$	$-a_{11}$

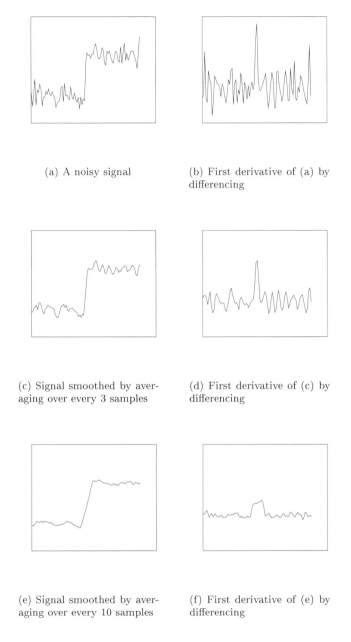

(a) A noisy signal

(b) First derivative of (a) by differencing

(c) Signal smoothed by averaging over every 3 samples

(d) First derivative of (c) by differencing

(e) Signal smoothed by averaging over every 10 samples

(f) First derivative of (e) by differencing

Figure 7.14: The left column shows a noisy signal and two smoothed versions of it. On the right is the result of estimating the first derivative of the signal by simple differencing. The edge manifests itself with a sharp peak. However, the more the noise is left in the signal, the more secondary peaks will be present, and the more the smoothing, the more blunt the main peak becomes.

5. In the case of a smooth signal, and as we differentiate in the direction of columns, we expect each column to produce 0 output. Therefore, $a_{21} = 0$:

a_{11}	a_{12}	a_{11}
0	0	0
$-a_{11}$	$-a_{12}$	$-a_{11}$

We can divide these weights throughout by a_{11} so that finally, this mask depends only on one parameter:

1	K	1
0	0	0
-1	$-K$	-1

$$(7.23)$$

What is the best value of parameter K?

It can be shown that the orientations of edges which are almost aligned with the image axes are not affected by the differentiation, if we choose $K = 2$. We have then the *Sobel* masks for differentiating an image along two directions:

1	2	1
0	0	0
-1	-2	-1

-1	0	1
-2	0	2
-1	0	1

Note that we have changed convention for the second mask and subtract the values of the pixels "behind" the central pixel from the values of the pixels "in front". This is intentional so that the orientation of the calculated edge, computed by using the components of the gradient vector derived using these masks, is measured from the horizontal axis anticlockwise (see Box **B7.2**).

B7.2: Derivation of the Sobel masks

Edges are characterized by their strength and orientation defined by:

$$\text{Strength} = E(i,j) = \sqrt{[\Delta f_x(i,j)]^2 + [\Delta f_y(i,j)]^2}$$

$$\text{Orientation: } a(i,j) = \tan^{-1}\frac{\Delta f_y(i,j)}{\Delta f_x(i,j)}$$

The idea is to try to specify parameter K of mask (7.23) so that the output of the operator is as faithful as possible to the true values of E and a which correspond to the non-discretized image:

$$E = \sqrt{\left(\frac{\partial f}{\partial x}\right)^2 + \left(\frac{\partial f}{\partial y}\right)^2}$$

$$a = \tan^{-1}\frac{\frac{\partial f}{\partial y}}{\frac{\partial f}{\partial x}}$$

Consider a straight step edge in the scene, passing through the middle of a pixel. Each pixel is assumed to be a tile of size 1×1. Suppose that the edge has orientation θ and suppose that θ is small enough so that the edge cuts lines AB and CD as opposed to cutting lines AC and BD ($0 \le \theta \le \tan^{-1}(\frac{1}{3})$) (see *Figure 7.15*).

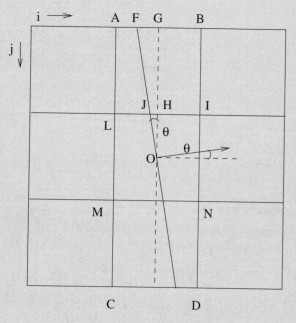

Figure 7.15: **Zooming into a 3×3 patch of an image around pixel (i, j).**

Also, assume that on the left of the edge we have a dark region with grey value G_1 and on the right a bright region with grey value G_2. Then clearly the pixel values inside the mask are:

$$f(i-1, j-1) = f(i-1, j) = f(i-1, j+1) = G_1$$
$$f(i+1, j-1) = f(i+1, j) = f(i+1, j+1) = G_2$$

The pixels in the central column have mixed values. If we assume that each pixel is like a tile with dimensions 1×1 and denote the area of a polygon by the name of the polygon inside brackets, then pixel $ABIL$ will have value:

$$f(i,j-1) = G_1(AFJL)+G_2(FBIJ) = G_1[\frac{1}{2}-(FGHJ)]+G_2[\frac{1}{2}+(FGHJ)] \quad (7.24)$$

We must find the area of trapezium $FGHJ$. From the triangles OJH and OFG we have:

$$JH = \frac{1}{2}\tan\theta, \qquad FG = \frac{3}{2}\tan\theta$$

Therefore, $(FGJH) = \frac{1}{2}(JH + FG) = \tan\theta$, and by substitution into equation (7.24) we obtain:

$$f(i,j-1) = G_1(\frac{1}{2} - \tan\theta) + G_2(\frac{1}{2} + \tan\theta)$$

By symmetry:

$$f(i,j+1) = G_2(\frac{1}{2} - \tan\theta) + G_1(\frac{1}{2} + \tan\theta)$$

Clearly

$$f(i,j) = \frac{G_1 + G_2}{2}$$

Let us see now what mask (7.23) will calculate in this case:

$$\begin{aligned}
\Delta f_x &= f(i+1,j+1) + f(i+1,j-1) + Kf(i+1,j) \\
&\quad - [f(i-1,j+1) + f(i-1,j-1) + Kf(i-1,j)] \\
&= (G_2 + G_2 + KG_2) - (G_1 + G_1 + KG_1) = (G_2 - G_1)(2 + K) \\
\Delta f_y &= -[f(i-1,j+1) + f(i+1,j+1) + Kf(i,j+1)] \\
&\quad + f(i-1,j-1) + f(i+1,j-1) + Kf(i,j-1) \\
&= -G_1 - G_2 - KG_2\left(\frac{1}{2} - \tan\theta\right) - KG_1\left(\frac{1}{2} + \tan\theta\right) + G_1 + G_2 \\
&\quad + KG_1\left(\frac{1}{2} - \tan\theta\right) + KG_2\left(\frac{1}{2} + \tan\theta\right) \\
&= -K(G_2 - G_1)\left(\frac{1}{2} - \tan\theta\right) + K(G_2 - G_1)\left(\frac{1}{2} + \tan\theta\right) \\
&= K(G_2 - G_1)\left(-\frac{1}{2} + \tan\theta + \frac{1}{2} + \tan\theta\right) \\
&= 2K(G_2 - G_1)\tan\theta
\end{aligned}$$

The magnitude of the edge will then be

$$E = \sqrt{(G_2 - G_1)^2(2 + K)^2 + (2K)^2(G_2 - G_1)^2\tan^2\theta}$$

$$= (G_2 - G_1)(2 + K)\sqrt{1 + \left(\frac{2K}{2+K}\right)^2 \tan^2\theta}$$

and the orientation of the edge:

$$\tan\alpha = \frac{\Delta f_y}{\Delta f_x} = \frac{2K\tan\theta}{2+K}$$

Note that if we choose $K = 2$:

$$\tan\alpha = \tan\theta$$

i.e. the calculated orientation of the edge will be equal to the true orientation.
    One can perform a similar calculation for the case $\tan^{-1}\frac{1}{3} \le \theta \le 45^o$. In that case we have distortion in the orientation of the edge.

Example 7.9

Write down the formula which expresses the output $O(i,j)$ at position (i,j) of the convolution of the image with the Sobel mask that differentiates along the i axis, as a function of the input values $I(i,j)$. (Note: Ignore boundary effects, i.e. assume that (i,j) is sufficiently far from the image border.)

$$\begin{aligned} O(i,j) = {} & -I(i-1, j-1) - 2I(i-1, j) - I(i-1, j+1) \\ & + I(i+1, j-1) + 2I(i+1, j) + I(i+1, j+1) \end{aligned}$$

Example 7.10

You have a 3×3 image which can be represented by a 9×1 vector. Construct a 9×9 matrix which, when it operates on the image vector, will produce another vector each element of which is the estimate of the gradient component of the image along the i axis. (To deal with the boundary pixels assume that the image is repeated periodically in all directions.)

Consider a 3×3 image:

$$\xrightarrow{i}$$

$$j\downarrow \begin{pmatrix} f_{11} & f_{12} & f_{13} \\ f_{21} & f_{22} & f_{23} \\ f_{31} & f_{32} & f_{33} \end{pmatrix}$$

Periodic repetition of this image implies that we have:

$$\begin{matrix} f_{33} & f_{31} & f_{32} & f_{33} & f_{31} \\ f_{13} & \begin{pmatrix} f_{11} & f_{12} & f_{13} \end{pmatrix} & f_{11} \\ f_{23} & \begin{pmatrix} f_{21} & f_{22} & f_{23} \end{pmatrix} & f_{21} \\ f_{33} & \begin{pmatrix} f_{31} & f_{32} & f_{33} \end{pmatrix} & f_{31} \\ f_{13} & f_{11} & f_{12} & f_{13} & f_{11} \end{matrix}$$

Using the result of Example 7.9, we notice that the first derivative at position $(1,1)$ *is given by:*

$$f_{32} + 2f_{12} + f_{22} - f_{33} - 2f_{13} - f_{23}$$

The vector representation of the image is:

$$\begin{pmatrix} f_{11} \\ f_{21} \\ f_{31} \\ f_{12} \\ f_{22} \\ f_{32} \\ f_{13} \\ f_{23} \\ f_{33} \end{pmatrix}$$

If we operate with a 9×9 *matrix on this image, we notice that in order to get the desired output, the first row of the matrix should be:*

$$0 \quad 0 \quad 0 \quad 2 \quad 1 \quad 1 \quad -2 \quad -1 \quad -1$$

The derivative at position $(1,2)$, *i.e. at pixel with value* f_{21}, *is given by:*

$$f_{12} + 2f_{22} + f_{32} - f_{13} - 2f_{23} - f_{33}$$

Therefore, the second row of the 9×9 *matrix should be:*

$$0 \quad 0 \quad 0 \quad 1 \quad 2 \quad 1 \quad -1 \quad -2 \quad -1$$

The derivative at position $(1,3)$ *is given by:*

$$f_{22} + 2f_{32} + f_{12} - f_{23} - 2f_{33} - f_{13}$$

Therefore, the third row of the 9 × 9 matrix should be:

$$0 \quad 0 \quad 0 \quad 1 \quad 1 \quad 2 \quad -1 \quad -1 \quad -2$$

Reasoning this way, we conclude that the matrix we require must be:

$$
\begin{pmatrix}
0 & 0 & 0 & 2 & 1 & 1 & -2 & -1 & -1 \\
0 & 0 & 0 & 1 & 2 & 1 & -1 & -2 & -1 \\
0 & 0 & 0 & 1 & 1 & 2 & -1 & -1 & -2 \\
-2 & -1 & -1 & 0 & 0 & 0 & 2 & 1 & 1 \\
-1 & -2 & -1 & 0 & 0 & 0 & 1 & 2 & 1 \\
-1 & -1 & -2 & 0 & 0 & 0 & 1 & 1 & 2 \\
2 & 1 & 1 & -2 & -1 & -1 & 0 & 0 & 0 \\
1 & 2 & 1 & -1 & -2 & -1 & 0 & 0 & 0 \\
1 & 1 & 2 & -1 & -1 & -2 & 0 & 0 & 0
\end{pmatrix}
$$

Example 7.11

Using the matrix derived in Example 7.10, calculate the first derivative along the i axis of the following image:

$$
\begin{pmatrix}
3 & 1 & 0 \\
3 & 1 & 0 \\
3 & 1 & 0
\end{pmatrix}
$$

What is the component of the gradient along the i axis at the centre of the image?

$$
\begin{pmatrix}
0 & 0 & 0 & 2 & 1 & 1 & -2 & -1 & -1 \\
0 & 0 & 0 & 1 & 2 & 1 & -1 & -2 & -1 \\
0 & 0 & 0 & 1 & 1 & 2 & -1 & -1 & -2 \\
-2 & -1 & -1 & 0 & 0 & 0 & 2 & 1 & 1 \\
-1 & -2 & -1 & 0 & 0 & 0 & 1 & 2 & 1 \\
-1 & -1 & -2 & 0 & 0 & 0 & 1 & 1 & 2 \\
2 & 1 & 1 & -2 & -1 & -1 & 0 & 0 & 0 \\
1 & 2 & 1 & -1 & -2 & -1 & 0 & 0 & 0 \\
1 & 1 & 2 & -1 & -1 & -2 & 0 & 0 & 0
\end{pmatrix}
\begin{pmatrix}
3 \\ 3 \\ 3 \\ 1 \\ 1 \\ 1 \\ 0 \\ 0 \\ 0
\end{pmatrix}
=
\begin{pmatrix}
4 \\ 4 \\ 4 \\ -12 \\ -12 \\ -12 \\ 8 \\ 8 \\ 8
\end{pmatrix}
$$

At the centre of the image the component of the gradient along the i axis is -12.

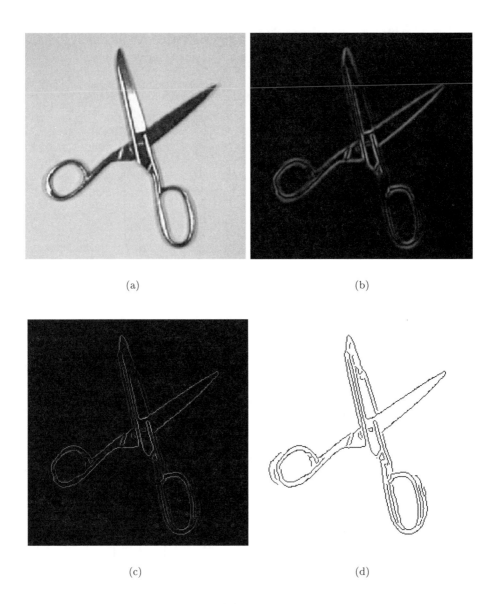

(a)   (b)

(c)   (d)

Figure 7.16: (a) Original image. (b) The gradient magnitude "image": the brighter the pixel, the higher the value of the gradient of the image at that point. (c) After non-maxima suppression: only pixels with locally maximal values of the gradient magnitude retain their values. The gradient values of all other pixels are set to zero. (d) The edge map: the gradient values in (c) are thresholded: all pixels with values above the threshold are labelled 0 (black) and all pixels with values below the threshold are labelled 255 (white).

In the general case, how do we decide whether a pixel is an edge pixel or not?

Edges are positions in the image where the image function changes. As the image is a 2D function, to find these positions we have to calculate the gradient of the function $\nabla f(x, y)$. The gradient of a 2D function is a 2D vector with components the partial derivatives of the function along two orthogonal directions. In the discrete case these partial derivatives are the partial differences computed along two orthogonal directions by using masks like, for example, the Sobel masks. If we convolve an image with these masks, we have a gradient vector associated with each pixel. Edges are the places where the magnitude of the gradient vector is a local maximum along the direction of the gradient vector (i.e. the orientation of the gradient at that pixel position). For this purpose, the local value of the gradient magnitude will have to be compared with the values of the gradient estimated along this orientation and at unit distance on either side away from the pixel. In general, these gradient values will not be known because they will happen to be at positions in between the pixels. Then, either a local surface is fitted to the gradient values and used for the estimation of the gradient magnitude at any interpixel position required, or the value of the gradient magnitude is calculated by interpolating the values of the gradient magnitudes at the known integer positions.

After this process of *non-maxima suppression* takes place, the values of the gradient vectors that remain are thresholded and only pixels with gradient values above the threshold are considered as edge pixels identified in the *edge map* (see *Figure* 7.16).

Example 7.12

The following image is given

0	1	2	0	4	5
0	0	1	1	4	5
0	2	0	4	5	4
0	0	5	4	6	6
0	0	6	6	5	6
5	4	6	5	4	5

Use the following masks to estimate the magnitude and orientation of the local gradient at all pixel positions of the image except the boundary pixels:

−1	0	1
−3	0	3
−1	0	1

−1	−3	−1
0	0	0
1	3	1

Then indicate which pixels represent horizontal edge pixels and which represent vertical.

If we convolve the image with the first mask, we shall have an estimate of the gradient component along the x (horizontal) axis:

	5	4	16	17
	6	11	19	6
	21	20	7	6
	24	23	−4	2

ΔI_x

If we convolve the image with the second mask, we shall have an estimate of the gradient component along the y (vertical) axis:

	1	−1	11	6
	4	15	15	10
	0	18	12	4
	18	8	2	−6

ΔI_y

The gradient magnitude is given by: $|G| = \sqrt{(\Delta I_x)^2 + (\Delta I_y)^2}$

	$\sqrt{26}$	$\sqrt{17}$	$\sqrt{377}$	$\sqrt{325}$
	$\sqrt{52}$	$\sqrt{346}$	$\sqrt{586}$	$\sqrt{136}$
	$\sqrt{441}$	$\sqrt{724}$	$\sqrt{193}$	$\sqrt{52}$
	$\sqrt{900}$	$\sqrt{593}$	$\sqrt{20}$	$\sqrt{40}$

(7.25)

The gradient orientation is given by:

$$\theta = \tan^{-1}\frac{\Delta I_y}{\Delta I_x}$$

	$\tan^{-1}\frac{1}{5}$	$-\tan^{-1}\frac{1}{4}$	$\tan^{-1}\frac{11}{16}$	$\tan^{-1}\frac{6}{17}$	
	$\tan^{-1}\frac{2}{3}$	$\tan^{-1}\frac{15}{11}$	$\tan^{-1}\frac{15}{19}$	$\tan^{-1}\frac{5}{3}$	
	$\tan^{-1}0$	$\tan^{-1}\frac{9}{10}$	$\tan^{-1}\frac{12}{7}$	$\tan^{-1}\frac{2}{3}$	
	$\tan^{-1}\frac{3}{4}$	$\tan^{-1}\frac{8}{23}$	$-\tan^{-1}\frac{1}{2}$	$-\tan^{-1}3$	

We know that for an angle θ to be in the range $0°$ to $45°$ $(0° \leq |\theta| \leq 45°)$ its tangent must satisfy $0 \leq |\tan\theta| \leq 1$. Also, an angle θ is in the range $45°$ to $90°$ $(45° < |\theta| \leq 90°)$ if $1 < |\tan\theta| \leq \infty$.

As we want to quantize the gradient orientations to vertical and horizontal ones only, we shall make all orientations $0° \leq |\theta| \leq 45°$ horizontal, i.e. set them to 0, and all orientations with $45° < |\theta| \leq 90°$ vertical, i.e. we shall set them to $90°$.

By inspecting the orientation array above we infer the following gradient orientations:

0	0	0	0	
0	90°	0	90°	
0	0	90°	0	
0	0	0	90°	

A pixel is a horizontal edge pixel if the magnitude of its gradient is a local maximum when compared with its vertical neighbours and its orientation is 0. A pixel is a vertical edge if its orientation is $90°$ and its magnitude is a local maximum in the horizontal direction.

By inspecting the gradient magnitudes in (7.25) and the quantized orientations we derived, we identify the following horizontal $(-)$ edges only:

			$-$		
		$-$			

Example 7.13

Are the masks used in Example 7.12 separable? How can we take advantage of the separability of a 2D mask to reduce the computational cost of convolution?

The masks used in Example 7.12 are separable because they can be implemented as a sequence of two 1D filters applied in a cascaded way one after the other:

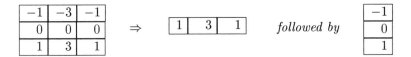

Any 2D $N \times N$ separable mask can be implemented as a cascade of two 1D masks of size N. This implementation replaces N^2 multiplications and additions per pixel by $2N$ such operations per pixel.

Are Sobel masks appropriate for all images?

Sobel masks are appropriate for images with low levels of noise. They are inadequate for noisy images. See for example *Figure* 7.17 which shows the results of edge detection using Sobel masks for a very noisy image.

How can we choose the weights of the mask if we need a larger mask owing to the presence of significant noise in the image?

We shall consider the problem of detecting abrupt changes in the value of a 1-dimensional signal, like the one shown in *Figure* 7.18.

Let us assume that the feature we want to detect is $u(x)$ and it is immersed in additive white Gaussian noise $n(x)$. The mask we want to use for edge detection should have certain desirable characteristics, called *Canny's criteria*:

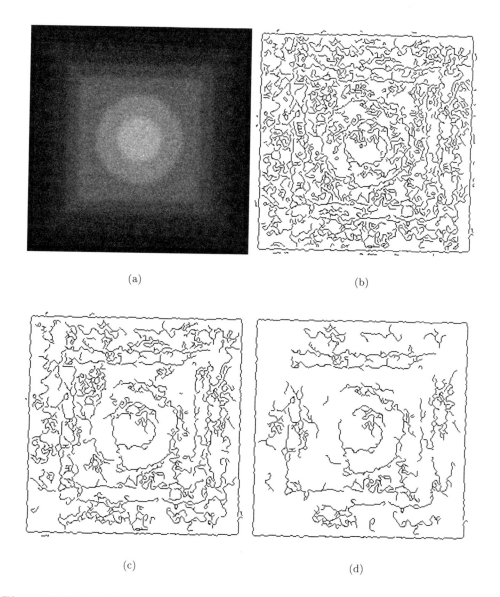

(a)   (b)

(c)   (d)

Figure 7.17: Trying to detect edges in a blurred and noisy image using the Sobel edge detector may be very tricky. One can play with different threshold values but the result is not satisfatory. The three results shown have been obtained by using different thresholds within the same edge detection framework, and they were the best among many others obtained for different threshold values.

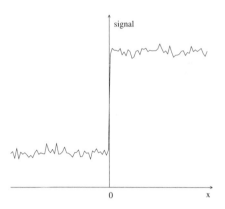

Figure 7.18: A noisy 1D edge.

1. Good signal to noise ratio.
2. Good locality; i.e. the edge should be detected where it actually is.
3. Small number of false alarms; i.e. the maxima of the filter response should be mainly due to the presence of the true edges in the image rather than to noise.

Canny showed that a filter function $f(x)$ has maximal signal to noise ratio if it is chosen in such a way that it maximizes the quantity:

$$SNR = \frac{\int_{-\infty}^{\infty} f(x)u(-x)dx}{n_0 \sqrt{\int_{-\infty}^{\infty} f^2(x)dx}} \tag{7.26}$$

where n_0 is the standard deviation of the additive Gaussian noise. As this is a constant for a particular image, when we try to decide what function $f(x)$ should be, we omit it from the computation. So, we try to maximize the quantity:

$$S \equiv \frac{\int_{-\infty}^{\infty} f(x)u(-x)dx}{\sqrt{\int_{-\infty}^{\infty} f^2(x)dx}}$$

Canny also showed that the filter function $f(x)$ detects an edge with the minimum deviation from its true location if it is chosen in such a way that its first and second derivatives maximize the quantity:

$$L \equiv \frac{\int_{-\infty}^{\infty} f''(x)u(-x)dx}{\sqrt{\int_{-\infty}^{\infty} [f'(x)]^2 dx}} \tag{7.27}$$

Finally, he showed that the output of the convolution of the signal with the filter $f(x)$ will contain minimal number of false responses if function $f(x)$ is chosen in such a way that its first and second derivatives maximize the quantity:

$$C \equiv \sqrt{\frac{\int_{-\infty}^{\infty} (f'(x))^2 dx}{\int_{-\infty}^{\infty} (f''(x))^2 dx}} \tag{7.28}$$

One, therefore, can design an optimal edge enhancing filter by trying to maximize the above three quantities:

$$S, \quad L, \quad C$$

Canny combined the first two into one performance measure P, which he maximized under the constraint that $C = const$. He derived a convolution filter that way, which, however, he did not use because he noticed that it could be approximated by the derivative of a Gaussian. So he adopted the derivative of a Gaussian as a good enough edge enhancing filter.

Alternatively, one may combine all three quantities S, L, and C into a single performance measure P:

$$P \equiv (S \times L \times C)^2 = \frac{\left[\int_{-\infty}^{\infty} f(x)u(-x)dx\right]^2 \left[\int_{-\infty}^{\infty} f''(x)u(-x)dx\right]^2}{\int_{-\infty}^{\infty} f^2(x)dx \int_{-\infty}^{\infty} (f''(x))^2 dx} \tag{7.29}$$

and try to choose $f(x)$ so that P is maximum. The free parameters which will appear in the functional expression of $f(x)$ can be calculated from the boundary conditions imposed on $f(x)$ and by substitution into the expression for P and selection of their numerical values so that P is maximum.

It must be stressed that the filters defined in this way are appropriate for the detection of **antisymmetric features**; i.e **step** or **ramp edges** when the noise in the signal is **additive**, **white** and **Gaussian**.

Can we use the optimal filters for edges to detect lines in an image in an optimal way?

No. Edge detection filters respond badly to lines. The theory for optimal edge detector filters has been developed under the assumption that there is a single isolated step edge. We can see that from the limits used in the integrals in equation (7.26), for example: the limits are from $-\infty$ to $+\infty$, assuming that in the whole infinite length of the signal there is only a single step edge. If we have a line, its profile looks like the one shown in *Figure* 7.19.

It looks like two step edges back to back. The responses of the filter to the two step edges interfere and the result is not satisfactory: the two steps may or may not be detected. If they are detected, they tend to be detected as two edges noticeably misplaced from their true position and shifted away from one another.

Apart from this, there is another more fundamental difference between step edges and lines.

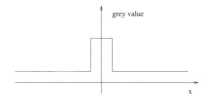

Figure 7.19: The profile of a line.

What is the fundamental difference between step edges and lines?

A step edge is scale invariant: it looks the same whether we stretch it or shrink it. A line has an intrinsic length-scale: it is its width. From the moment the feature we wish to detect has a characteristic "size", the size of the filter we must use becomes important. Similar considerations apply also if one wishes to develop appropriate filters for ramp edges as opposed to step edges: the length over which the ramp rises (or drops) is characteristic of the feature and it cannot be ignored when designing the optimal filter. Petrou and Kittler used the criterion expressed by equation (7.29), appropriately modified, to develop optimal filters for the detection of ramp edges of various slopes. *Figure* 7.20 shows attempts to detect edges in the image of *Figure* 7.17a, using the wrong filter; the edges are blurred, so they actually have ramp-like profiles and the filter used is the optimal filter for step edges. *Figure* 7.21 shows the results of using the optimal filter for ramps on the same image, and the effect the size of the filter has on the results. *Figure* 7.22 shows an example of a relatively "clean" and "unblurred" image for which one does not need to worry too much about which filter one uses for edge detection.

Petrou also modified Canny's criteria for good edge filters for the case of lines and used them to derive optimal filters for line detection that depend on the width and sharpness of the line.

Example 7.14(B)

Assume that $u(x)$ is defined for positive and negative values of x and that we want to enhance a feature of $u(x)$ at position $x = 0$. Use equation (7.26) to show that if the feature we want to enhance makes $u(x)$ an even function, we must choose an even filter, and if it makes it an odd function, we must choose an odd filter.

Any function $f(x)$ can be written as the sum of an odd and an even part:

$$f(x) \equiv \underbrace{\frac{1}{2}[f(x) - f(-x)]}_{f_o(x)(odd)} + \underbrace{\frac{1}{2}[f(x) + f(-x)]}_{f_e(x)(even)}$$

In general, therefore, $f(x)$ can be written as:

$$f(x) = f_e(x) + f_o(x)$$

Suppose that $u(x)$ is even. Then the integral in the numerator of S is:

$$\int_{-\infty}^{\infty} f(x)u(-x)dx \;=\; \int_{-\infty}^{\infty} f_e(x)u(-x)dx + \underbrace{\int_{-\infty}^{\infty} f_o(x)u(-x)dx}$$

odd intergand integrated over a symmetric interval: it vanishes

$$= \int_{-\infty}^{\infty} f_e(x)u(-x)dx$$

The integral in the denominator in the expression for S is:

$$\int_{-\infty}^{\infty} f^2(x)dx \;=\; \int_{-\infty}^{\infty} f_e^2(x)dx + \int_{-\infty}^{\infty} f_o^2(x)dx + \underbrace{2\int_{-\infty}^{\infty} f_e f_o(x)dx}$$

odd intergand integrated over a symmetric interval: it vanishes

$$= \int_{-\infty}^{\infty} f_e^2(x)dx + \int_{-\infty}^{\infty} f_o^2(x)dx$$

So, we see that the odd part of the filter does not contribute at all in the signal response, while it contributes to the noise response; i.e. it reduces the signal to noise ratio. Thus, to enhance an even feature we must use even filters. Similarly, to enhance an odd feature, we must use odd filters.

Example 7.15

What type of feature are the digital filter masks below appropriate for enhancing?

−1	2	−1
−1	2	−1
−1	2	−1

−1	−1	−1
2	2	2
−1	−1	−1

The first enhances vertical lines and the second horizontal lines, in both cases brighter than the background.

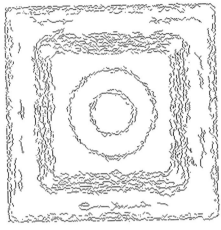

(a) Assuming a step edge and using a filter of size 17×17

(b) The same parameters as in (a) with the size of the filter increased to 21×21. Some detail is lost and no improvement has been achieved

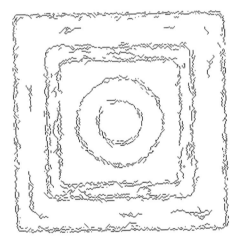

(c) Assuming a step edge and a filter of size 17×17 and increasing the threshold

(d) As in (c) with the threshold increased even further

Figure 7.20: Trying to deal with the high level of noise by using an optimal filter of large size and playing with the thresholds may not help, if the edges are blurred and resemble ramps, and the "optimal" filter has been developed to be optimal for step edges.

(a) Assuming a shallow ramp and using a filter of size 7×7

(b) Exactly the same as in (a) but setting the filter size to 13×13

(c) Assuming a ramp edge and filter of size 17×17

(d) Assuming a ramp edge and using a filter of size 21×21

Figure 7.21: Having chosen the right model for the features we wish to detect, using the right filter size for the level of noise may prove crucial.

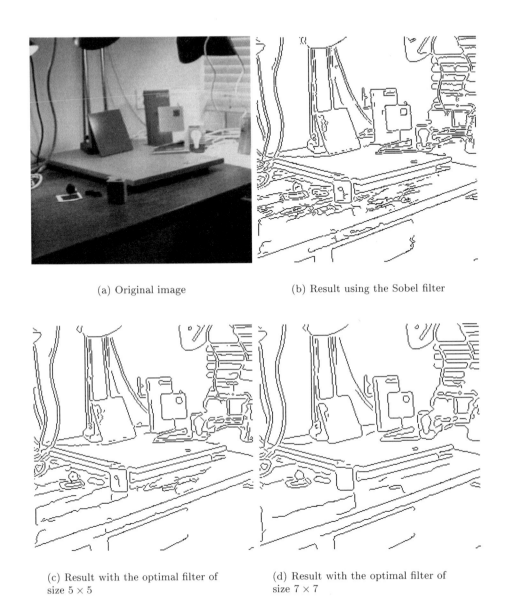

(a) Original image   (b) Result using the Sobel filter

(c) Result with the optimal filter of   (d) Result with the optimal filter of
size 5 × 5   size 7 × 7

Figure 7.22: Everyday life images are relatively clean and good results can be obtained by using relatively small filters. Note that the smaller the filter, the more details are preserved.

Example 7.16

Use the masks of Example 7.15 to process the image below (do not process the border pixels).

1	1	5	3	0
0	1	4	1	0
1	1	3	2	1
0	2	5	3	0
1	0	4	2	0

Then by choosing an appropriate threshold for the output images calculated, produce the feature maps of the input image.

The result of the convolution of the image with the first mask is:

	-8	15	-1
	-5	14	-1
	-8	14	1

The result of the convolution of the image with the second mask is:

	-2	-3	-7
	-2	-4	-1
	4	8	4

(We do not have output for the border pixels.)
Note that the outputs contain values that can easily be divided into two classes: positive and negative. We choose therefore as our threshold the zero value $t = 0$. This threshold seems to be in the largest gap between the two populations of the output values (i.e. one population that presumably represents the background and one that represents the features we want to detect). When thresholding we shall give one label to one class (say all numbers above the threshold will be labelled 1) and another label to the other class (all numbers below the threhold will be labelled 0). The two outputs then become:

0	1	0
0	1	0
0	1	0

0	0	0
0	0	0
1	1	1

These are the feature maps of the given image.

B7.3: Convolving a random noise signal with a filter.

Suppose that our noise signal is $n(x)$ and we convolve it with a filter $f(x)$. The result of the convolution will be:

$$g(x_0) = \int_{-\infty}^{\infty} f(x)n(x_0 - x)dx \qquad (7.30)$$

As the input noise is a random process, this quantity is a random process too. If we take its expectation value, it will be 0 as long as the expectation value of the input noise is 0 too. We may also try to characterize it by calculating its variance, its mean square value. This is given by:

$$E\{[g(x_0)]^2\} = E\{g(x_0)g(x_0)\} \qquad (7.31)$$

This is the definition of the autocorrelation function of the output calculated at argument 0. So we must calculate first the autocorrelation function of g, $R_{gg}(0)$. We multiply both sides of equation (7.30) with $g(x_0 + \tau)$ and take expectation values:

$$E\{g(x_0)g(x_0 + \tau)\} = E\left\{\int_{-\infty}^{\infty} f(x)g(x_0 + \tau)n(x_0 - x)dx\right\}$$

$$\Rightarrow R_{gg}(\tau) = \int_{-\infty}^{\infty} f(x)E\{g(x_0 + \tau)n(x_0 - x)\}dx$$

$$\Rightarrow R_{gg}(\tau) = \int_{-\infty}^{\infty} f(x)R_{gn}(x_0 + \tau - x_0 + x)dx$$

$$\Rightarrow R_{gg}(\tau) = \int_{-\infty}^{\infty} f(x)R_{gn}(\tau + x)dx \qquad (7.32)$$

where $R_{gn}(a)$ is the cross-correlation function between the input noise signal and the output noise signal at relative shift a. We must calculate R_{gn}. We multiply both sides of (7.30) with $n(x_0 - \tau)$ and then take expectation values. (We multiply with $n(x_0 - \tau)$ instead of $n(x_0 + \tau)$ because in (7.32) we defined the argument of R_{gn} as the difference of the argument of g minus the argument of n.)

$$E\{g(x_0)n(x_0 - \tau)\} = \int_{-\infty}^{\infty} f(x)E\{n(x_0 - x)n(x_0 - \tau)\}dx$$

$$\Rightarrow R_{gn}(\tau) = \int_{-\infty}^{\infty} f(x)R_{nn}(x_0 - x - x_0 + \tau)dx$$

$$\Rightarrow R_{gn}(\tau) = \int_{-\infty}^{\infty} f(x)R_{nn}(\tau - x)dx \qquad (7.33)$$

where $R_{nn}(a)$ is the autocorrelation function of the input noise signal at relative shift a. However, $n(x)$ is white Gaussian noise, so its autocorrelation function is a delta function given by $R_{nn}(\tau) = n_0^2\delta(\tau)$ where n_0^2 is the variance of the noise. Therefore:

$$R_{gn}(\tau) = \int_{-\infty}^{\infty} f(x)n_0^2\delta(\tau - x)dx = n_0^2 f(\tau) \qquad (7.34)$$

Therefore, R_{gn} with argument $\tau + x$ as it appears in (7.32) is:

$$R_{gn}(\tau + x) = n_0^2 f(\tau + x) \qquad (7.35)$$

If we substitute in (7.32) we have:

$$R_{gg}(\tau) = n_0^2 \int_{-\infty}^{\infty} f(x)f(\tau + x)dx \Rightarrow R_{gg}(0) = n_0^2 \int_{-\infty}^{\infty} f^2(x)dx \qquad (7.36)$$

If we substitute in equation (7.31) we obtain:

$$E\left\{[g(x_0)]^2\right\} = n_0^2 \int_{-\infty}^{\infty} f^2(x)dx \qquad (7.37)$$

The mean filter response to noise is given by the square root of the above expression.

B7.4: Calculation of the signal to noise ratio after convolution of a noisy edge signal with a filter.

Suppose that $f(x)$ is the filter we want to develop, so that it enhances an edge in a noisy signal. The signal consists of two components: the deterministic signal of interest $u(x)$ and the random noise component $n(x)$:

$$I(x) = u(x) + n(x)$$

The response of the filter due to the deterministic component of the noisy signal, when the latter is convolved with the filter, is:

$$s(x_0) = \int_{-\infty}^{\infty} f(x)u(x_0 - x)dx \tag{7.38}$$

Let us assume that the edge we wish to detect is actually at position $x_0 = 0$. Then:

$$s_0 = \int_{-\infty}^{\infty} f(x)u(-x)dx \tag{7.39}$$

The mean response of the filter due to the noise component (see Box **B7.3**) is $n_0\sqrt{\int_{-\infty}^{\infty} f^2(x)dx}$. The signal to noise ratio, therefore, is:

$$SNR = \frac{\int_{-\infty}^{\infty} f(x)u(-x)dx}{n_0\sqrt{\int_{-\infty}^{\infty} f^2(x)dx}} \tag{7.40}$$

B7.5: Derivation of the good locality measure.

The position of the edge is marked at the point where the output is maximum. The total output of the convolution is the output due to the deterministic signal $u(x)$ and the output due to the noise component $n(x)$. Let us say that the noisy signal we have is:

$$I(x) = u(x) + n(x)$$

The result of the convolution with the filter $f(x)$ will be:

$$O(x_0) \equiv \int_{-\infty}^{\infty} f(x)I(x_0 - x)dx = \int_{-\infty}^{\infty} f(x_0 - x)I(x)dx$$

$$= \int_{-\infty}^{\infty} f(x_0 - x)u(x)dx + \int_{-\infty}^{\infty} f(x_0 - x)n(x)dx \equiv s(x_0) + g(x_0)$$

where $s(x_0)$ is the output due to the deterministic signal of interest and $g(x_0)$ is the output due to noise.

The edge is detected at the local maximum of this output; i.e. at the place where its first derivative with respect to x_0 becomes 0:

$$\frac{dO(x_0)}{dx_0} = \frac{ds(x_0)}{dx_0} + \frac{dg(x_0)}{dx_0} = \int_{-\infty}^{\infty} f'(x_0 - x)u(x)dx + \int_{-\infty}^{\infty} f'(x_0 - x)n(x)dx \quad (7.41)$$

We expand the derivative of the filter, about point $x_0 = 0$, the true position of the edge, and keep only the first two terms of the expansion:

$$f'(x_0 - x) \simeq f'(-x) + x_0 f''(-x) + \ldots$$

Upon substitution into $\frac{ds(x_0)}{dx_0}$ we obtain:

$$\frac{ds(x_0)}{dx_0} \simeq \int_{-\infty}^{\infty} f'(-x)u(x)dx + x_0 \int_{-\infty}^{\infty} f''(-x)u(x)dx \quad (7.42)$$

The feature we want to detect is an antisymmetric feature; i.e. an edge has a shape like ⌐͟ .

According to the result proven in Example 7.14, $f(x)$ should also be an antisymmetric function. This means that its first derivative will be symmetric and when it is multiplied with the antisymmetric function $u(x)$ it will produce an antisymmetric integrand which upon integration over a symmetric interval will make the first term in equation (7.42) vanish. Then:

$$\frac{ds(x_0)}{dx_0} \simeq \underbrace{x_0 \int_{-\infty}^{\infty} f''(-x)u(x)dx}_{\text{set } \tilde{x} \equiv -x} = x_0 \int_{-\infty}^{\infty} f''(\tilde{x})u(-\tilde{x})d\tilde{x} \quad (7.43)$$

The derivative of the filter response to noise may be written in a more convenient way:

$$\frac{dg(x_0)}{dx_0} = \int_{-\infty}^{\infty} f'(x_0 - x)n(x)dx = \int_{-\infty}^{\infty} f'(x)n(x_0 - x)dx \quad (7.44)$$

The position of the edge will be marked at the value of x that makes the sum of the right hand sides of equations (7.43) and (7.44) zero:

$$\frac{ds(x_0)}{dx_0} + \frac{dg(x_0)}{dx_0} = 0 \Rightarrow$$

$$x_0 \int_{-\infty}^{\infty} f''(x)u(-x)dx + \int_{-\infty}^{\infty} f'(x)n(x_0 - x)dx = 0 \Rightarrow$$

$$x_0 \int_{-\infty}^{\infty} f''(x)u(-x)dx = -\int_{-\infty}^{\infty} f'(x)n(x_0 - x)dx \quad (7.45)$$

The right hand side of this expression is a random variable, indicating that the location of the edge will be marked at various randomly distributed positions around the true position, which is at $x_0 = 0$. We can calculate the mean shifting away from the true position as the expectation value of x_0. This, however, is expected to be 0. So, we calculate instead the variance of the x_0 values. We square both sides of (7.45) and take the expectation value of:

$$E\{x_0^2\}\left[\int_{-\infty}^{\infty} f''(x)u(-x)dx\right]^2 = E\left\{\left[\int_{-\infty}^{\infty} f'(x)n(x_0 - x)dx\right]^2\right\}$$

Note that the expectation value operator applies only to the random components.

In Box **B7.3** we saw that if a noise signal with variance n_0^2 is convolved by a filter $f(x)$, the mean square value of the output signal is given by:

$$n_0^2 \int_{-\infty}^{\infty} f^2(x)dx$$

(see equation (7.37)). The right hand side of equation (7.45) here indicates the convolution of the noise component by filter $f'(x)$. Its mean square value therefore is $n_0^2 \int_{-\infty}^{\infty} [f'(x)]^2 dx$. Then:

$$E\{x_0^2\} = \frac{n_0^2 \int_{-\infty}^{\infty} [f'(x)]^2 dx}{\left[\int_{-\infty}^{\infty} f''(x)u(-x)dx\right]^2} \qquad (7.46)$$

The smaller this expectation value is, the better the localization of the edge. We can define, therefore, the good locality measure as the inverse of the square root of the above quantity. We may also ignore factor n_0 as it is the standard deviation of the noise during the imaging process, over which we do not have control. So a filter is optimal with respect to good locality, if it maximizes the quantity:

$$L \equiv \frac{\int_{-\infty}^{\infty} f''(x)u(-x)dx}{\sqrt{\int_{-\infty}^{\infty} [f'(x)]^2 dx}} \qquad (7.47)$$

Example 7.17 (B)

Show that the differentiation of the output of a convolution of a signal $u(x)$ with a filter $f(x)$ can be achieved by convolving the signal with the derivative of the filter.

The output of the convolution is:

$$s(x_0) = \int_{-\infty}^{\infty} f(x)u(x_0 - x)dx$$

or

$$s(x_0) = \int_{-\infty}^{\infty} f(x_0 - x)u(x)dx \tag{7.48}$$

Applying Leibnitz's rule for differentiating an integral with respect to a parameter (Box B7.1) we obtain from equation (7.48):

$$\frac{ds(x_0)}{dx_0} = \int_{-\infty}^{\infty} f'(x_0 - x)u(x)dx$$

which upon changing variables can be written as:

$$\frac{ds(x_0)}{dx_0} = \int_{-\infty}^{\infty} f'(x)u(x_0 - x)dx$$

B7.6: Derivation of the count of false maxima.

It has been shown by Rice that the average density of zero crossings of the convolution of a function h with Gaussian noise is given by:

$$x_{av} = \pi \left[\frac{-R_{hh}(0)}{R''_{hh}(0)} \right]^{\frac{1}{2}} \tag{7.49}$$

where $R_{hh}(\tau)$ is the spatial autocorrelation function of function $h(x)$, i.e:

$$R_{hh}(\tau) = \int_{-\infty}^{\infty} h(x)h(x + \tau)dx \tag{7.50}$$

Therefore:

$$R_{hh}(0) = \int_{-\infty}^{\infty} (h(x))^2 dx$$

Using Leibnitz's rule (see Box B7.1) we can differentiate (7.50) with respect to τ to obtain:

$$R'_{hh}(\tau) = \int_{-\infty}^{\infty} h(x)h'(x+\tau)dx$$

We define a new variable of integration $\tilde{x} \equiv x + \tau \Rightarrow x = \tilde{x} - \tau$ and $dx = d\tilde{x}$ to obtain:

$$R'_{hh}(\tau) = \int_{-\infty}^{\infty} h(\tilde{x} - \tau)h'(\tilde{x})d\tilde{x} \qquad (7.51)$$

We differentiate (7.51) once more:

$$R''_{hh}(\tau) = -\int_{-\infty}^{\infty} h'(\tilde{x} - \tau)h'(\tilde{x})d\tilde{x} \Rightarrow R''_{hh}(0) = -\int_{-\infty}^{\infty} (h'(x))^2 dx$$

Therefore, the average distance of zero crossings of the output signal when a noise signal is filtered by function $h(x)$ is given by:

$$x_{av} = \pi \sqrt{\frac{\int_{-\infty}^{\infty} (h(x))^2 dx}{\int_{-\infty}^{\infty} (h'(x))^2 dx}} \qquad (7.52)$$

From the analysis in Box **B7.5**, we can see that the false maxima in our case will come from equation $\int_{-\infty}^{\infty} f'(x)n(x_0 - x)dx = 0$ which will give the false alarms in the absence of any signal $(u(x) = 0)$. This is equivalent to saying that the false maxima coincide with the zeroes in the output signal when the noise signal is filtered with function $f'(x)$. So, if we want to reduce the number of false local maxima, we should make the average distance between the zero crossings as large as possible, for the filter function $f'(x)$. Therefore, we define the good measure of scarcity of false alarms as:

$$C \equiv \sqrt{\frac{\int_{-\infty}^{\infty} (f'(x))^2 dx}{\int_{-\infty}^{\infty} (f''(x))^2 dx}}$$

What is the "take home" message of this chapter?

This chapter dealt with the reduction of the information content of an image so that it can be processed more easily by a computer vision system. It surveyed the two basic approaches for this purpose: region segmentation and edge detection. Region segmentation tries to identify spatially coherent sets of pixels that appear to constitute uniform patches in the image. These patches may represent surfaces or parts of surfaces of objects depicted in the image. Edge detection seeks to identify boundaries between such uniform patches. The most common approach for this is based on the estimate of the first derivative of the image. For images with low levels of noise the

Sobel masks are used to enhance the edges. For noisy images the Canny filters should be used. Canny filters can be approximated by the derivative of a Gaussian; i.e. they have the form $xe^{-\frac{x^2}{2\sigma^2}}$. Although parameter σ in this expression has to have a **specific** value for the filters to be optimal according to Canny's criteria, often people treat σ as a free parameter and experiment with various values of it. Care must be taken in this case when discretizing the filter and truncating the Gaussian, not to create a filter with sharp ends. Whether the Sobel masks are used or the derivatives of the Gaussian, the result is the enhancement of edges in the image. The output produced has to be further processed by non-maxima suppression (i.e. the identification of the local maxima in the output array) and thresholding (i.e. the retention of only the significant local maxima).

Edge detectors consist of all three stages described above and produce fragmented edges. Often a further step is involved of linking the fragments together to create closed contours that identify uniform regions. Alternatively, people may bypass this process by performing region segmentation directly and, if needed, extract the boundaries of the regions afterwards. Region-based methods are much more powerful when both attribute similarity and spatial proximity are taken into consideration when deciding which pixels form which region.

Bibliography

The material of this book is based largely on classical textbooks. The method described in example 6.10 was taken from reference 3. Chapter 7 is based mainly on research papers. From the references given below, numbers 13, 15 and 23 refer to thresholding methods. References 5, 6, 13, 18–22 and 28 refer to linear edge detectors. Paper 17 refers to non-linear edge detectors with an application of them presented in 10. Reference 21 is an extensive review of the Canny-related approaches to edge detection.

1. M Abramowitz and I A Stegun (eds), 1970. *Handbook of Mathematical Functions*, Dover Publications, ISBN 0-486-61272-4.

2. K G Beauchamp, 1975. *Walsh Functions and their Applications*, Academic Press, ISBN 0-12-084050-2.

3. C R Boukouvalas, 1996. *Colour Shade Grading and its Applications to Visual Inspection*. PhD thesis, University of Surrey, UK.

4. R N Bracewell, 1978. *The Fourier Transform and its Applications*, McGraw-Hill, ISBN 0-07-007013-X.

5. J Canny, 1983. *Finding Edges and Lines in Images*. MIT AI Lab Technical Report 720.

6. J Canny, 1986. "A computational approach to edge detection". *IEEE Transactions on Pattern Analysis and Machine Intelligence*, Vol PAMI-8, pp 679–698.

7. R A Gabel and R A Roberts, 1987. *Signals and Linear Systems*, John Wiley & Sons, ISBN 0-471-83821-7.

8. R C Gonzalez and R E Woods, 1992. *Digital Image Processing*, Addison-Wesley, ISBN 0-201-50803-6.

9. I S Gradshteyn and I M Ryzhik, 1980. *Table of Integrals, Series and Products*, Academic Press, ISBN 0-12-294760-6.

10. J Graham and C J Taylor, 1988. "Boundary cue operators for model-based image processing". *Proceedings of the Fourth Alvey Vision Conference, AVC88, University of Manchester, 31 August–2 September 1988*, pp 59–64.

11. H P Hsu, 1970. *Fourier Analysis*, Simon & Schuster, New York.

12. T S Huang (ed), 1979. "Picture processing and digital filtering". In *Topics in Applied Physics*, Vol 6, Springer-Verlag, ISBN 0-387-09339-7.

13. J Kittler, 1983. "On the accuracy of the Sobel edge detector". *Image and Vision Computing*, Vol 1, pp 37–42.

14. J Kittler and J Illingworth, 1985. "On threshold selection using clustering criteria". *IEEE Transactions on Systems, Man, and Cybernetics*, Vol SMC-15, pp 652–655.

15. N Otsu, 1979. "A threshold selection method from gray level histograms". *IEEE Transactions on Systems, Man, and Cybernetics*, Vol SMC-9, pp 62–66.

16. A Papoulis, 1965. *Probability, Random Variables and Stochastic Processes*, McGraw-Hill Kogakusha Ltd, Library of Congress Catalog Card number 64-22956.

17. I Pitas and A N Venetsanopoulos, 1986. "Non-linear order statistic filters for image filtering and edge detection". *Signal Processing*, Vol 10, pp 395–413.

18. M Petrou and J Kittler, 1991. "Optimal edge detectors for ramp edges". *IEEE Transactions on Pattern Analysis and Machine Intelligence*, Vol PAMI-13, pp 483–491.

19. M Petrou and A Kolomvas, 1992. "The recursive implementation of the optimal filter for the detection of roof edges and thin lines". In *Signal Processing VI, Theory and Applications*, pp 1489–1492.

20. M Petrou, 1993. "Optimal convolution filters and an algorithm for the detection of wide linear features." *IEE Proceedings: Vision, Signal and Image Processing*, Vol 140, pp 331–339.

21. M Petrou, 1994. "The differentiating filter approach to edge detection". *Advances in Electronics and Electron Physics*, Vol 88, pp 297–345.

22. M Petrou, 1995. "Separable 2D filters for the detection of ramp edges". *IEE Proceedings: Vision, Image and Signal Processing*, Vol 142, pp 228–231.

23. M Petrou and A Matrucceli, 1998. "On the stability of thresholding SAR images". *Pattern Recognition*, Vol 31, pp 1791–1796.

24. W K Pratt, 1978. *Digital Image Processing*, John Wiley & Sons, ISBN 0-471-01888-0.

25. R M Pringle and A A Rayner, 1971. *Generalized Inverse Matrices with Applications to Statistics*, No. 28 of Griffin's Statistical monographs and courses, edited by A Stuart, ISBN 0-85264-181-8.

26. S O Rice, 1945. "Mathematical analysis of random noise". *Bell Systems Tech. J.*, Vol 24, pp 46–156.

27. A Rosenfeld and A C Kak, 1982. *Digital Picture Processing*, Academic Press, ISBN 0-12-597301-2.

28. L A Spacek, 1986. "Edge detection and motion detection". *Image and Vision Computing*, Vol 4, pp 43–53.

29. P H Winston, 1992. *Artificial Intelligence*, Addison-Wesley, ISBN 0-20-153377-4.

30. T Y Young and K S Fu (eds), 1986. *Handbook of Pattern Recognition and Image Processing*, Academic Press, ISBN 0-12-774560-2.

Index